FOUNDATIONS OF THE EARTH

Foundations of the Earth

Global Ecological Change and the Book of Job

H. H. Shugart

COLUMBIA UNIVERSITY PRESS

NEW YORK

Columbia University Press
Publishers Since 1893
New York   Chichester, West Sussex
cup.columbia.edu

Library of Congress Cataloging-in-Publication Data
Shugart, Herman H. (Herman Henry), 1944–
    Foundations of the earth : global ecological change and the book of Job /
H. H. Shugart.
        pages cm
    Includes bibliographical references and index.
    ISBN 978-0-231-16908-0 (cloth : alk. paper) —ISBN 978-0-231-53769-8
(e-book)
    1. Human ecology—Religious aspects. 2. Environmental degradation—
Religious aspects. 3. Natural history—Religious aspects. 4. Religion and
science. 5. Bible. Job—Criticism, interpretation, etc. I. Title.

    GF80.S348   2014
    304.2—dc23   2013045871

Jacket design: Noah Arlow

Contents

Preface   ix

1. Introduction   1
   On Job and the Whirlwind Speeches   2
   The Whirlwind Speech   4
   The Whirlwind Speech as an Account of Planetary Creation
      and Function   6
   The Antiquity of the Job Text   8
   Questions to Job from the Whirlwind   10

2. Laying the Foundation of the Earth   13
   The Origin of the Earth   15
   Laying the Foundation of the Earth   20
   Concluding Comments   31

3. Taming the Unicorn, Yoking the Aurochs: Animal and
   Plant Domestication and the Consequent Alteration of
   the Surface of the Earth   35
   The Unicorn   36
   The Wild Ox   39
   The Domestication of Animals   41
   The Dog as the First Domesticated Animal   42
   Yoking the Aurochs   50
   Changing Regional Land Cover   58
   Concluding Comments   69

4. Freeing the Onager: Feral and Introduced Animals   71

   The Onager   72
   The Onager as a Draft Animal   73
   Humans as a Keystone Species   77
   Introduced Species   87
   Ecosystems and Anthro-Ecosystems on a Human-Dominated
      Planet   98
   Concluding Comments   100

5. Bounding the Seas, Freezing the Face of the Deep:
   When the Sea Is Loosed from Its Bonds   103

   The Tides   104
   Past Sea Levels   112
   Future Sea Level Rise   116
   Concluding Comments   120

6. The Ordinances of the Heavens and Their Rule on Earth:
   Adaptation and the Cycles of Life   123

   The Ordinances of the Heavens and Their Rule on Earth   127
   Biological and Ecological Timing   132
   Knowing When the Mountain Goats Give Birth   140
   Migration: When Birds Turn Their Wings Toward
      the South   142
   Concluding Comments   149

7. The Dwelling of the Light and the Paths to Its Home:
   Winds, Ocean Currents, and the Global Energy Balance   153

   Föhns and Chinooks: Physical Processes Behind
      the Pattern   154
   Tropical Cyclones, Hurricanes, and Typhoons   155
   The Hadley Circulation: The Paths to Home   163
   Winds and Ocean Gyres   171
   Light's Pathway Home: The Global Radiation Balance   178
   Concluding Comments   180

8. Making the Ground Put Forth Grass: The Relationship
   Between Climate and Vegetation   183
       Climate and Vegetation   187
       Climate-Vegetation Relationships in a Changing World   199
       Concluding Comments   218

9. Feeding the Lions: The Conservation of Biological Diversity
   on a Changing Planet   221
       On the Conservation of the Bright and Beautiful,
          Big and Dangerous   223
       The Megafaunal Extinctions at the End of the Pleistocene   229
       Conserving Diversity on a Changing Planet   251
       Concluding Comments   254

10. Making Weather and Influencing Climate: Human Engineering
    of the Earth   257
        Making Weather   261
        Geoengineering: Climate Modification   273
        Concluding Comments   279

11. Conclusion: Comprehending the Earth   283

    Notes   289
    Index   355

Preface

Foundations of the Earth intends to demonstrate the intrinsic connectedness of the Earth's systems, their dynamic change, and their interactions with humans. The book emphasizes environmental synthesis at large scales—regional to global scales in space, centuries to millennia to even longer scales in time. The issues in the chapters, all on our changing and human-altered planet, are interwoven. The overarching themes involve planetary complexity, connectedness, and dynamism. These are large themes for a small book, and the diversity of disciplines considered is substantial. The mutual interactions among different Earth systems provide a unity to the text.

The book uses a set of questions from a small section of the Book of Job as an overarching theme to provide additional connectedness. These biblical questions frame chapter discussions on such topics as, "Where did the solar system come from? How were animals domesticated? How do changes in the greenhouse gases in the atmosphere imply global warming? How does climate and its change alter the world's vegetation and vice versa?" The intention here is neither to substantiate the questions, which coalesce into a creation account starting in the thirty-eighth chapter of Job. Nor is the intent to reconcile the Bible with science somehow. Rabbi Shelomo Ben Isaac, or "Rashi" (France, 1040–1105 CE), distinguished what a biblical text "says" from what it "means" and separated the literal translation from traditional interpretation.[1] In considering the whirlwind questions, this book considers the "what-it-says" part of Rashi's distinction with literal translation from a standard source (the New Revised Standard Version of the Bible) and largely interprets what it "means" as if it were a direct question to a scientist.

The verses starting with Job 38 through the first part of Job 42 are a composite of divine questions called the "whirlwind speech." In the speech, God interrogates Job concerning the workings of the Earth, its creation, and nature. Some Jewish and Christian environmentalist writers see the whirlwind speech as a "green" creation story, particularly in its representation of the human role on the Earth and its natural systems.

Despite the gravitas of pointed questions posed by the deity that made and controls the Earth, the questions from the whirlwind speech considered in *Foundations of the Earth* are direct and easily understood. They ask Job and humankind in general if they understand the workings of the planet upon which they live.

Foundations of the Earth will appeal to some readers because it discusses the environmental implications of verses from the Bible. These were important questions in their ancient context. They are certainly no less important today. *Foundations of the Earth* is framed in a religious topic, but it is simultaneously about global systems science, particularly global ecology. It presupposes no religious background or even religious interest on the part of the reader. However, readers with a religious bent should find the scientific discussions of the whirlwind speech's questions significant. Certainly, the divine questions in this text are quintessentially religious. After all, they are an account of the directly spoken word of God to a person.

I came upon the whirlwind speech when asked by a minister to find Bible verses appropriate to be read at my mother's funeral. After a longish night of compulsive searching, my eyes fell upon Job 39:26–27: "Is it by your wisdom that the hawk soars, and spreads its wings toward the south? Is it at your command that the eagle mounts up and makes its nest on high?" These two verses seemed appropriate to Kathryn Luvois Rich Shugart, an avid, keen-eyed birdwatcher with a steady hand that could beam sixteen-power binoculars on everything from small warblers to eagles, dead at ninety on November 7, 2006. I periodically reread these and the surrounding Joban verses, which I eventually learned was the whirlwind speech. The questions in these verses spun out to issues of cosmic creation, Earth's planetary function, rain, storms, birds, and mammals.

This book, *Foundations of the Earth*, responds to some of the questions in the whirlwind speech. The response comes from a scientist with a research interest in the dynamics of ecosystems attempting to comprehend and explain some remarkably good and very ancient questions. The text is annotated with many hundreds of backnotes for those who would like to take up the issues in more detail, as might be done in a college seminar or discussion group. This note format is used to give access to these sources without simultaneously chopping the narrative into bits. For those unfamiliar with the story of Job, the first chapter provides an abbreviated version of the Book of Job. The other chapters elaborate specific whirlwind questions, which provide epigraphs to each chapter as Joban quotes.

I have enjoyed a lot of help from friends in the process of developing this book. Several read the entire text through different stages of its ontogeny. My debt to them is great as is my gratitude. These stalwarts were Lyndele von Schill, my longtime friend and coworker for these many years; Ramona K. Shugart, my wife and soulmate even after forty-seven years of marriage; Rachel Most, my coteacher to anthropologists and ecologists; Erika C. Shugart, my elder daughter and talented science communicator; G. Carleton Ray and Jerry McCormick Ray, dedicated marine biologists and conservationists; Megan McGroddy, tropical ecologist and soil scientist par excellence; Alan Crowden, a remarkably helpful and consistent supporter of my writing endeavors; two broad-thinking anonymous readers that Alan managed to entice to read the manuscript; and four anonymous reviewers from Columbia University Press. Patrick Fitzgerald, Bridget Flannery-McCoy, Kathryn Schell, and Robert Fellman at Columbia University Press worked with patience, contagious enthusiasm, and professionalism, each in their own way, to push this project to a satisfactory conclusion. Thank you all!

Several colleagues took on one or more chapters and really dug in to try to help with a diverse array of technical issues, details, interpretations, and general directions: Simon Bickler, Michael Pace, Bruce Hayden, Robert Dolan, Fred Damon, Scott Schnee, Howard Epstein, David E. Smith, and Amato Evan. My special appreciation goes out to Martien Halverson-Taylor and the members of her Job class (RELJ 5559—New Course in Judaism, Department of Religious Studies, University of Virginia). Martien and her students willingly tolerated a scientific interloper untrained in ancient Greek, Aramaic, and Hebrew and were the participants in engrossing discussions on the Book of Job and its roots. Finally, thanks to friends who were willing to listen to bits and pieces of Joban lore: Bruce Hayden, Bob Dolan, Todd Scanlon, Pat Allinson, David and Cyndy Martin, Dwayne and Yvonne Osheim, Brian Walker, and many others. My own scientific research often would spill over into my thinking during writing of this book. This was supported by several grants from NASA.[2]

FOUNDATIONS OF THE EARTH

"Then the Lord answered Job out of the Whirlwind." *Source*: Plate 13 from William Blake, *Illustrations of the Book of Job Invented and Engraved by William Blake* (London: James Lahee, No. 3, Fountain Court, Strand, 1826). Collection of Robert N. Essick; Copyright © 2013 William Blake Archive. Used with permission.

I

Introduction

There was once a man in the land of Uz whose name was Job. That man
was blameless and upright, one who feared God and turned away from evil.

—Job 1:1 (NRSV)

This book treats the omnipresence of global environmental change and the role of humans as agents for this change. We are significantly altering our planet in ways that are patently obvious. When one looks from an airplane window, one sees the hand of humanity upon the land: fields where there were once forests, fragmented landscapes and grasslands that were once continuous, muddy rivers that were once clear. A walk on a beach, anywhere, reveals the flotsam and jetsam of things we have thrown in the seas. More subtle but also easily demonstrated are the changes we have wrought upon the ocean's fauna and chemistry. Similarly, it is straightforward to show that human actions have changed the chemistry of the air. We know from basic physics that a change in atmospheric dynamics follow from a change in its chemistry—namely, an increase in global temperature.

The "whirlwind speech" from the Book of Job, a set of divine questions on the functioning of the planet and its natural systems, is used in this book to provide a starting point to discuss the current knowledge of global environmental issues. The Joban whirlwind questions connect the consequences of human actions to the state of our planetary home. The global changes, which we have wrought, operate in physical, chemical, and biological domains. The poetry of Job and the power of the whirlwind speech are the skeleton of a more unified synthesis across the diverse global issues that Earth and its peoples now face.

ON JOB AND THE WHIRLWIND SPEECHES

Job's home, the biblical land of Uz, is a land of grazing pastoralists whose wealth is measured in numbers of sheep, oxen, camels, and donkeys. Regretfully, this pastoral scene is marred by murderous bands of marauding nomads, the Sabeans and the Chaldeans, who sweep across the land, burning houses, killing families, and stealing livestock. Patriarchs such as Job try to appease the wrath of God in private sacrificial ceremonies.

The Book of Job begins with a brief prose narrative describing Job's situation (Job 1–2). He is wealthy; his seven sons and three daughters enjoy reciprocal feasts at one another's houses. Job makes regular burnt offerings to God out of concern that his adult children may have sinned. In a celestial conversation with the tone of a friendly wager, God's positive opinion of Job, "There is no one like him on the earth, a blameless and upright man who fears God and turns away from evil," is countered by Satan, who asserts that Job's piety is a result of his great prosperity: "But stretch out your hand now and touch all that he has, and he will curse you to your face" (Job 1:11). Satan is given permission to take away Job's possessions.

Immediately, raiding barbarians arrive to destroy Job's wealth. Sabeans kill his servants and steal Job's oxen and donkeys; Chaldeans raid his camels and put more servants to the sword. Fire falls from the sky and destroys more servants and Job's sheep; a terrible wind blows down his eldest son's house during a feast, and all his sons and daughters perish. Nonetheless, Job is undaunted in his faith: "The Lord gave and the Lord has taken away; blessed be the name of the Lord" (Job 2:21). Satan doubles down on his wager and tells God that Job's faith will waver if he is physically afflicted: "But stretch out your hand now and touch his bone and flesh and he will curse you to your face" (Job 2:5). God grants permission for Satan to inflict infirmities on Job practically to the point of killing him.

Job is covered in sores from head to toe. His wife, who has suffered the same terrible losses, tells Job to "curse God, and die" to end his suffering.[1] As Job sits in ashes scraping his sores with a potsherd, three friends, Eliphaz the Temanite, Bildad the Shuhite, and Zophar the Namathite, gather to console him. They sit with him on the ground for seven days and seven nights without speaking. When they do begin to speak (Job 3:1), the text shifts from prose to poetry.

Job breaks the silence by cursing the day he was born. In the verses that follow, Job's three friends opine on his terrible situation, and Job responds. This continues through three cycles, with the three friends becoming progressively more insistent that Job has sinned in some way.[2] Because they believe God rewards good and punishes evil, they surmise that sin must be at the root of Job's calamitous fall. Each time Job answers that he has not sinned and that his suffering is unjustified. Eventually a younger man, Elihu son of Barachel the Buzite, begins to speak. Elihu has a different message from the three others. The creation of the world is clear testimony to God's incomprehensible power. Further, God is under no obligation to treat Job well for Job's good behavior—Job's self-righteousness does not constrain God's actions by some mortal moral imperative.

God then speaks to Job from a whirlwind. In a series of direct questions, God asks Job what he knows of the creation and the functioning of the planet. Job knows no answer to these questions and humbles himself before the Almighty: "I know that you can do all things, and that no purpose of yours can be thwarted" (Job 41:2). The writing shifts back to prose at Job 41:7. God rebukes Job's three friends. He tells them to contribute seven bulls and seven rams as a burnt offering and to ask Job to pray for God to forgive them. Job's brothers and sisters and all his friends gather to comfort him, and each of them gives him a ring and a coin. His fortune is doubled to "fourteen thousand sheep, six thousand camels, a thousand yoke of oxen, and a thousand donkeys" (Job 42:12). Seven sons and three beautiful daughters (Jemimah, Keziah, and Keren-happuch) are born to Job. He lives another 140 years and sees "his children, his children's children, four generations" (Job 42:16).

THE WHIRLWIND SPEECH

The "whirlwind speech" frames the present book, *Foundations of the Earth*. Many of the questions out of the whirlwind are posed in a straightforward, understandable manner. Others are somewhat more difficult to understand, and a few are incomprehensible because of our inability to appreciate the Joban context. The whirlwind questions largely concern the formation, functioning, and dynamics of our planet and its ecosystems. These are questions whose answers are critical to our future. We must understand the planet's function as our actions change its land surface, atmosphere, and oceans. Job professed that he did not have an answer to any of the whirlwind questions. But can we answer the whirlwind questions about the functioning of the planet any better than Job could?

That said, *Foundations of the Earth* does not simply tally some sort of "scorecard" of modern technological society versus the grazing patriarchal society of Job's day. The religious understanding found in the Book of Job operates in a different dimension from the scientific implications of questions in the whirlwind speech. For example, when God asks the first question, "Gird up your loins like a man, I will question you, and you shall declare to me. 'Where were you when I laid the foundation of the earth? Tell me if you have understanding'" (Job 38:3–4), it is posited by the same Almighty who created the Earth. Clearly there is a correct answer, which he uniquely knows. This is the case for all of the whirlwind questions. One either has the knowledge to answer the question or one does not.

Job testified that he did not know how to "lay the foundation of the earth." A modern scientist could produce the statement that a current estimate is that the Earth's age is 4.54 billion years, plus or minus 45.4

million years, based on radiometric dating.[3] The inexactitude of this response to the Joban God might well provoke a divine demonstration of why one might need to gird up one's loins when speaking to the Almighty. If Job or a modern scientist were asked to produce the *absolutely true* age of the Earth, "I don't know," would be the wisest answer from either.

Science pursues truth, but its findings are not represented as the truth. Science as a way of knowing is a process. Scientific understanding develops with observations, experiments, and analyses, but this understanding is subject to change, radical revision, or even abandonment when new information and new understanding develop. This continual evolution of scientific wisdom can be, and often is, used to discredit scientists in public, political, and legal forums.

In the current politicization of global sciences, particularly concerning the implications of a human-augmented global warming, and with tensions between cosmologists and evolutionists with some religious fundamentalists, questions on the "truth" of scientific findings often erupts as a fatuous discussion point. This is not a new tactic. Early in the evaluation of the environmental impact assessment of public works projects, lawyers for the industry asked questions of scientist witnesses such as, "Do you believe without a shadow of a doubt that the heat from this power plant will destroy the fish populations in the river?" Today, a scientist might be asked, "Do you believe in global warming?" These questions are rhetorical devices, intentional or not, that cloud scientific issues. Scientists do not deal in absolute certainties that are beyond a shadow of a doubt—they do not deal with believed truths. Belief is central to religion but is one of the weakest of arguments in support of a scientific position. The following discussions of the "answers" to the whirlwind questions are not truths. Hopefully, they are our current best understanding of the world and its functioning.

Foundations of the Earth focuses on the first two-thirds of the whirlwind speech. The last part of the whirlwind speech, which concerns the Behemoth (a giant semiaquatic creature that some see as patterned after the hippopotamus) and the Leviathan (a fire-breathing dragon of the abyssal depths that could be an amplified account of a crocodile) is not treated here. Both of these mythic creatures were created in the time of chaos. They are the playthings of God but are enormous, horrific, and dangerous to humans. The creatures are significant theologically but it is difficult to see how they relate to meaningful scientific questions.[4]

THE WHIRLWIND SPEECH AS AN ACCOUNT OF PLANETARY CREATION AND FUNCTION

What the Book of Job clearly presents in the whirlwind speech is an account of the creation of the Earth along with examples of the way it functions. Under divine interrogation, Job clearly does not pass muster in his knowledge. When compared to the whirlwind speech, other creation stories in the Bible differ significantly on the issue of the importance of humans and whether the Earth serves them or vice versa.

The best known of the seven biblical creation stories are the two accounts of creation in Genesis.[5] The "priestly" creation story in Genesis 1 describes, among other things, a wet world where the land must be separated from the sky and the seas,

> And God said, "Let there be a dome in the midst of the waters, and let it separate the waters from the waters." So God made the dome and separated the waters that were under the dome from the waters that were above the dome. And it was so. God called the dome Sky. And there was evening and there was morning, the second day. And God said, "Let the waters under the sky be gathered together into one place, and let the dry land appear." And it was so. God called the dry land Earth, and the waters that were gathered together he called Seas. And God saw that it was good.
> (Genesis 1:6–10)

The "Yahwistic account" of creation in Genesis 2 and 3 describes an arid land watered by an artisan spring:

In the day that the Lord God made the earth and the heavens, when no plant of the field was yet in the earth and no herb of the field had yet sprung up—for the Lord God had not caused it to rain upon the earth, and there was no one to till the ground; but a stream would rise from the earth, and water the whole face of the ground.

(Genesis 2:4–6)

Both of these accounts make humans the focus of creation, as does the creation account in Psalms 8:4–5, which says, "What are human beings that you are mindful of them, mortals that you care for them? Yet you have made them a little lower than God, and crowned them with glory and honour."[6] The same can be said for Psalms 104:14–15, which, while it has many parallels with the Joban creation story, sees humanity in a positive light and assures that the happiness of humankind is a divine objective: "You cause the grass to grow for the cattle, and plants for people to use, to bring forth food from the earth, and wine to gladden the human heart, oil to make the face shine, and bread to strengthen the human heart." In contrast to this, the creation story in the Book of Job sees man's position in the creation and in the resultant world to be unimportant—in fact, insignificant. The Joban creation story places humanity as mere part of a created world and certainly not as its centerpiece.

For readers interested in the religious or devotional aspects of the whirlwind questions, three recent books provide insights into these issues. *The Comforting Whirlwind: God, Job, and the Scale of Creation* by Bill McKibben provides a discussion of the environmental implications of the whirlwind speech as a paradigm for environmental living. Kathryn Schifferdecker's scholarly but readable text, *Out of the Whirlwind: Creation Theology in the Book of Job*, is a theological analysis of the whirlwind speech. William Brown's *Seven Pillars of Creation: The Bible, Science, and the Ecology of Wonder* presents the seven different creation accounts in the Bible and uses them to interweave theology and science at different scales of time and space. All three of these books identify the difference in the creation account in Job from the other biblical accounts and discuss its implications for a more environmentally conscious behavior for humans on the planet.[7]

THE ANTIQUITY OF THE JOB TEXT

We are changing our planet and our local environments. The whirlwind speech in the context of this book asks environmental questions from antiquity that have only gained in importance in modern times. How old are the questions from the whirlwind speech? How long have people been concerned with these questions? One indication of the antiquity of biblical sources comes from the words, use of words, and phraseology in a text. From these analyses, the Book of Job is generally thought to have been composed sometime in the sixth century BCE[8] or perhaps slightly later, after the Babylonian exile.[9] This said, there is a wide range of opinions by religious scholars as to the time of its composition.[10] The sixth century BCE was a dynamic time in the history of the Jewish people, with the fall of Jerusalem to King Nebuchadnezzar II of Babylon in 599 BCE,[11] the eventual capture of Judah by the Babylonian Empire,[12] and the destruction of Jerusalem and the exile of its religious and political infrastructure to Babylon.[13] With this social trauma came a time of reformulation of concepts of state and religion. The Babylonian exiles produced significant portions of the Hebrew Bible, including the Book of Job.[14]

The Book of Job reads of a different time, a time more ancient than the apparent time of its composition.[15] It was at a time when the Chaldeans, who early in the first millennium BCE formed a kingdom in what today is Yemen,[16] and the Sabeans, who became the eleventh dynasty of Babylon in the sixth century BCE, were nomadic raiders; when wealth was measured in herds and slaves; and when religion involved private sacrifice with neither a shrine nor a priest.[17] The concept that sacrifices assuage the anger of God is a primeval religious concept. Job lived to an extremely old age, another 140 years after the whirlwind speech. This is an age appropriate to the Patriarchs from the earliest books of the Bible. It is not as old as Methuselah's 969 years but is certainly older than people normally live, by a substantial margin. The use of the ancient Hebrew word קשיטה (or qesitah) for the monetary unit given to Job by his friends and relatives after his ordeal is otherwise found only in the early books of Genesis and Joshua, in connection to the story of Jacob.[18] If the temporal setting of the Book of Job is ancient, the composition of the narrative likely is more recent, perhaps 2,500 years ago.

The Book of Job also fuses elements from older religious texts from various other peoples. One obvious indication is that even though it deals with the God of Israel, no one in the Book of Job is actually an Israelite: Job is an Edomite, and his friends (Eliphaz the Temanite, Bildad the Shuhite, Zophar the Namathite, and Elihu son of Barachel the Buzite) are from locations to the east. At the story's end, Job divides his property among his daughters and sons. The Mosaic code would allocate inheritance to daughters only in cases in which there were no surviving sons; this implies some different sort of cultural milieu.[19]

The whirlwind speech has imbedded in its essential questions the elements of a creation myth.[20] This myth, as the first of the two creation stories in Genesis, is of a wet world. Lands must be separated from the waters. The seas are divinely constrained: "Or who shut up the seas with doors when it burst out of its womb?—when I made the clouds its garment, and thick darkness its swaddling band, and prescribed bounds for it, and set bars and doors, and said, 'Thus far you shall come, and no farther, and here shall your proud waves be stopped?'" (Job 38:8–11). The watery-universe creation from Genesis 1 shares elements with a Babylonian creation epic known as the *Enŭma Elish*.[21] The *Enŭma Elish* version of creation reflects the annual flooding of the Tigris and Euphrates Rivers and the resultant chaos in Babylon. It is likely that the Babylonian exiles, who shaped the story of Job, would have also been familiar with the *Enŭma Elish*.[22]

There also seem to be elements of Ugaritic legends in the Book of Job. Ugaritic, a Semitic language, is known only from a large library of clay tablets unearthed in 1928 from the ancient city of Ugarit, which was located on the shores of the Mediterranean near modern Ras Shamra in Syria.[23] Ugaritic was an unknown Semitic language thought to have become extinct in the twelfth century BCE after Ugarit's destruction. It is written using a cuneiform *adjab* (alphabet without vowels). The cryptology of translating Ugaritic challenged archeology and linguistics; it took decades. Some of the literary texts that emerged involve the religion of the Ugaritic people, who in the Bible are referred to as Canaanites. These include *The Legend of Keret*, *The Aqhat Epic* (or *Legend of Danel*), and *The Baal Cycle* (*The Myth of Baal* combined with *The Death of Baal*). Similarities with the Book of Job are seen, for example, in the use of literary permutations based on the number seven (six events happen followed by a climax on the seventh event), the number of sons and daughters of Job (and Baal), and the naming of Job's (and Baal's) daughters in the story but not the sons.[24]

The Book of Job is beautiful literature that is made more compelling by its ancient tone. The effect of the text is felt through the obscurations of time and translation. It is set in a deeper past than the already deep past in which it was written. Compiled 2,500 years ago, it dates at least another six hundred years to Ugarit and is likely much older. As one peels away time, one also peels away technology. Push the origins of some of the elements back to the time of the Canaanites, a Late Bronze Age people, and one has moved at least six hundred years into the past.[25] The Job story could have roots well beyond this;[26] one is left to wonder just how long the questions in the whirlwind speech have been on the minds of humans.

QUESTIONS TO JOB FROM THE WHIRLWIND

In this book, specific quotations from the Joban whirlwind speech (using the New Revised Standard Version of the Bible) begin each discussion of ecology and environmental change. The lone exception to this rule is in chapter 3, where both the King James Bible and the NRSV are quoted to illustrate some of the subtle effects of word choices by biblical translators. This chapter discusses the effects of the domestication of animals and plants on terrestrial ecosystems.

Chapter 2 discusses current concepts of the creation of the universe and, more specifically, the solar system. The story of whirling stellar dust clouds under the triggering event of a nearby exploding star is no less fantastic than any number of creation myths (including that in Job). The issue of how a rich array of observations implies the scientific account of creation provides a platform to discuss science as a way of thinking. Chapter 4 deals with the effect of the release of feral species and the importation of invasive animals on the Earth's biotic diversity; chapters 5 and 7 discuss the interaction among the components of the Earth systems of atmosphere, land, water, and ice on our dynamic planet. The lack of environmental constancy and the naturalness of change in the planetary history is a basic theme in both of these chapters. The physical interactions discussed in chapters 5 and 7 also provide background to the chapters that finish this book. Chapter 6 investigates the periodicities in the environment at multiple scales and the phenology and biological timing of plants and animals. Chapter 8 discusses the interaction between vegetation and climate at the global scale along with the possible consequences of human modification of climate. The simultaneous effects of climate change and human interactions with animals frame the global change–related challenge to conservation biologists treated in chapter 9. Chapter 10 discusses the possibility of human planetary engineering now being considered as a possible remedy to the negative effects of humanity's changes to our planet.

Even in the face of a long geological history of change, humankind and its systems are now a major force shaping the nature and the future of the planet. The issues in all of these chapters act in concert on our changing and human-altered planet. The whirlwind questions, posed to Job by the deity that made and controls the Earth, ask humans if they understand the workings of the planet upon which they live. These were important questions in their ancient context. They are no less important today.

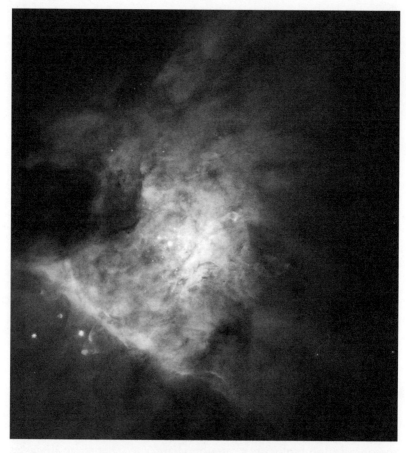

Black-and-white rendering of a color mosaic of the Orion nebula, in the constellation of Orion, from images taken with NASA's Hubble Space Telescope. Orion shows prominently in the early winter evenings at northern latitudes. The Orion nebula is visible to the naked eye and is located to the south (or north, in the southern hemisphere) of the three bright stars that form Orion's belt. Embedded in this image are at least 153 glowing embryotic solar systems called protoplanetary disks. Each of these will eventually form a system of planets orbiting a central star. Object names: Orion Nebula, M42, NGC 1976. Image Type: Astronomical. *Source*: NASA (http://hubblesite.org/ and the Hubble Heritage Team), C. R. O'Dell and S. K. Wong (Rice University). Material credited to Space Telescope Science Institute on this site was created, authored, and/or prepared for NASA under Contract NAS5-26555.

2

Laying the Foundation of the Earth

Where were you when I laid the foundation of the earth? Tell me, if you have understanding. Who determined its measurements—surely you know! Or who stretched the line upon it? On what were its bases sunk, or who laid its cornerstone when the morning stars sang together and all the heavenly beings shouted for joy? Or who shut in the sea with doors when it burst out from the womb?—when I made the clouds its garment, and thick darkness its swaddling band, and prescribed bounds for it, and set bars and doors, and said, "Thus far shall you come, and no farther, and here shall your proud waves be stopped"?

—Job 38:4–11 (NRSV)

In the so-called priestly account of the origin of the Earth found in Genesis 1:1–10, a universe of water exists, and the formation of light initiates Earth's creation. "In the beginning when God created the heavens and the earth, the earth was a formless void and darkness covered the face of the deep, while a wind from God swept over the face of the waters. Then God said, 'Let there be light'; and there was light." The water is partitioned into two parts on the second day. "And God said, 'Let there be a dome in the midst of the waters, and let it separate the waters from the waters.'" One part of this partitioned universe of waters then is separated into sky and sea. "So God made the dome and separated the waters that were under the dome from the waters that were above the dome. And it was so. God called the dome Sky." On the third day, land and sea are separated from the other beneath the dome of the sky. "God said, 'Let the waters under the sky be gathered together into one place, and let the dry land appear. And it was so. God called the dry land Earth, and the waters that were gathered together he called Seas.'" This sky-domed earth and sea sat under a sky-lined bubble in the watery universe.

The creation accounts of other cultures in the Middle Eastern region have similarities with the biblical priestly account. They postulate a creation that involves a primordial ocean often represented as a goddess. In these creation stories, this ocean deity is often somehow divided into components—the land, seas, and sky of Earth. For example, in the Babylonian creation mythology *Enūma Elish*, Marduk, the storm god, slays Tiamat, a primordial sea goddess and mother of the first generation of Babylonian deities. Marduk then divides her corpse to create the sea and the land.[1] Nearby in time and space on the Mediterranean coast of what is now Syria, Yamm, whose name in the Ugaritic Baal legend means "the Sea," must be defeated before cosmic order and subsequent creation can be established. In Sumerian mythology, Nammu, also a primordial sea goddess, gives birth to the heaven and earth. She later creates humankind.[2] In the priestly creation in Genesis and as in the Baal and *Enūma Elish* legends, a water-filled universe precedes divine creation.[3]

The creation account in Job differs from the watery priestly account and of the Ugaritic, Babylonian, and Sumerian "wet creation" accounts. In the Joban creation, an initial divine construction project with foundations and cornerstones produces land. Then the sea bursts forth from the womb. The infant sea is wrapped in clouds and swaddled in darkness. The oceanic part of creation is not a dividing event about the separation of primordial waters or the dissection of sea goddesses. It is a birth with God as a divine midwife to a powerful infant whose seething waves pulsate with tides and storms.[4] God instructs this newborn where its boundaries are and where it should not go. Joban creation involves elements of birth and childrearing, elements that match the other procreation-centered aspects in the whirlwind questions about animals that follow in the text.

THE ORIGIN OF THE EARTH

Behind all the creation accounts, biblical, scientific, or otherwise, lies an essential and difficult question: "Where did all of this come from?" The current scientific explanation is that the Earth coalesced from an eddy in a vast swirling flat disk of space dust whirling ever faster around a forming star, a star we call the Sun. This creation seems remarkable when expressed as a single statement. Indeed, it is no less fantastic than creation stories from other cultures—that the Earth was made by a water beetle bringing up the mud that then spread to form the land,[5] that it was coughed up from the stomach of a solitary god named Bumba,[6] or that it was constructed from the corpse of Ymir, a frost giant.[7] One significant aspect of the whirling-disk-of-space-dust creation account is that this process can actually be observed taking place across the universe in different stages of completion by using modern instruments such as the Spitzer and the Hubble space telescopes. A second significant aspect is that it explains some of the regular features of our solar system. What are these patterns, and what do they tell us about the origins of the major Earth systems, the land, the seas, and the atmosphere?

PATTERNS IN THE SOLAR SYSTEM

The formation of our solar system began about 4.6 billion years ago with a collapse in a cloud of celestial dust like the ones currently forming protoplanetary disks in the Orion Nebula and shown in the illustration at the front of this chapter. There are regularities in our solar system that are consistent with such a whirling disk coalescing to form the Sun and its planets. Before discussing the processes that formed our solar system, it is useful to point out some of the planetary patterns generated by the process.[8]

Regularities in the Planets

The planets in the solar system are sorted according to their composition. The planets in orbits near the Sun, called the terrestrial planets (Mercury, Venus, Earth, and Mars), are relatively small, rocky, and dense. Their atmospheres represent a minor fraction of their mass. They rotate slowly and have no or few moons. The planets that are further from the Sun, the Jovian planets (Jupiter, Saturn, Uranus, and Neptune), are large and are almost all atmosphere—they are primarily made of gases. They spin rapidly and have many moons; they also all have rings.

Regularities in Orbits

Planets do not clump together; each planet is relatively isolated in space. They are also spread with regularity. Each planet from Mercury out to Saturn (skipping the asteroid belt) is roughly twice as far from the Sun as its next inward neighbor. The orbits of the planets are ellipses, but they are close to being circles, except for Mercury's strongly eccentric orbit, which is likely caused by this innermost planet's interaction with the nearby Sun. The orbits of the planets all lie in nearly the same plane. The orbital plane for the Earth is called the ecliptic. The plane of each planet's orbit aligns with the others to within a few degrees (except for Mercury, at 7°); thus the planetary orbits in the solar system all lie on a flat disk.

Regularities in Rotations

The planets orbit the Sun in the same direction as the Sun's rotation on its axis—counterclockwise, if viewed from the direction of north on Earth. Further, the direction of the rotation of most planets is the same as the Sun's spin (counterclockwise). The exceptions are Venus, which spins in the opposite direction (retrograde), and Uranus, which has its axis tipped to lie in line with the plane of the Uranian orbit around the Sun.

Regularities in Moons

Most of the moons revolve about their planet in the same direction as their planet rotates on its axis. Some moons, such as those of Jupiter, resemble the arrangement seen in the solar system and revolve about their planet on roughly the same plane as the planet's equator.

These regular patterns are seen in the eight planets, Mercury, Venus, Earth, Mars, Jupiter, Saturn, Uranus, and Neptune, listed in their order from the Sun. Their regularities remind one of a set of Chinese carved ivory balls with an intricate nest of carved globes, one inside the other, connoting pattern within pattern created at the hand of a remarkable craftsman.

When Galileo Galilei observed the rotation of the moons around Jupiter, he realized that the observable patterns for the solar system suggested that the other planets did not rotate about the Earth, a position strongly held by the Catholic Church at the time. His initial glimpse of the remarkable order of the solar system eventually led Galileo to a sentence of heresy and a life of house arrest under the Inquisition in Rome on July 1633.[9]

———

Now demoted from its status as a planet to that of a dwarf planet, Pluto does not conform to the aforementioned planetary regularities. Its orbit is well off the plane of the eight planets (over 17° off) and is highly eccentric (noncircular). Unlike the other "gas giant" planets that lie beyond Mars's orbit, Pluto is rocky. Named in 1930, Pluto was demoted from planetary status at the twenty-fifth General Assembly of the International Astronomical Union meeting in Prague in August 2006. The assembly defined a planet as a celestial body that orbits the Sun, that is not a moon, that is massive enough that its gravity has compressed it to a nearly spherical shape, and *that has cleared the neighborhood around its orbit*. Pluto fails in this last regard, because its gravity has not captured all of the material in its vicinity. Thus, Pluto is considered a dwarf planet.

So is Ceres, a spherical asteroid and dwarf planet found in the asteroid belt. Its gravity has not cleared the ice and rock asteroids in its orbit. The asteroid belt is a ring of primordial material from the nebula that produced our solar system. Found between Mars and Jupiter, its components seem to be composed of three broad categories of material.[10] About 75 percent of the asteroids are carbon rich, 17 percent are rock-like and rich in silicates,[11] and the remaining metallic asteroids appear to be rich in iron and nickel.[12] Ceres was discovered by Giuseppe Piazzi of the Academy of Palermo, Sicily, on January 1, 1801.[13] It is made of rock (at its core) and ice. It is about 950 kilometers in diameter. More will be learned about this dwarf planet when the NASA *Dawn Discovery* mission reaches it in 2015.[14]

There are other dwarf planets, and likely more will be discovered, particularly given our increasing observational ability. Officially, the International Astronomical Union lists three dwarf planets: Ceres (the only known round asteroid), Pluto (the now-demoted planet), and Eris (a dwarf planet far from the Sun that seems a bit larger than Pluto).[15] Eris and Pluto, along with forty-two objects with a size large enough to be round (having over a 400-kilometer diameter), have been found in a region called the Kuiper belt, which is located beyond the orbit of Neptune at a distance equal to thirty to fifty times the distance from Earth to the Sun. One dwarf planet–sized object, Sedna, has been identified in the region beyond the Kuiper Belt. Mike Brown, the discoverer of Eris, reckons that the number of dwarf planets could be around two hundred for the region in the Kuiper belt but that even more, perhaps around two thousand, will be found when the region beyond the Kuiper belt is explored.[16]

The Early Formation of the Solar System and Earth

So what processes explain the patterns we see in the planets, the chemical composition of the primordial material in the asteroid belt, the Kuiper belt, and other regions, along with the collage of planets and dwarf planets and the cosmic menagerie of comets, larger comets called centaurs, meteorites, and asteroids? The starting point for the current explanation is a large cloud of stellar dust like the Orion nebula.

Stars have life cycles. They form, they grow, they ignite with internal thermonuclear reactions, they burn up their nuclear fuel, they collapse, and they explode. The patterns of these cycles are observable in the sense that we can piece together the typical star life-cycle sequence by looking at different stars at different stages of the process. The stellar explosions at the end of this cycle form vast clouds of space dust called nebulas, which are the starting point of the next cycle. The patterns are regular, but the details of the specific path and duration of a star's life cycle vary with a star's size.

The nebula that produced our Sun was a cloud of dust produced from the dying explosions of a previous generation of stars. With the stardust were gases, mostly hydrogen and helium, left from the initial creation of the universe. The solar nebula that spawned our Sun was a large rotating interstellar cloud in orbit around the center of the Milky Way galaxy. From the isotopes of elements found in meteorites, the dust cloud appears to have been relatively well mixed.[17] About 4.54 billion years ago, something caused the solar nebula to collapse from its gravity. This cause appears to have left at least some "fingerprints"—notably an unusual radioactive isotope of iron (^{60}Fe), which is produced in the fierce conditions of a supernova, a large stellar explosion.[18] Material with this isotopic signature seems to have been blown from a nearby supernova into the dust cloud that produced our solar system. The shock from that stellar explosion initiated a collapse of the dust cloud and the formation of the protostar that would become the Sun.[19]

The supernova-triggered collapse was intensified by positive feedback—the increased mass of the developing Sun increased its gravitational pull, and the increased gravity allowed the embryonic Sun to capture more material. Eventually, the Sun captured over 99 percent of the matter in our solar system. When any spinning mass is pulled toward a center, conservation of angular momentum increases the number of rotations per unit time. An example is the increased rate of rotation of a spinning figure skater as she pulls her extended arms into her body. Conservation of angular momentum caused the material spinning into the embryonic Sun to rotate at greater and greater speeds as it was drawn into the Sun. The rapidly rotating dust cloud whirled into a flattened disk spinning about the Sun's equator.

Eventually, the Sun gained enough mass that the internal pressure and heat from the gravitational compression of the material forming it ignited a thermonuclear reaction that generated heat from the fusion of hydrogen atoms into helium. This lit the Sun, and it became a star emitting sunlight and the stream of high-velocity particles, mostly electrons and protons, called the solar wind. The geomagnetic storms that can interfere with radios and communication satellites and even knock out power grids on Earth are consequences of solar wind. Solar wind also causes the tails of comets always to point away from the Sun. The dramatic aurora borealis and aurora australis (the northern and southern lights, respectively) originate through interactions between the Earth's magnetic field and the solar wind.

LAYING THE FOUNDATION OF THE EARTH

In the solar protoplanetary disk, like the disks seen today in the Orion Nebula, specks of dust bump into one another and form larger bits of dust. The bits of stardust over time collide to form grit, then grains, then pebbles, then cobbles, boulders, hills, mountains . . . After tens or hundreds of thousands of years into this process, many ten-kilometer objects spin and collide in the disk forming around the developing Sun.[20] Eventually, the growth of these ever-larger pieces of matter form embryonic planets with enough gravity to pull in materials at higher rates than could be obtained from random collisions alone. As these merge to form planets, their gravity increases, which "sweeps" other material in their orbits. The increasing gravity in an accreting planet whirls up a smaller version of the protoplanetary disk that surrounds the Sun (and of which the accreting planet is a part). Eddies in these smaller whirling disks then become moons, which orbit the planet. The moons' orbits have the same spin as the planet, which has the same spin as the direction of its own orbit around the Sun. Our planet Earth formed by this process in the protoplanetary disk of our Sun.

As the Earth's gravity pulled matter from its surroundings, its growing gravity compressed it into a sphere. The heat from this compression, along with the heat generated by the decay of radioactive isotopes and from the heat from the bombardment of meteors upon the Earth, turned the planet into a glowing molten ball. Needless to say, molten rock is hot, and many elements (for example, zinc, lead, and sodium) were vaporized by the heat. The gases in this first Earth atmosphere were whisked away by the solar wind simply because the small Earth did not have enough gravity to hold them. A similar process can be seen today, 4.5 billion years after the solar system's formation, in the ongoing loss of Venus's atmosphere.[21] Venus has a tail observable by space probes. Venus's tail, pushed by the solar wind, almost reaches out as far as Earth's orbit.

When matter is heated to high temperatures, the patterns in the wavelengths of radiation that are emitted are called a spectrum. Different elements emit different spectra. Thus, the light radiation emitted from the Sun spectrographically reveals the proportions of the elements in this gigantic repository of the primordial dust that originally formed our solar system. The chemical composition of the protoplanetary disk that formed the solar system also can be estimated from chemical analysis of meteorites and by analyzing the spectra from comets. This allows an estimate of the composition of the material that originally whirled up to become the Sun and solar system.

As it consolidated into a spherical shape, the molten Earth also began to separate into different components. The heavier molten iron sank to the center of the forming planet, and the lighter melted rock floated on top of this molten iron core. This process changed the chemical makeup to what would become the crust of the Earth. Molten iron, like more familiar liquids such as water, alcohol, or gasoline, dissolves some substances but not others. The elements that are soluble in molten iron are called siderophilic (iron-loving) elements. These, including gold, cobalt, and nickel, among many others,[22] dissolved into the iron of the molten Earth and ended up concentrated in the Earth's metallic core. Other elements, such as magnesium, calcium, aluminum, and the rare earth elements, are lithophilic (rock loving)[23] and remained in the rocky crust of the Earth.[24]

Elements determined from the spectrography of the Sun's emitted radiation can be compared to the nonvolatile, rock-loving elements found in the oldest rocks on the Earth's surface. Similar determinations can be made by analyzing meteorites that have fallen to Earth and from the spectra emitted from light reflected off of comets. The proportions of these lithophilic elements and the ratios of their isotopes in these indicate that Earth, the Sun, meteorites, and comets all derive from the same whirling stellar dust cloud. The missing elements that boiled away to be whisked off by the solar wind, as well as the other siderophilic elements dissolved in molten iron to be carried to the Earth's core, do not match—just as one would expect.

The proportions of volatile, lithophilic, and siderophilic elements (and their isotopes) in the ancient rocks of the Earth's crust provide a unique signature—a fingerprint reflecting the details of Earth's formation processes. Some rocky meteorites, thought to have been blasted into space by a Martian collision with asteroids, have a different elemental signature for these elements than do ancient Earth rocks. In addition, the ratios of the isotopes of other elements indicate that they are made from the same stardust as Earth. An ancient rock from Mars would have a different elemental signature from ancient Earth rock, because the details of a Martian rock's formation would reflect Mars's planetary formation, which would yield subtle chemical differences from an Earth rock.

Earth is the first planet from the Sun to have a moon, and Earth's moon is an unusually large one. Moon rocks brought to Earth by the U.S. *Apollo* astronauts, by the Soviet robotic *Luna* mission, and from lunar meteorites knocked from the Moon have an Earth fingerprint.[25] The Moon somehow came from the Earth's surface. How did this happen?

THE BIRTH OF THE MOON

The formation of Earth was a violent process. The "clearing its orbit of large objects" part of the definition of a planet sounds much, much gentler than the actual process, which features a violent bombardment of the forming planet by objects kilometers across hurtling in from space. As material on the protoplanetary disk coalesced, the collisions between these objects became less frequent, but the sizes of the objects became larger. About 40 million years into the formation of the solar system, the forming Earth collided with a Mars-sized planet (about half the diameter of Earth) named Theia.[26] In Greek mythology, Theia was the titan who was the mother of Selene, the Moon. This is an apt name: the collision between Theia and Earth is thought to have produced the Moon.

Theia is postulated to have overtaken the Earth's orbit from a more-or-less shared orbit. Theia hit the Earth with a glancing blow, which knocked it off kilter and gave its axis a 23.5° tilt. The collision was relatively slow in astrophysical terms but fast by normal human reckoning, around 14,400 kilometers per hour. The core of Theia would have eventually merged with the Earth's core.[27] The impact would have pulverized the surface of both planets to form a dust cloud around the Earth. Eventually this cloud would have spun up into a disk around the Earth. The Moon coalesced from this whirling dust cloud. Thus, the Moon was made of rock from the Earth's surface. This source material was depleted of iron-loving elements, which already would have been collected into the molten iron cores of Earth and Theia before their impact.

Questions remain to be answered in this theory of the Moon's origin. The surfaces of the Moon and of the Earth actually are too similar—computer simulations of the collision between Earth and Theia predict that the Moon mostly would be composed of material from the Theian surface material.[28] At the time of writing this book, two gravity-sensing GRAIL satellites are in a tandem orbit around the Moon mapping its internal structure.[29] They will map the fine-scale variations in the Moon's gravitational field to determine its internal structure. Hopefully, this mission will reveal more about the Moon's core and help us better understand the details of the Moon's origin.

A BRIEF ACCOUNT OF THE ORIGINS OF MAJOR PHYSICAL SYSTEMS OF THE EARTH

The first 700 million years of the Earth's history is called the Hadean eon, after the Greek god Hades, god of the underworld. Hades is an appropriate eponym for this period in the history of our then hellish planet. The Hadean is the interval between the agglomeration of the Earth into a great molten ball up until the end of the time of the heavy meteorite bombardments.[30] If one looks at the full moon, the signature of the so-called Late Heavy Meteorite Bombardment is easily seen on the Moon's scarred, cratered face. The geological record of this period is sparse because the crust of the planet was undergoing active change both from geological dynamics and meteorite impacts.

The Oceans and the Lands

Zircons, prism-shaped mineral crystals of zirconium silicate ($ZrSiO_4$), are among the scant surviving relics of the Hadean. Zircons have some uranium embedded in their crystalline matrix. The radioactive decay of the isotopes of this uranium produces isotopes of lead and provides the dates of the formation of zircons far back into the past. This uranium-lead clock provides what turns out to be an estimate of the minimum age of the oceans.[31] Zircons found in marine sediments from Greenland date back to 3.85 billion years ago. Even older are a few samples of zircons found in Western Australia, which appear to have been deposited in liquid water even earlier in the Hadean.[32] These imply an origin of the Earth's ocean as early as 4.4 billion years ago.[33] They also indicate a continental crust existing at the same time.[34] This ancient crust demonstrated plate tectonics, the moving of continents, whose collisions are associated with earthquakes, volcanoes, and the building of mountains and ocean trenches on modern Earth.[35]

Whence came the oceans? There has been and remains considerable scientific discussion on the source of the Earth's water. The first issue in these debates contrasts two fundamental hypotheses regarding the accumulation of water on Earth. Was Earth's water delivered at the initial accretion of the planet? Or has water continued to be delivered to the Earth right up to the present?[36] The hypothesis of continued water delivery derives from satellite measurements, which indicates a considerable influx of water from small comets.[37] This implies an ever-growing Earth ocean. If the input of water observed in the falls of "microcomets" now observed also occurred across all of Earth's history, this accumulation adds up to about three times the mass of the water found in the current oceans.

However, that there was early formation of deep ocean water is indicated by the existence of minerals (over three billion years old) that could have only formed in water over two miles deep.[38] The most accepted hypothesis is that the planetary inventory of water, which amounts to about 0.02 percent of the Earth's mass, was available soon after the Earth formed.[39] Like today's ocean, this early ocean was salty.[40]

But one still needs to explain where the early ocean water might have come from. Currently there are several theories on the origin of Earth's water. These indicate several possible sources and potential combinations of these sources. Prominent among these sources:

1. The water could have originated in the very material from which the Earth originally formed. The heat generated by the gravitational compression of primordial Earth produced a very hot, molten planet. As the heat radiated into space, the Earth cooled and an atmosphere

formed from volatile chemicals cooked out of material that made up the coalescing planet. This atmosphere apparently had enough mass to provide a surface pressure large enough to allow liquid water to exist on the Earth's surface. Recall that water boils at lower temperatures in the lower air pressures of high altitudes (and vice versa), so one needs a relatively dense atmosphere to hold water in liquid form. On the one hand, some calculations of how much water could have come from the Earth's crust indicate that not enough water would have been produced to fill the oceans.[41] On the other hand, some studies suggest that volcanoes do produce enough water and gases, indicating that volcanism could have produced both the ocean and the atmosphere. The problem is that the ratios of stable hydrogen ions in the oceans do not match those for the material thought to have whirled up originally to form the Earth, based on the ratios of hydrogen isotopes in the Sun[42] and the atmospheres of the larger planets.[43] It seems that the part of our solar system with isotope signatures that match the oceans should occur in planetesimals formed in the region of Jupiter and Saturn and in the region of the outer part of the asteroid belt.[44] This suggests the next possible source for ocean water.

2. Comets; other objects from beyond the orbit of Neptune, now the outermost planet after the demotion of Pluto; or water-rich meteorites originating in the outer reaches of the main asteroid belt could have brought water to supply the world's oceans. Measurements of the ratio of the two hydrogen isotopes[45] found in the carbon-rich chondritic asteroids resemble the ratios found in the oceans, but this is not the case for hydrogen ratios measured in comets and trans-Neptunian objects. This indicates these asteroids as a source of the ocean's water. We only really have measurements for a small number of comets (Hykutake, Hale-Bopp, and Halley).[46] If these are indicative of all the comets past and present, then comets did not contribute much water to the seas, perhaps 10 percent of the total.[47] However, if the isotope ratios of the early comets were different, then the contribution from comets could be different as well.[48]

The current dominant view is that the waters of the oceans originated from small chondritic asteroids from the outer reaches of the asteroid belt and were violently delivered near the end of the accretion of the Earth.[49]

The Atmosphere and Life

What of our atmosphere? The Earth lost its initial atmosphere to the solar winds, which stole it from the growing Earth's weak gravity. It may have lost another atmosphere after its collision with Theia. Very significantly, Earth's current atmosphere bears the signature of life. It is full of oxygen—an element that, without its constant replenishment by the photosynthesis of plants, would combine with other elements comprising the Earth's surface and be lost from the atmosphere.

With hot land surfaces and boiling seas, the hellish conditions of the Hadean eon are hostile to life as we think of it. Even when the surface cooled to an average surface temperature below the boiling point of water, the occasional large asteroid impact could generate enough heat to once again boil the oceans.[50] A logical question regarding the origin of life on Earth is to ascertain when the planet began to have the conditions that would allow life to survive. Life is fragile, but it is also tougher than one might first think. The current "record" for a lifeform surviving at high temperatures is above the boiling point of water at 122°C for a strain of microorganisms brought up from a hydrothermal vent found in the abyssal depths of the Indian Ocean and grown in a laboratory.[51] The microorganism, *Methanopyrus kandleri* strain 116, is an archaeon. Archaeons, members of domain Archaea, once were considered a form of bacteria named the archaebacteria. After further research, they seem to have a long and separate evolutionary history.[52] They also have a quite different biochemical makeup from bacteria.[53]

Archaeons are ancient microorganisms. These one-celled organisms are very simple—they do not have a cell nucleus or any other membrane-covered structures inside their cells. They are ubiquitous organisms found in a wide range of conditions[54] and may represent as much as 20 percent of the Earth's living mass.[55] Significantly for the current discussion, they can live in extremely hostile environments. Some of the first archaeons described were isolated from hot springs, salt lakes, geysers, hydrothermal vents in the deep ocean, and oil wells. A new species of the genus *Picrophilus* has been found in sulfuric geothermal springs with a pH between 0 and 2,[56] *Carnobacterium pleistocenium* were found to be alive after being frozen for 32,000 years in permafrost in Alaska,[57] and *Thermococcus alkaliphilus* can grow in strong alkali conditions with a pH of 10.5.[58] Archaeons could live on present-day Mars, and they also could have lived early in the Hades of the Hadean.

Archaeons may not have been the "last common ancestor" for all life on Earth. Indeed, life may not have even developed and evolved on Earth, although this is hard to prove and generally thought unlikely.[59] When did the Earth cool into the habitable range for life? If one expands the range of life-supporting environments to include the wide range of conditions that can be tolerated by various species of archaea, it could have been quite early in the Earth's history but also quite briefly.[60] Another complexity involves the degree to which the young Earth cooled. Based on astronomical observations of younger stars otherwise like the Sun, the early Sun should have been about 25 to 30 percent less bright than today. The eventual cooling of the Earth to balance the heat input from this dimmer young Sun should be at a temperature below freezing. The early Earth may have been warmed by the greenhouse effects of water vapor, carbon dioxide, and methane in its atmosphere or by a change in the amount of solar radiation reflected by the Earth into space. Or it may have frozen in the dim sunlight. Debate remains.

Early fossil evidence of mats of microorganisms referred to as stromolites, as well as of microscopic fossils of living organisms, date from 3.496 billion years ago.[61] Life could have developed and evolved in a hyperthermophile Eden with high-heat-loving organisms developing during a brief period, perhaps a million years in hot ocean water following an asteroid impact, or in a hydrothermal zone of the ocean that was hot but deep enough to be shielded from the effects of asteroid sea-boiling and the associated planetary steambath.[62] Or perhaps initial life was not heat loving at all. Perhaps life diversified into all sorts of environments but then was wiped out by an asteroid impact—except for the forms that had evolved to tolerate hot conditions.[63] This "Noah hypothesis" could produce multiple locations where life survived—perhaps bacteria evolving in one location and archaeons in another to found two lineages of descendants.[64] The sea-boiling collisions with asteroids could have selected life for tolerance of high heat (the Noah hypothesis) as could the hot-but-cooling Earth associated with the formation of liquid water (the Eden hypothesis). A third alternative is that life evolved on Mars, given its (at the time) slightly different and more beneficent conditions than those on Earth.[65] Bits of material knocked off from Mars by asteroid impacts could then seed the Earth with life. Of course, life has not been found on Mars, but should it be found its nature could reveal much about the development of life.

The U.S. National Aeronautics and Space Administration (NASA) planned missions to land unmanned probes on Mars starting in 1966 under the name Voyager Mars Program. The program was cancelled in 1968 but rose from its ashes as the highly successful Viking missions, which orbited Mars with satellite sensors and landed robotic rovers on the surface. The Viking mission had the objectives of mapping the Martian surface in high resolution, analyzing Mars's atmosphere, and searching for signs of past or present life.[66] From speculations as early as 1958, details of a mission to Mars and the question of how one might detect "life" in the broadest sense captured the imagination of researchers from a variety of fields. How would you detect life on another planet? James Lovelock, an inventor, sensor designer, and systems thinker, directed: "Search for the presence of compounds in the planet's atmosphere, which are incompatible on a long-term basis. For example, oxygen and hydrocarbons co-exist in Earth's atmosphere."[67] This insight led Lovelock to a global view that the biogeochemical cycles of materials at the planetary scale produce a holistic feedback system analogous to the physiological feedbacks in an organism: the Gaia hypothesis.[68]

Indeed, our air carries the signature of life.[69] It is 78 percent nitrogen, with a concentration influenced by bacteria,[70] archaeons, and more and more by human activities. It is 21 percent oxygen, the product of plant photosynthesis. Carbon dioxide, used by green plants in photosynthesis, is rare, composing only around 0.039 percent of the atmosphere. Our air has a trace amount of methane, generated by microbes. Mars and Venus, our nearest neighbors, have atmospheres that are 96.5 percent carbon dioxide in the dense atmosphere of Venus and 95 percent carbon dioxide in the thin atmosphere of Mars. Nitrogen is rare in the atmospheres of both planets (3.5 percent of the Venusian atmosphere, 2.7 percent of Mars's). Mars has 0.13 percent oxygen and a whiff of methane that does not demonstrate life's presence there but is tantalizing, nonetheless.

The signature of life has probably been writ upon Earth's atmosphere for a long time.[71] The Hadean eon ended about 3.8 billion years ago, when the Archean eon (sometimes called the Archeozooic) began. Understanding the proportions of gases in the early atmosphere requires a complicated observational, experimental, and theoretical synthesis. Significant in these determinations is reconciling the apparent warmth of the Earth despite the dim Sun.[72] The early Earth's atmosphere until well toward the end of the Archean was high in carbon dioxide but also methane, the signature of the metabolic wastes of anaerobic archaeons and bacteria.[73] The oxygen-producing photosynthesis process removes carbon dioxide from the atmosphere and replaces it with oxygen. The atmosphere also loses some of its methane from its reaction with the increased atmospheric oxygen. This removes greenhouse gases from the atmosphere and possibly induces glaciation, even a frozen tropics, on a "snowball Earth." The increase in atmospheric oxygen also promotes the formation of ozone in the upper atmosphere, which shields living cells on the Earth's surface from ultraviolet sunlight—an additional feedback system. There may have been a considerable length of time during the Archean when Earth flipped between being a snowball and a warm planet, the latter stage perhaps triggered by volcanoes and other geological events.[74]

At the end of the Archean, around 2.45 billion years ago, the forces for oxygen scored a major victory, captured the control of the makeup of the atmosphere for photosynthetic organisms, and likely threw the planet into an ice age.[75] This Earth has retained its oxygenated atmosphere despite even the occasional asteroid impact, the movement of the continents, and volcanic eruptions. If Earth has moved into a relatively stable condition with respect to oxygen in its atmosphere, even given the regular increase in the heat influx from a gradually warming Sun, the climate continues to vary in ways that strongly affect life when there are changes in the land, atmosphere, and oceans.

The interaction of these large systems will be taken up in chapters 5 and 7, which discuss the interactions of land, air, and ocean. It is important to stress here that these examples are but a glimmer of the complexity of interactions that occur on different time scales from decades, to centuries, to millennia, to millions and multiple millions of years—a marvelously complex system that comprises the clockwork of our planet's functioning.

CONCLUDING COMMENTS

The creation account in the whirlwind speech is one of seven biblical accounts of creation.[76] Each of these reflects a different aspect of the creation. Prior to the Babylonian captivity, YHWH, the God of Israel, was an example of monolatry, the worship of a single god but without the claim that it is the only god.[77] One sees this from Deuteronomy 32:7–9 (NRSV): "Remember the days of old, consider the years long past; ask your father, and he will inform you; your elders, and they will tell you. When the Most High apportioned the nations, when he divided humankind, he fixed the boundaries of the peoples according to the number of the gods; The Lord's own portion was his people, Jacob his allotted share." By the end of the Babylonian captivity, YHWH took on the images, authority, and attributes of other deities from the region. In absorbing the qualities of these other gods, Israel's God also displaced them. The former belief system of monolatry became a monotheistic system.[78] One of the significant aspects of this transition is the role of God as creator.[79] The theological importance of God as a creator is clearly evident from the first questions of the whirlwind speech, "Where were you when I laid the foundation of the earth? Tell me, if you have understanding" (Job 38:4, NRSV). At the same time, understanding creation and the Earth processes that follow is central to scientific inquiry. This is all the more so today, when the actions of humans appear as agents of planetary change. Both science and religion need to understand the process of cosmic creation. Hopefully, each can learn from the other.

In chapter 1, a point was made that belief is a weaker basis for argument in science. Our solar system's cosmic origins, when separated from the direct evidence for its plausibility, would seem as hard to imagine at face value as other alternative accounts from a diversity of other religious traditions. The scientific narrative of the origins of our solar system is a ripping yarn: supernova detonations, colossal cosmic vortices, thermonuclear star ignition under the crush of gravity, and interplanetary collision. It is a tale with a rich filigree of details: rare isotopes of elements that can only be cooked in an exploding star's death, symmetries in planetary orbits, atmospheres forming and boiling away, celestial organization arising from seeming chaos. It is still a tale not completely told, and there is much more to be learned. If humans are the focus of this particular story, they show up very late in the last paragraph of the last chapter of a very long epic, an epic that beggars belief.

Certainly, we live in an exceedingly exciting time for astronomers, cosmologists, and astrophysicists. New technologies (satellites, space missions, and the capability to measure the isotopes of elements) meld with innovative theorists and clever interpreters of observational data. This synthesis produces a heady mixture of creativity, argument, and discovery. There are riddles to be solved and questions to be asked. And this is essentially what science is about.

The Unicorn in Captivity tapestry (1495–1505). Wool warp, wool, silk, silver, and gilt wefts. 368×251.5 cm. South Netherlandish. Gift of John D. Rockefeller Jr., 1937 (37.80.6). The Metropolitan Museum of Art, New York, NY. *Source*: Image copyright © The Metropolitan Museum of Art. Art Resource, NY.

3

Taming the Unicorn, Yoking the Aurochs

Animal and Plant Domestication and the Consequent Alteration of the Surface of the Earth

Will the unicorn be willing to serve thee, or abide by thy crib? Canst thou bind the unicorn with his band in the furrow? or will he harrow the valleys after thee? Wilt thou trust him, because his strength is great? or wilt thou leave thy labour to him?

—Job 39:9–11 (King James Bible)

Is the wild ox willing to serve you? Will it spend the night at your crib? Can you tie it in the furrow with ropes, or will it harrow the valleys after you? Will you depend upon it because its strength is great, and will you hand over your labor to it?

—Job 39:9–11 (NRSV)

S o what is it, the unicorn or the wild ox? These two translations of Job 39:9 differ as to the creature that is the topic of God's question from the whirlwind. These two creatures, one now extinct and one that never existed, are used by translators for the Hebrew word ראם (*re'em*, a wild, untamable animal) in the Hebrew Job text. Unicorn is also translated from *re'em* in the Greek Septuagint (as *monokeros*) and in the Latin Vulgate (as *unicornis*).[1] The unicorn and the wild ox are part of God's question to Job, "Do you know how to domesticate fierce, wild animals?" The unicorn and the wild ox illustrate different facets of this question. These creatures are discussed below, followed by a narrative on the large topic of plant and animal domestication, and then of the eventual environmental consequences of such domestication. The larger issue here really is not the proper translation of *re'em*; rather, it is, "How are animals domesticated, and what are the implications of such domestication for human activity?" But first, a discussion of the creatures, the unicorn and the aurochs.

THE UNICORN

Nowadays, we relegate the unicorn to imaginary or mythological status, but it was originally reported as a real animal by Ctesias of Cnidus (fifth century BCE), the physician to Artaxerxes Mnemon (Xerxes) in Persia. Ctesias enthusiastically reported on his experiences in Persia in his book *Persica*. He also compiled what was reported to him about India in his book *Indica*.[2] The unicorn, as described by Ctesias in *Indica*, was a one-horned wild ass, a creature extant in India.[3]

> In India there are wild asses as large as horses, or even larger. Their body is white, their head dark red, their eyes bluish, and they have a horn in their forehead about a cubit in length. The lower part of the horn, for about two palms distance from the forehead, is quite white, the middle is black, the upper part, which terminates in a point, is a very flaming red. Those who drink out of cups made from it are proof against convulsions, epilepsy, and even poison, provided that before or after having taken it they drink some wine or water or other liquid out of these cups.[4]

It is thought Ctesias's unicorn was derived from a garbled description of the rhinoceros. Nonetheless, a century later, Megasthenes, who actually traveled to India, reported "horses with a single horn, and a head like a deer" found there.[5] Probably elaborating from these more ancient chroniclers, Aristotle, Strabo, and Pliny the Elder, among others, reported the existence of the unicorn and its complex attributes.

With the endorsement of famed ancient scholars, the unicorn became part of the repertoire of medieval creatures. It was described in the *Physiologus*,[6] a Greek text from fourth-century CE Alexandria. The oldest surviving versions are in Latin. The *Physiologus* collected ancient animal stories and other accounts from Aristotle, Pliny, and other natural philosophers into forty-eight or forty-nine chapters. The beasts, birds, and stones described in the chapters illustrated aspects of early Christian dogma. The *Physiologus* was copied; translated into Ethiopian, Armenian, Syrian, and Latin; retold; and elaborated over the next several centuries. Most of these copies were lost, but several bestiaries based on the *Physiologus* were developed in the twelfth century.[7] Some of these still exist. One is the *Aberdeen Bestiary*.[8] The book was part of the Old Royal Library, a library assembled at Westminster Palace in 1542 by Henry VIII. It now resides in Aberdeen University's library. Along with an illustration of a unicorn (the *monoceros*), the Aberdeen Bestiary describes:

> The monoceros is a monster with a horrible bellow, the body of a horse, the feet of an elephant and a tail very like that of a deer. A magnificent, marvellous horn projects from the middle of its forehead, four feet in length, so sharp that whatever it strikes is easily pierced with the blow. No living monoceros has ever come into man's hands, and while it can be killed, it cannot be captured.

Of course, one of the most celebrated medieval unicorn legends is the concept that the unicorn can only be captured by a young woman. Leonardo da Vinci sketched a *Young Woman Seated in a Landscape with a Unicorn*, now housed in the Ashmolean Museum at the University of Oxford. The sketch depicts a young woman holding the leash of a quieted unicorn on the ground near her side. Leonardo's unicorn is a smallish animal, perhaps the size of a donkey, with an equine face and a straight horn with a spiral twist. Leonardo wrote of the unicorn in his notebooks, "Because of its intemperance, not knowing how to control itself before the delight it feels towards maidens, forgets its ferocity and wildness, and casting aside all fear it will go up to the seated maiden and sleep in her lap, and thus the hunter takes it."[9]

With complex elaborations, the hunt for the unicorn is the theme of several series of tapestries from the late sixteenth century. The central theme from this great collection of unicorn lore is that the creature is untamable or that it can only be tamed by special, mystical procedures. These latter procedures involve the unique power of virgins over the otherwise untamable unicorn and serve as allegories of the Virgin Mary, Mother of Jesus.

The accounts of ancient authorities and the incorporation of the unicorn into spiritual teaching and striking religious art would help convince the translator that the unicorn is, indeed, an excellent word for the Hebrew re'em. Additionally, the existence of what seemed to be actual "unicorn horns" would have been a compelling line of evidence that would incline a medieval translator (or the later translator of the King James Version) to use "unicorn" for re'em.

Unicorn horns were part of the power paraphernalia of royal dynasties and popes. These "horns" were the long (two meters or more), straight, spiral tusks of a small whale called the narwhal, *Monodon monoceros*. The males of the species have a single ivory tusk, which actually is an elongated left tooth.[10] Thus, the scientific name of the species translates to "one-tooth one-horn." The trading pathway of these helical ivory tusks from the seas above the Arctic Circle through Viking intermediaries to the courts of Europe—as well as the derring-do of the medieval confidence men who were bold enough to sell these remarkable objects to the medieval power elite—is the stuff of an action movie. Narwhal tusks are part of the royal jewels of several nations. A narwhal tusk still resides at the Schatzkammer (Imperial Treasury) in the Hapsburg's Hofburg Palace in Vienna. King Christian V of Denmark was anointed in 1670 on a throne made of narwhal tusks.

The unicorn is commonly displayed in coats of arms and other heraldic emblems. Two unicorns stand as the supporters of the Royal Coat of Arms of Scotland.[11] The beasts are leashed with golden chains, indicating that the creatures cannot be tamed and, by extension, neither can Scotland. In these representations, as in the Scottish coat of arms, the unicorn is horselike, with a long spiraled single horn, a goat's beard, a lion's tail, and cloven hooves. Cloven hooves would zoologically remove this representation of the unicorn from any close taxonomic relationship with the single-hoofed horses. It would similarly remove it from any close taxonomic relation with the rhinoceros, which often is posited as the source animal for the unicorn's description.[12]

Through its long and fictitious history, the unicorn represents an untamed and untamable creature. Medieval legends notwithstanding, it symbolizes an intrinsically wild thing that cannot be made otherwise.

THE WILD OX

The second creature used by translators for the *re'em* is the wild ox or aurochs (*Bos primigenus*), the species from which domestic cattle were derived. The aurochs stood as much as 180 centimeters tall at the shoulder and weighed a thousand kilograms.[13] The males were black-brown with a pale stripe down the spine. The females and calves were reddish-brown. The long aurochs horns pointed forward and were lyre shaped—going outward at the base and then curving inward at the tips. Evolutionarily, the aurochs originated in what is now India in the Pleistocene epoch between 1.5 and two million years ago.[14] The Pleistocene began about two million years ago and featured an alternation of intense glacial periods ("ice ages") with relatively milder interglacials. We are living in an interglacial we call the Holocene epoch, which began about 12,000 years ago.

Aurochs spread through Asia and eventually to northern Africa and Europe.[15] The species' range contracted southward during geological intervals of more intense glaciation. Aurochs were large, mean-tempered, dangerous beasts. These were the stereotypic attributes of the aurochs perhaps well into prehistory. The beautiful Paleolithic cave paintings in Lascaux and Liveron, France, estimated to be 17,300 years old, are illustrated with striking images of aurochs. The Hall of Bulls in the Lascaux caves features paintings of four black aurochs, including one that is 5.2 meters long.[16] It is inappropriate to speculate on the "meaning" of the aurochs images at Lascaux and Liveron, but we do have a long history of the aurochs as a symbol of power. This symbolic imagery certainly sharpens as the time reference becomes closer.

A striking piece of ancient architecture, the Ishtar Gate, named for the Babylonian goddess Ishtar, has a blue-tile mosaic incorporating alternate rows of dragons and aurochs in bas-relief. This beautiful part of the Ishtar Gate, fourteen by thirty meters, has been reconstructed and is displayed in the Pergamon Museum in Berlin. The Ishtar Gate, the eighth gate to the inner city of Babylon, was built about 575 BCE— the time of Nebuchadnezzar II and of the Babylonian captivity of the people of Judah. This is also contemporaneous with the compilation of the Book of Job during the Babylonian captivity.[17] This reinforces the wild ox as an appropriate translation for the Hebrew *re'em*, as does the possible cognate *rimu*, Assyrian for wild ox.

The fierceness of the aurochs made them a trophy animal for hunters wishing to demonstrate their prowess against a worthy foe. This practice appears to have been widespread, and there is a vivid account from the first century BCE in Julius Caesar's *Gallic Wars* (6:28):

> These are a little below the elephant in size, and of the appearance, color, and shape of a bull. Their strength and speed are extraordinary; they spare neither man nor wild beast which they have espied. These the Germans take with much pains in pits and kill them. The young men harden themselves with this exercise, and practice themselves in this kind of hunting, and those who have slain the greatest number of them, having produced the horns in public, to serve as evidence, receive great praise. But not even when taken very young can they be rendered familiar to men and tamed. The size, shape, and appearance of their horns differ much from the horns of our oxen. These they anxiously seek after, and bind at the tips with silver, and use as cups at their most sumptuous entertainment.[18]

Prowess demonstrated by success in hunting fierce beasts such as aurochs continued through medieval times in Europe. One might argue that today's hunting clubs or safaris to Africa continue this historically deep tradition into the present.

For the aurochs, hunting pressure with habitat change and the transfer of diseases from domesticated cattle restricted the range of the animal to eastern Europe by the thirteenth century.[19] Even with the eventual cessation of aurochs hunting by noblemen and attempts to conserve the aurochs, the species remained in irreversible decline. The last aurochs, a female, died in the Jaktorów Forest in Poland in 1627. Even though it is extinct, the aurochs remains a symbol of power. It serves in this context as in the national symbols of Romania and Moldavia as well as in the coats of arms of several European provinces and municipalities.

Aurochs are the source animals for domestic cattle. It is hard to imagine the docility seen in a modern Holstein dairy cow emerging from the menacing aurochs, although the Holstein bull remains a formidable animal. If the *re'em* in the whirlwind question is the wild ox or the aurochs, then the answer to the whirlwind question should be that people can and have domesticated the *re'em*. Job, although he owned five hundred yoke of oxen (Job 1:3), apparently did not know this (Job 40:3–5). One must sympathize with Job on this issue. It is quite surprising that a belligerent, powerful creature such as the aurochs could be tamed at all. This is also the case for the next topic animal. How was the first domesticated animal, the dog, derived from a pack-hunting fierce predator, the wolf?

THE DOMESTICATION OF ANIMALS

The unicorn illustrates that there are some animals that intrinsically cannot be domesticated. The other member of the pair of animals we have discussed, the aurochs, reflects that some remarkably fierce creatures can be tamed and domesticated. Francis Dalton, Charles Darwin's cousin, wrote an early essay on domestication and summarized its necessary conditions. His list of conditions is quoted here and is followed by a more modern disambiguation derived from the work of Juliet Clutton-Brock:

"1, they should be hardy"—"The young animal has to be physically tough enough to survive the trauma of removal from its own mother (probably before it is weaned). Further, it must be hardy enough to adapt to a new diet and environment." "2, they should have an inborn liking of man;"—"The species' behavior must match human social structure. This implies a social animal with dominance hierarchies that will accept humans as leaders and that will remain imprinted on humans for life." "3, they should be comfort loving"—"The animals should not bolt into instant flight (like antelopes or deer). They should be amenable to close contact, and to being penned or herded together." "4, they should be found useful to the savages"—"They should be useful to humans in some way. For many captive animals, this stems from their role as an easily maintained, non-spoiling source of food that can supply meat when required." "5, they should breed freely"—"The breeding behaviors of the animals should not require complex or hard-to-create conditions." "6, they should be easy to tend."—"Particularly for herded animals, they should be reasonably placid, versatile in their feeding habits and gregarious so that herds will stay together to be herded."[20]

For Galton, animal domestication results from symbiotic interactions between humans and the to-be-domesticated species. One element of this symbiosis includes the companionship afforded by pets.

THE DOG AS THE FIRST DOMESTICATED ANIMAL

The first domesticated animal was the dog (*Canis familiaris*), which from several different lines of evidence was obtained from the wolf (*Canis lupus*) at least 15,000 years ago and perhaps but debatably earlier.[21] The domestication of tamed wolves did not have as much of the intrinsic nature of the animal to overcome as the domestication of some other animal might have had. Wolves conform well to Galton's conditions, but most animals do not. But how could wolves have been tamed, given their ferocious natural state? Wolf taming would seem a risky task for anyone to undertake, particularly the first time it was tried.

Taming the Wolf

Some have conjectured that wolves might have been tamed by becoming accustomed to humans because they were cleaning up after the kills made by human hunters or scavenging in Stone Age human waste dumps.[22] One certainly finds ample supplies of feral dogs associated with the refuse heaps of modern cities, but these are domestic dogs transitioning back to a wild state. Domestication works in the opposite direction—from wild to tame to domesticated.

An alternate method of wolf taming is revealed in historical as well as recent records. The first are from accounts from explorers and travelers; the latter from reports of anthropologists. As an initial example of an early anthropological account of animal taming consider the report on the "Guiana Indians" living in the upper reaches of the rivers of British Guiana and reported by W. E. Roth.[23] This compilation includes a remarkable inventory of how to make the crafts and material of these people, ranging from how to weave a hammock, ink tattoos, build an alligator hook, and make patterned baskets. Providing a hint of his thoroughness, Roth described over seventy "cat's cradles" (patterns made of a loop of string with the fingers) and their transitional sequences used to illustrate stories and games. His chapter on the topic "Animals under domestication and captivity" begins: "Women will often suckle young mammals just as they would their own children; e.g. dog, monkey, oppossum-rat, labba, acouri, deer, and few, if any are the vertebrate animals which the Indians have not succeeded in taming."[24]

Roth's report of women nursing baby animals in the hinterlands of British Guiana is not a one-off, weird observation from the wilds of the South American rainforest. There are legions of such cases collected from all over the world for a range of indigenous peoples. Suckling baby animals is sufficiently alien to a European explorer's expectation that one would expect it to be reported. Observations come from all the populated continents and Oceania. Along with pigs, bears, and menagerie of other creatures, puppies receive frequent mention in these accounts.[25]

One can easily appreciate that nursing baby animals would reduce their fear of humans, behaviorally imprint them on humans, and ultimately tame them. Hunting people in their pursuit of food encounter baby animals, often after killing their mother in her lair. When these babies are brought back to the hunters' base camp, the nursing and feeding of the pups, along with the inevitable play with young children afterward, is widespread as a method of taming wild animals for companionship and potentially for food in difficult times.[26]

Is this the first step in the transformation of the wolf into the dog? The interrelation between Australian Aboriginal people and the Australian dingo serves as a useful example in a hunting-and-gathering society somewhere between the taming process and the ultimate domestication of dogs.

The Australian dingo (*Canis lupus dingoensis*) is classified as a subspecies of the wolf, even though it is often referred to as the aboriginal "dog." To tame dingoes, aboriginal women suckle dingo pups removed from their mothers. From observations of wildlife biologists working with dingoes, pups taken from their mothers when their eyes are still shut (up to thirteen days old) and nursed with milk from eyedroppers imprinted on humans and grew up to become good pets. In the same study, dingo pups taken from their mothers only a few days older, but after their eyes had opened, were all difficult or even impossible to train.[27]

The dingo is a relative newcomer to Australia.[28] Human habitation in Australia dates from at least forty to sixty thousand years ago and is possibly even earlier, but archaeological evidence of dingoes shows up only about four thousand years ago.[29] The skull structure of dingoes resembles that of South Indian pariah dogs and Thai dogs.[30] Their skull measurements overlap in these same skull measurements with wolves, particularly Asian wolf subspecies (the Indian wolf, *Canis lupus pallipes*; and the Arab wolf, *C. lupus arabs*).[31]

The dingo's arrival to Australia was almost indubitably as passengers on boats from a location in southern Asia. The people who supplied the dingo to Australia were probably the Lapita people, seafarers who originated somewhere near Taiwan about six thousand years ago.[32] Based on the practices of the voyaging Polynesians, a later group of ocean-exploring people whose languages imply they were derived from the Lapita, the dingo may have been on board as a portable food item on its boat trip to Australia.[33]

Studies on the genetics of dogs place the dingo in a group with several older dog breeds.[34] These include ancient breeds such as the New Guinea singing dog, the African basenji, and the greyhound. Also grouped with the dingo by this analysis are a number of more modern breeds of dogs. Subsequent studies support a close relation between the dingo and Asian dogs and imply a separation of the dingo from the Asian parent stock at approximately six thousand years ago. Given the variability in such estimates, this date is consistent with the archeological evidence of the species' presence.[35] At the about same time in Australian prehistory one finds of remains of dingoes, and contemporaneously there is a widespread change in aboriginal weapons (for example, a switch to the use of smaller spear points) and a disappearance of the largest marsupial carnivores, *Thylacinus* (sometimes called the Tasmanian wolf or Tasmanian tiger) and *Sarcophilus* (the Tasmanian devil) from mainland Australia.

The dingo provides an example how the process of converting wolves to dogs might proceed. The suckling of baby animals, including wolf pups, has ample representation as a human practice from cultures all over the word and represents an effective method of animal taming. The Australian aboriginal example is significant because the imprinting and taming of the animals by breast feeding the pups is still part of the bonding process between animal and human. When this bond is established, dingoes function much like domestic dogs. Today, the greater convenience of the dog versus the dingo has produced a shift in favor of dogs for Aboriginal people. Also, because of their hybridization with dogs, purebred dingoes are disappearing.[36]

TURNING TAME WOLVES INTO DOGS

If nursing the pups provides a way to tame wolves, the other necessary step for domestication involves the selecting of the wolf's genes to produce more docile animals that eventually become dogs. How do tamed animals differ from their domesticated relatives? Tamed animals are dependent on people and are willing stay either with or close to them. Domesticated animals are raised in captivity. Humans control the species' breeding, territorial organization, and food supply.[37] The control of the animals' breeding is the key in the domestication process.

Consider the hypothetical case in which a human hunting group manages to tame wolf pups. When hard times reduce the food supply, a selection process arises if the less docile members of the tamed wolves are the first to be put on the menu. If the people in this hypothetical case can carry the tamed wolves over several generations, then human control of the animals' breeding and genetic destiny becomes an artificial selection process.

The phrase "artificial selection" was used by Darwin to describe processes that produced domesticated animals and plants and also to indicate the power of natural selection to produce new species:

> Slow though the process of selection may be, if feeble man can do much by his powers of artificial selection, I can see no limit to the amount of change, to the beauty and infinite complexity of the coadaptations between all organic beings, one with another and with their physical conditions of life, which may be effected in the long course of time by nature's power of selection.[38]

Although artificial selection was described by Darwin, it is a concept with deep historical roots.[39]

An intriguing long-term experiment on the results of selection for tameness and against aggression in foxes (*Vulpes vulpes*) was developed in Novosibirsk, Russia, in 1959 by Dmitry K. Belyaev.[40] The larger intent of the study was to understand better the process of dog domestication using another canid species, the fox. One aspect of the experiment was to observe the change in other physical features in animals that had been selected only for tameness. There are several features that domesticated animals share: the development of dwarf and giant varieties, piebald coat color, wavy or curly hair, shortened tails and fewer vertebra, floppy ears, and changes in the reproductive cycles of the animals. Belyaev spent the last twenty-six years of his life on this experiment, and his colleagues have continued up to the present.[41]

Belyaev's experiment bred silver foxes (a fox variety used in the fur trade) and then selected successive generations of offspring for tameness. The foxes spent their time in cages and were only allowed brief contact with humans. When they reached the age of sexual maturity at about seven or eight months, the foxes were graded on their friendliness toward humans based on a standard scoring system. For each new generation of foxes, 4 or 5 percent of the tamest males and 20 percent of the tamest females were allowed to breed to produce the next generation.

The tamest class of foxes in the scoring system, the "elite domesticated" class, were animals that were eager to establish human contact. They whimpered to attract human attention, wagged their tails, and licked the experimenters, just as dogs might. Ten fox generations into the experiment, the domesticated elites made up 18 percent of the population, by the twentieth generation 35 percent, and after forty years about 70 to 80 percent of the fox population.[42] Other changes also were occurring with the elites: reduction in the fear response, floppy ears, shorter legs, rolled tails, shorter tails. The genetics of these traits, which also show up in other domesticated animals, and the selection for behavior (tameness) are still being explored.[43]

The foxes from this study were selected favoring the tamest individuals. The results after a surprisingly short period of time are "domesticated" foxes that enjoy human companionship. These domesticated foxes are enthusiastically affectionate toward people. Videos of them playing with people are fun to watch and bring a smile to the observer. The affectionate creatures resulting from this study are currently being sold as luxury pets on the Internet.

One of the results of Belyaev's study was an illustration of how fast domestication could develop. Selection toward docile tamed animals by humans eliminating the fiercer ones in a breeding population of tamed wolves should work in a similar fashion to produce the dog.

Dog Domestication

Several different lines of evidence indicate that dogs were the first animals domesticated by humans, probably more than 15,000 years ago. This time is indicated from three rather different lines of analysis. First, there are several examples of fossil dog bones found in conjunction with human settlements. Early dates for fossil dogs include the earliest date from Germany from 14,000 years ago and from Israel and a cave in Iraq thought to be around 12,000 years old.[44] There are other finds of dogs that are only slightly later than this date from widely dispersed locations, including North America.[45]

The use of nuclear DNA in the blood, semen, or saliva of perpetrators in criminal proceedings for identification purposes is commonly reported in the news. Similar application of DNA technology determines the relatedness of animal species and produces a second line of evidence on the time of origin of dogs. Mutations cause variations in the DNA or the genes of animals and plants. These genetic mutations accumulate over generations, and the DNA of two related species becomes progressively more different. This provides an evolutionary "clock." The greater the numbers of different mutations, the longer the time since two species were derived from a common ancestor. By such analyses, the domestic dog appears to have derived from the wolf about 15,000 years ago. Given the precision of such calculations, this estimate matches the archeological evidence.

Further insight comes from a third line of evidence involving the analysis of a special type of DNA called mitochondrial DNA. Mitochondria are organelles inside cells that drive the cells' energy metabolism. Like the nucleus of the cell, they also contain DNA. Unlike nuclear DNA, half of which is from the organism's mother and the other half is from its father, mitochondrial DNA is passed from generation to generation *only* from the mother—egg cells have it; sperm cells don't. The mitochondrial DNA's mutations also can be used as a genetic "clock." This clock counts the time since two individuals' ancestors shared a common mother.

A recent mitochondrial DNA study by Jun-Feng Pang and a large team of molecular geneticists inspected the genetics of well over a thousand dogs.[46] Data analyses produced results consistent with the dog arising as a domesticated animal from a location in southern China from multiple tamed wolves (perhaps several hundred) sometime between 11,500 to 16,300 years ago. A core of genetic material appears to be shared by all dogs. Dog population genetic diversity drops away as one moves away from a region of China south of the Yangtze, which has the maximum genetic diversity found. When "founding populations" with a subset of the total population of genetic diversity spread out from the location of origin, some of the genes are lost, and the genetic diversity decreases. This is same rationale is applied in human genetics to support an "out of Africa" hypothesis for human dispersal from the genetically diverse Africa.

The most straightforward explanation of this Chinacentric pattern of dog mitochondrial DNA would be that there was a domestication "event" involving several different animals in China south of the Yangtze River, the place of the dog's origin. The combination of the genes in the mitochondrial DNA implies that this domestication of wolves involved at least fifty-one females. Presumably somewhere in the region of China south of the Yangtze River, a tribe or tribes of sedentary hunter-gathers (or early agricultural) peoples traded the wolves they had tamed. Eventually, dogs were domesticated from these tamed wolves. Pang and his colleagues speculate that the antiquity of the still current regional practice of eating dogs implies that the first dogs were primarily domesticated as a meat animal.[47]

Much remains to be learned about the evolution of the dog as a product of wolf domestication. The determination of the timing of the initial domestication of dogs remains a scientific challenge to unravel.[48] However, it is clear that dogs were the first domesticated

animal and were likely domesticated by a hunting and gathering people, probably in East Asia sometime in the late Pleistocene around
15,000 years ago.

YOKING THE AUROCHS

If the first domesticated animal, the dog, was initially tamed from the
wolf by the succor of women, how might the aurochs, the wild ox from
Job, be tamed? Given the strength of the animal and its potential to
harm its captors, it is difficult to understand the utility of a semitamed
aurochs to a Neolithic people. The logistics of containing potentially
large creatures such as an aurochs immediately implies a more settled
situation than one normally expects of the hunter-gather lifestyle—it is
enough of a challenge to contain the creatures in the first place; moving
them about with a change of hunting grounds is a substantial additional
difficulty. The taming of the aurochs, as well as a number of other large
grazing and browsing animals, developed along with the invention of
crop agriculture in the Fertile Crescent region of western Asia.[49]

The Fertile Crescent was a region of rich soils and complex topography that arched from Mesopotamia (with the watersheds of the lower
Tigris and Euphrates rivers) to the Levant (modern Israel, Palestine,
and the western part of Jordan). It was bounded on the south by the
Syrian Desert and the Anatolian highlands to the north. On a modern
map, the Fertile Crescent region would include Iraq, the western edge
of Iran, Syria, the southeastern edge of Turkey, Jordan, Lebanon, and
Israel. Ecologically, this was a region with a rich mixture of plant and
animal species from both Africa and Asia. This biotic diversity con-
tained the progenitors for early crops, which would eventually sustain
complex human cultures: emmer wheat (*Triticum dicoccum*), einkorn
(*T. monococcum*), barley (*Hordeum vulgare*), flax (*Linum usitatissi-
mum*), the chickpea (*Cicer arietinum*), pea (*C. arietinum*), lentil (*Lens
culinaris*), and bitter vetch (*Vicia ervilia*).

The Fertile Crescent was also in a zone of overlap of the ranges of
the wild progenitors of the four of the five important animals that

compose the Western domesticated animal "kit" (cows, goats, sheep, and pigs). The fifth member of this kit, the horse, was found nearby. In parts of the Fertile Crescent, the ecosystem's productivity was great enough that hunting-and-gathering societies were able to prosper and eventually construct cities. This more sedentary lifestyle allowed plant domestication and the domestication of the meat animals of Western civilization—goats, sheep, pigs, and the topic of this section, cattle.

One proposed mode of aurochs taming would maintain wild herds of aurochs near a village by providing strategic supplies of salt (or water) to the herd. Plant protoplasm differs from that of animals by the virtual absence of sodium in most plants. Large herbivores excrete large amounts of sodium but feed on plants that have very little sodium in their tissue. To replace the excreted sodium, these animals are drawn to any available supply of salt. For a living example of this practice from the Assam Hills of northern India, free-ranging bovids called mithans (*Bos frontalis*), a domesticated wild gaur (*Bos gaurus*), can be entreated to take salt from people's hands.[50] The resultant tamed creatures are not used as food or for labor. Instead, they are used in ritual sacrifices—usually with a sequence of increasingly larger sacrificial animals, building to the crescendo of a mithan as the ultimate sacrifice. Mithan are also bartered as part of the bride price by the hill tribes of Assam.

As an intermediate stage of the process, the mithan represents a potential model of the domestication of aurochs (much as the dingo was proposed as an intermediate case of wolf–dog domestication). Such a practice on aurochs would provide an available herd close at hand should a meat supply be needed. Further, if the mithan can be taken as a broader model, one expects that the animals were also used for ritual practices.[51] Such appears to be the case in early archaeological sites of people who had cattle. Eventually, aurochs taming would involve rearing young aurochs calves in some sort of enclosed conditions. The first available sign of domestication in aurochs is smaller animals. From Neolithic times to the Iron Age, cattle shrunk to only a meter tall at the withers (the area between the shoulders). Today, these sizes are found in dwarf breeds.[52]

The time of the "first" domesticated population of a species is intrinsically difficult to ascertain in the spectrum of hunting and capturing, taming and domestication.[53] There is some evidence of domesticated cattle from the oases of western Egypt as early as 7700 BCE.[54] A significant and very early archeological record of domestic cattle comes from the site of a settlement in Anatolia (in modern Turkey) called Çatal Hüyük,[55] an ancient city with of population of around five thousand to six thousand people.[56]

Located on the bed of a dried-out ancient lake, the Konya Plain in central Anatolia, Çatal Hüyük survived for a millennium after its origin between eight and nine thousand years ago, from the middle of the seventh millennium BCE. Cattle bones from the site date from 5800 BCE, and there is the strong likelihood that cattle were domesticated animals there five hundred years earlier than that. Joining these dates would place domestic cattle in Anatolia originating at about 8,300 years ago.

Çatal Hüyük was a remarkable place, particularly for its time. It is the largest known preliterate site in Asia. It had a complex culture that compares to other Near East centers of developing civilization. Ongoing investigations for several decades have revealed much of the life of the city. Their diet resembled a shopping list from a modern health-food store: grains (naked six-row barley, emmer wheat, einkorn, along with other grains), legumes (peas and vetch), nuts (pistachios, acorns, and almonds), fruits and berries (apple, blackberry, and juniper), seeds of crucifers for oil, and leafy herbs. From the bones left in their rubbish, they were a cattle-raising people who also owned dogs. Ninety percent of the meat in their diet was from cattle.[57] For the other 10 percent of their meat protein, they hunted wild animals, mostly wild sheep and onagers.[58] While the town was centered on cattle raising, many of the frescos and molded figurines from the city featured leopards and leopard-skin-draped figures (no leopard bones have been found in the city) and images of red deer hunting (very rarely found among the bones left in the city).

Çatal Hüyük was a large, densely inhabited city with spectacular painted and incised mud-brick walls. It was filled with young people.

The life expectancy for males was slightly over thirty-four years and for women slightly under thirty years. There were many adult women, over 60 percent of the population being female. It was a trading city for obsidian and animal hides, along with other goods such as baskets, bowls, textiles, and foodstuffs.[59] Indeed, the variability in the appearance of the city's inhabitants indicates the mixture of people that might be expected in a trade center. Analysis of the skeletons of the inhabitants indicated that they were strong, limber, and able to climb and bend easily.[60]

In this vibrant place inhabited by a diverse people with artistic creativity and trading acumen, there was darker side to life: periodic outbreaks of the deadliest form of malaria—cerebral malaria caused by the mosquito-dispersed protozoan *Plasmodium falciparum*. Presence of malaria selects the human population for genetically based malaria resistance, but this resistance has a price. From evidence seen in the bones of people buried at Çatal Hüyük, 41 percent of the people there had severe anemia (either sickle-cell anemia or thalassaemia from abnormal blood hemoglobin). This is the genetically caused byproduct of strong selection for malaria resistance in a location with deadly malaria outbreaks.[61]

The city was first excavated by James Mellaart between 1961 and 1965, and it continues to be explored today.[62] It is a remarkable example of the transformative power of plant and animal domestication on a human society. From such locations, cattle and other plant and animal domesticates eventually spread across the known world.

THE SPREAD OF THE DOMESTICATED AUROCHS

As summarized in the poetry of Book of Job, cattle are willing to serve, they can be yoked to furrow and harrow land for crop planting, their strength is great, and they can replace grueling human labor. They also provide meat, milk, blood, bone, horn, and hides for human use. Cattle were the engine driving agricultural production and, with this production, the development of civilization. Cattle would be a valuable commodity to acquire if one did not have them. If one did have cattle, then obtaining better animals to breed into one's herds would be a high priority. It is no surprise that cattle-raising people in many locations have a deep history of cattle raiding, cattle trading, and cattle dowering—practices that among other things move cattle about a regional landscape.

Scientists are studying the DNA of cattle to unravel the complex patterns of the spread of cattle across Eurasia and Africa. Analyses using mitochondrial DNA, which is transmitted to subsequent generations only from the mothers, have already been discussed as tools for understanding the origin and dispersal of the dog. It has similar potential for the revealing aspects of the domestication of cattle. Cattle were domesticated at least twice in two well-separated locations.[63]

The first domestication was of the taurine cattle (*Bos taurus*). These were the sorts of beasts maintained by the Çatal Hüyük people. Mehrgarh, another ancient city in what is now Pakistan, also was a cattle city. Cattle herding there dates from as early as 5000 BCE—not as early as the 6300 BCE date for Çatal Hüyük but a very early date nonetheless. These Mehrgarh cattle were the product of a second domestication of the aurochs from the subspecies that occupied the Indian subcontinent. The domesticated animals, zebu cattle (*Bos indicus*), had prominent humps across their shoulders, and so do the modern breeds derived from them. Along with these two domestication events that produced the taurine and the zebu cattle, two other centers for independent domestication of the aurochs have been proposed: one in the Nile valley and one in northeastern Asia.[64]

The portage of large animals in boats is a topic of religious epics, the *Epic of Atra-Hasis* from seventeenth-century BCE Sumer and its adaptation to the *Epic of Gilgamesh* in seventh-century BCE Babylon, or the Book of Noah in the Hebrew Bible. One significant movement of large animals appears to be the shipping of cattle from locations in northern Africa (from Egypt, Tunisia, Algeria, and Morocco) to points in Spain, Italy, and Greece.[65] The genes from these imported African taurine cows (specifically cows because this is a study of the female-based mitochondrial DNA of cattle) appear to have moved overland across northern Africa through Egypt from the Fertile Crescent. The northward spread of their genes stops when they reach the major European mountain ranges. The African cows shipped to Spain have genetically spread over Spain and Portugal, but their genes are not found in European cattle breeds north of the Pyrenees. Similarly, the Alps restrict the northward movement of cows' genes shipped to Italy. In the case of cows shipped to Greece, there was little flow of genetic material beyond the northern Rhodope Mountains that lie on Greece's border with Bulgaria.

North of these mountain boundaries, European cows from their domestication in the Fertile Crescent came overland across the Bosporus into Bulgaria and down into Albania. These cows swept across central and northern Europe into Germany, France, and England. During this expansion of the range of domesticated cattle, there is also evidence that there was some degree of interbreeding with wild aurochs, which presumably were well adapted to the local environment.[66] The result is a very complex multiple origin of European cattle with a considerable mixing of genetic material from distant and nearby locations.

One sees this same complex mixing in African cattle. Taurine cattle sweep west across the Mediterranean coast of Africa and then southward.[67] The Sahara at this time was grassland, and there may have been movements southward simultaneous with the push westward. Coming from the second locus of cattle domestication, zebu cattle were transported by ships, arrived in the Horn of Africa, and spread southward down the east side of Africa. The resultant complex genetic mixture is the source of genetic diversity for the varied cattle breeds found across Africa.

Mitochondrial DNA analysis reveals patterns in the movement and genetics of cows. In cattle breeding, the transport of breeding bulls often is used as an efficient way to change the genetic makeup of a cattle herd. Studies of the genes of Y-chromosomes, which are transferred only from father to son, also reveal the considerable historical mixing of cattle over Europe[68] and presumably elsewhere. This complex genetic mélange has allowed the breeding and creation of a diverse array of breeds of modern cattle with a remarkable capacity to prosper in a wide range of environments.

HANDING YOUR LABOR TO THE DOMESTICATED AUROCHS (AND THEIR COUSINS)

Cattle provide humans with a remarkable array of goods and services. One indication of the value of such an animal to an agricultural people is the variety of aurochs-like creatures—mithans, bantengs, yaks, and water buffaloes, all of which have been domesticated. Domesticated species in the same genus (*Bos*) as aurochs (and cattle) are:

- Mithan (*Bos frontalis*). Gigantic dark-colored beasts found in forests in India (mostly) and across southeast Asia. As discussed earlier, the mithan seems a creature midway between a wild animal and a domesticated animal. The wild source of the mithan is the gaur (*Bos gaurus*), which is one of the largest land animals. Gaurs are exceeded in size only by elephants, rhinoceroses, and hippopotami. Gaur horns resemble the large crescent-shaped horns of the water buffalo; the horns of the mithan are flatter but equivalently imposing. Mithans are gaurs tamed by tribal people, who reward the free-ranging animals with salt. Mithans are used for ritual sacrifices, and they also are bartered for bridal dowries by the indigenous tribes living in the Assam Hills region of India.
- Banteng (*Bos javanicus*), also known as Bali cattle or tembadau. Well over a million domesticated banteng are found in Java, Borneo, Myanmar, Thailand, Cambodia, Laos, and Vietnam. They are similar in size to cattle but have a more slender neck and a smaller head. They are used as draft animals as well as for meat. As was the case for aurochs, bantengs differ in color by gender. The males are darker (black or dark chestnut) than either females or young, both of which are chestnut. They have white "stockings" on all four feet. The source animal, the wild banteng, is an endangered species, largely because of the illegal trade of banteng horns.[69]
- Yak (*Bos grunniens*). Yaks are wooly, long-haired bovids found in Central Asia, the Tibetan plateau, Mongolia, and Russia. They are

agile climbers, as might be expected for mountain animals. They are capable of carrying pack loads or a person at altitudes as high as six thousand meters for days and still remain in good condition.[70] Along with their remarkable capability as a high-altitude pack animal, they are also used for meat and milk. Yellow-colored yak milk has a very high butter content and is a staple food for the Sherpa people. Because of its high fat content, it also is used by them for lighting campfires.[71] The wild yak (*Bos mutus*), the source of the domesticated animal, is considered a vulnerable species because its population has declined 30 percent in the past thirty years.[72]

These creatures are closely related to aurochs and hence to cattle. They are all in the genus *Bos* and can potentially interbreed across species, often yielding sterile hybrids. Yak can be crossed with cattle to produce a hybrid (males sterile, females fertile) called the *dzo*. Further hybridization of female *dzo* usually produces inferior animals and is not normally done.[73] Mithan can be hybridized with cattle to produce a draft animal in Bhutan.[74] Bantengs can interbreed with zebu cattle.[75]

Another only slightly less related animal is:

- Water buffalo (*Bubalus bubalis*). Water buffalo are the plow animals that propel the traditional production and threshing of rice. The wild progenitor (*Bubalus arnee*) is the second largest wild bovid and is exceeded in size only by the gaur. Wild water buffalo are an endangered species, with fewer than four thousand individuals and possibly as few as two hundred animals lacking some domesticated water buffalo genes.[76] The domesticated animals are in two main collectives of breeds: the swamp water buffalo of China and Burma and the river water buffalo of India.[77] River water buffaloes prefer clear running water and are better milk producers than the swamp water buffalo breeds, which are the draft animals for paddy-rice farming. The animals existed in a domesticated form in the Mohenjo Daro civilization of the upper Indus valley, in what is now Pakistan, as early as 2500 BCE.[78] By the seventh century CE, water buffalo had been introduced to Italy,[79] and the familiar mozzarella cheese in its various forms originates with a traditional Italian cheese, *mozzarella di bufala campana*, made from water buffalo milk.

When one takes this collection of domesticated bovids as a whole, there are significant similarities. They range over many contrasting habitats. They are very large animals with even larger, obviously dangerous, wild progenitors. Today, the wild sources are often endangered species and heading toward the same sad extinction that befell the aurochs.

CHANGING REGIONAL LAND COVER

The domestication of wild animals, the animals discussed here as well as the myriad of other creatures that have been put into the service of humanity, gave people the ability to change much of the surface of the planet. Various kinds of cattle, along with donkeys and horses, provided the power to plow land and transport produce. Goats, sheep, pigs, reindeer, and other manmade animals provided a way to turn the inedible grass and other vegetation found in some very harsh settings into milk, meat, and useful products.

THE EFFECTS OF DOMESTICATION ON THE LAND COVER

By intelligent manipulation of the types of herds and their densities and time of movement from one location to another, grazers can manipulate the cover of the landscape to great advantage. Conversely, poor management of herds can destroy a landscape and make it subject to erosion and covered by weeds. Today, one sees the deforested eastern Mediterranean landscape with its depauperate plant communities as the product of ever-increasing agriculture, declining forests, and continual soil erosion, the consequences of the rise of urbanization and an increasingly demanding civilization.[80] The changes in the landscape can be reconstructed over the past by inspecting a variety of paleoecological evidence. The change in landscapes in the region reveals itself in bits of material preserved over time.

For example, Petra, located in modern Jordan, is a remarkable archeological site. It is a city carved from stone, perhaps best known for its use as a set in the movie *Indiana Jones and the Last Crusade*. In the cracks and crevices in and around Petra, animals about the size of a large rabbit, called hyraxes (*Procavia capensis*), construct their protective stick nests. Hyraxes are unique animals whose nearest living relatives are elephants. Petra's dry environment preserves hyrax nests from as long as two thousand years into the past. The changes in the kinds of sticks and other material used at different times in hyrax nest building reflect the vegetation near the nest site and, thus, record the vegetation history of the Jordanian landscape.[81]

In the region of Petra, material from progressively younger hyrax nests indicate that the landscape around the city degraded from an original Mediterranean forest of oaks (*Quercus*), pistachio (*Pistacia*), olive (*Olea*), pine (*Pinus*), and juniper (*Juniperus*) trees to a *maquis* (a degraded forest) and then to a *garigue* (an even further degraded shrubby forest) by the second century CE. After the collapse of Byzantine Petra, intensified grazing by livestock eventually reduced this vegetation to a Mediterranean *batha* (very open shrubby vegetation) and, in some cases, to sparse grassland.

This collapse of Petra's vegetation under land abuse and particularly from overgrazing certainly demonstrates the power of domesticated range animals to alter a landscape radically. Is this record of the past two thousand years, featuring a linear and sad decline of the land, the necessary path of agriculture for the Fertile Crescent and the agricultural systems that developed there? There are other records from the Jordan Valley that go deeper into the past to provide a more complex historical and prehistorical rendition of change. This change goes from before the dawn of agriculture to the present.

The so-called Neolithic Revolution is considered the final stage of the Stone Age.[82] It is associated with the development of agriculture, the transition from nomadic hunters to settled villages, and the use of domesticated animals and plants. The plant domestication process and a more settled village life produce a positive feedback—one encourages the other and vice versa. Domestication of plants is perfected in association with sedentary village life; sedentary villages are increasingly possible when agricultural crops have been developed. Extremely fertile locations are able to sustain sizable sedentary hunting-and-gathering communities and, thus, catalyze the positive feedback leading to the Neolithic lifestyle. The Fertile Crescent, with its early domestication of plants and animals, underwent this transition first.

Understanding the dynamics and evolution of the Neolithic Revolution involves collation of different sorts of prehistoric information. Archaeological studies can reveal the possible numbers of people at a location from the extent of ancient ruins and the materials that people used. Fragments of foods and scraps people leave behind provide clues to changes in farming and nutrition. Human bones can reveal the health of the people, their nutrition, and their life expectancy. Artifacts and art reflect cultural and religious practices. These analyses assembled over time capture the human condition as continuing technological innovation carries forward from the Neolithic to the Chalcolithic (Copper) Age, to the Bronze Age, to the Iron Age, and to the near present. Archaeologists assemble these bits and pieces to reconstruct how the people at a location functioned. If these are analogous to a forensic evaluation of what has happened in a "room" at a time in the past, the analyses of pollen in lake sediments reveals the view through the window in this ancient, reconstructed room. One can "see" the surrounding vegetation. The change in the pollen record demonstrates the vegetation changing over time.

Dustlike grains of pollen provide major clues about the vegetation of the past. Wind pollination in plants, the bane of hay fever sufferers, is a relatively inefficient process. A great number of the pollen grains are lost on the way to the recipient flower. The "rain" of pollen from the plants covering a landscape can be preserved in lakes. Pollen analysis (palynology) reconstructs the past vegetation and reveals vegetation change. There has been much work done to reconstruct vegetation dynamics associated with the climate variations that have occurred over the past ten to twenty thousand years (and in many locations much longer).

Pollen grains recovered from the Sea of Galilee and the Dead Sea in the Jordan Valley reveal the vegetation changes in the lands of the Levant region of the Fertile Crescent as agriculture spread across the surrounding landscapes.[83] Long cylindrical cores of mud taken from the lake bottoms provide clues to changes in the past.[84] From the layers of sediment laid down in the lake bottoms over the millennia, these pollen records document the surrounding vegetation for as far back as the past 17,000 years or more.[85]

In this swing and sway, land-cover composition is pushed by human history (and impelling the course of history in return). One significant pattern is clear evidence of a second agricultural revolution, the Secondary Products Revolution.[86] Some five thousand years after the initial advent of agriculture, innovations in animal husbandry allowed production of wool, leather, and dairy products. The development of widespread fruit trees in the Fertile Crescent also was part of a second wave of agricultural land-cover change. This change was fueled by the power of harnessed domesticated animals to convert land to agricultural use. In the Fertile Crescent, orchards with plants that could be propagated from cuttings (grape, figs, sycamore figs, and pomegranate), knobs from the base of the tree (olives), or transplanted offshoots (dates) covered the hillsides.[87] The first agricultural revolution in this region featured annual grains and meat animals; orchard crops developed somewhat later. People could move after harvest time and take their herds to better pastures. The second agricultural revolution included perennial plants. There is evidence that figs were domesticated quite early near Jericho,[88] but these and other orchard-type food sources greatly increase during the Secondary Products Revolution. Also at this time, animals attained value beyond the meat that their carcasses could produce. Yoked oxen and perennial plants encouraged more sedentary agricultural production systems.

A second important pattern suggested in the pollen record and corroborated from archaeological evidence is that the development of civilization is not a linear forward progression with technological advances and levels of urbanization increasing inexorably over the years. Even before agriculture, the region supplying pollen to the Sea of Galilee and the Dead Sea saw ancient "prototowns" such as Jericho, 'Ain Ghazal, Es-Sifiya, Ghwair, and others, spring up and prosper.[89] At these and other places like them, the more sedentary people began to develop and use domesticated plants. The prototowns prospered with the productivity of these early domesticated plants and then eventually were abandoned, perhaps as their local landscapes were deforested and their surrounding farmland degraded.[90] In the sediment cores, a reduction of the amount of pollen from deciduous oak trees indicates that these urban collapses may have arisen after local deforestation to obtain building timbers for multistory houses and firewood for heating and cooking. Following the collapse of the prototowns, almost three thousand years passed (until the Middle Bronze Age) before significant urbanization reoccurred in the region.

There are other recurring signals in the pollen record. Weed pollens increase when land is cleared, abandoned, or heavily grazed. Cities form, and the types of crop pollens shift to the cash crops sold in urban markets; cities fail, and pollen records show an agricultural shift to subsistence crops that provide food to their farmers.

Anthropogenic Landscapes

The history the Levant is deep. The implication for modern times of the domestications of plants and animals from the general region is great. The major religions incubated there touch a sizable portion of the Earth's modern population. However, another significant feature of its history, the sweeping landscape change driven by human culture, is not uniquely a phenomenon of the Levant. The Secondary Products Revolution is seen in Central American cities as well. Aztec and Mayan city-states had a significant effect on their surrounding rural environments.[91] The pre-Columbian cities produced a market for specialized orchard products (cacao, *Theobroma cacao*; avocado, *Persea americana*), which led to agricultural landscapes across the surrounding region.[92]

Humans have a remarkable capability to alter their surroundings. Human technology in all its manifestations provides increased capability to alter the regional landscape. Nonetheless, hunter-gatherers also have a remarkable capacity to alter land, notably by setting fires to obtain game and change the grazing grounds for hunted animals. The aboriginal people of Australia are famous for their use of fire as a game-management tool. "Fire-stick farming" is an example of extensive land alteration by these people.

Land Alteration by Hunter-Gatherers:
Martu of the Western Desert of Australia

With regard to the aboriginal people of Australia, it is generally thought that their anthropogenic fires produce a mosaic of landscape types; the "tiles" of the mosaic represent tracts of land burned at different times. Aboriginal fires are smaller, and the resultant fire mosaics of burned and regenerating vegetation are more finely grained.[93] Their fires increase local nutrient availability and enhance the short-term productivity of herbaceous plants typical of the early stages of recovering vegetation.[94] The fires set are of lower intensity but occur more frequently than Australia's natural, lightning-caused fires. The manmade fires often select particular types of vegetation for ignition.[95]

One recent well-studied example of these practices in action involves the traditional hunting by a contemporary Aboriginal people (the Martu) in the Western Desert of Australia.[96] The Martu sometimes set fires to "clean up" remote regions and to attract large turkey-like birds, bustards (*Ardeotis australis*), which forage in burned areas. However, most of their fires are set by women hunting the large monitor lizards called goannas (*Varanus gouldii*).[97] Martu fire hunts involve lighting up a suitable tract, walking behind the fire line searching for signs of fresh lizard burrows, and then using a specialized digging stick to root the animals from their dens. Fire control is a significant consideration. Individual hunters whose fire threatens a significant sacred site are subject to ritualized physical punishment and monetary payments. Martu fires are set to take advantage of the wind direction, and fire-breaks control the direction of the fire's spread. The Martu light their hunting fires in the winter; winter fires are more effective for lizard hunting, and the fires are more controllable.

In the case of the Martu, the women who hunt goannas and other small game using fires basically reconstruct the landscape ecosystem— their hunting creates a landscape different from what would occur without them, one more favorable to their activities. This is a positive outcome.[98] From their fires, the vegetation on the land is rearranged into relatively small patches. This fine-scale patch structure creates more biological diversity at the same spatial scales as the foraging ranges of the hunters. The immediate effect of the Martu women's use of fire on their landscape creates greater numbers of animals of the species that they hunt. They are "farming" small game through fire management.[99]

Neolithic Landscape Change: The Voyaging Polynesians

We have already discussed the extensive regionwide land changes that attended the Neolithic development of agriculture in the Fertile Crescent and particularly in the Levant. The same sorts of major land cover changes occurs with the Neolithic Revolution in other parts of the world as well. For example, the spread of the Polynesian people across the Pacific provides some excellent examples of just how effective people equipped with stone tools can be as agents for landscape change. The navigational skills used in these remarkable voyages will be discussed in chapter 6.

Polynesian arrival at a new island is a discrete event. Thus, their effect on a landscape is relatively more straightforward to determine archeologically via an examination of the change of an island's fauna and flora. The Polynesians sailing across the Pacific brought with them a "portmanteau biota"[100] of transportable plants and animals, which sustained the Polynesian agricultural base. These included several domesticated animals (dogs, pigs, and chickens)—along with the Polynesian rat. A diverse collection of food-producing plants were transported as well. In the Polynesian colonization of Hawaii, these included coconut (*Cocos nusifera*), breadfruit (*Artpcarpus altilis*), candlenut (*Aleurites molluccana*), taro (*Colocasia esculenta*), sweet potato (*Ipoemoea batatas*), and banana (*Musa acuminata*). In colonizing other islands, these and several other food plants were either brought by the Polynesians or were already part of a particular island's flora.[101] Along with food crops, other plants with a range of other uses also were either actively transported or utilized from the local flora: hibiscus (*Hibiscus tiliaceus*), with leaves used for sealing earth ovens; acute (*Broussonetia papyrifera*), with bark that can be pounded into tapa cloth; and ironwood (*Casuarina equisetifolia*), used for war clubs as well as house and boat parts.[102]

On the island of Mangaia in the Cook Islands, an interdisciplinary investigation documented the changes to the island before and after Polynesian habitation.[103] Mangaia is a small island (fifty-two square kilometers) with a central volcano. Shifts in the Earth's crust raised the island, and what was a surrounding reef was elevated out of the ocean to become a collar around the central volcano. This raised coral is called *Te maketa* (in Mangaian) or makatea (in English). The makatea traps the sediment from the streams that drain the central volcano. It also provides refuge caves for the living and burial places for the dead.

About 1000 CE, the Polynesian people who would become the Mangaians arrived at the island. They modified the island landscape by clearing and cultivating the upland forests on the slopes of the central volcano. They farmed the uplands with an agricultural practice called swidden agriculture, also called cut-and-burn or shifting agriculture. Small patches of forest are cleared and planted to crops. These small plots are abandoned and returned to forest cover when the soil fertility becomes depleted locally and the crop productivity lessens. The forest in the former garden plots eventually recovers, and the land is cleared and farmed again.

On Mangaia, as often occurs with this farming system, when the human population densities became higher, the time allowed for recovery of the cropped patches was shortened to leave more land in cultivation. This eventually degraded the soil and produced soil erosion. The eroded soil from the uplands began to fill the area behind the makatea. On the heavily used upland, a soil process known as laterization converted what remained of the soil to laterite. Lateritic soil occurs when high rainfall washes out most of the more soluble soil compounds, leaving behind iron and aluminum oxides. Dry laterite resembles brick. With soil laterization, the upland zone of the island largely went out of agricultural production.

In as little as five hundred years, Mangaian agriculture had dramatically changed the island's character. The lateritic soils of the uplands were covered with ferns, not forests as before. With the collapse of their upland agricultural production, the Mangaians shifted their effort to work the sedimentary deposits of soil, which had eroded from the uplands. In the sedimentary basins behind the makatea, complicated hydraulic systems flooded and drained highly productive fields of taro, a starchy root crop that then sustained the people. Such highly engineered and remarkably productive taro systems are found throughout Polynesia and impressed many historical observers. They are entirely a human construction and the byproduct of upland land degradation.

The changes that one sees in Mangaia are echoed across the Polynesian expansion across the Pacific. Mangaia's change is a logical consequence of the interactions among humans, crops, and the island's vegetation and its soil. The changes reconstructed by an interdisciplinary team of scientists for Mangaia and the kinds of landscapes produced there are found throughout the South Pacific, now and in the past. Charles Darwin on the HMS *Beagle* noted the predictable occurrence of fern-covered uplands across the Pacific Ocean and likened them to the heathlands of Britain. These fern-heathlands and the taro-growing systems, which are planted in the sedimented, eroded upland soil systems, both are products of human land alteration.

Neolithic people with neither draft animals nor metal tools can manifest landscape changes in a surprisingly short period of time. The implications from these changes to the power of technological human society to change the planet's landscapes are obvious. In modern Polynesia, the change in landscapes continues. The technologies are different (guns instead of snares, chain saws instead of stone axes), but fire is used as a tool for clearing land nowadays just as in the past.[104] This continued change is the case not just in the Polynesian islands but worldwide.

Frost Followed the Plow

The domestication of crop plants and the nutrition they ultimately provided fueled a stupendous change in the numbers and capabilities of humans. The question, "How much has the conversion of human labor to animal labor and the agricultural revolution changed our planet?" presages the current question for our technological society, "How much will the continued conversion of animal labor to fossil fuel–driven, mechanical labor and the industrial revolution change our planet in the future?" Let us consider the first question before taking on the latter in the chapters that follow.

There has been a scientific discussion over the past two hundred years on the topic of what the human alteration of the land has done to various regional climates. The relatively rapid growth, land clearing, and installation of European agriculture in the United States represents a compression of land-change history. Whole regions were converted from forest to agriculture within the lifetimes of colonial observers. Iconic American figures endorsed two different points of view. Thomas Jefferson in his *Notes on the State of Virginia* opined that "A change in our climate, however, is taking place very sensibly. Both heats and colds are become more moderate within the memory even of the middle-aged."[105] At about the same time, Noah Webster reported to the Connecticut Academy of Sciences:

> It appears that all the alterations in a country, in consequence of clearing and cultivation, result only in making a different distribution of heat and cold, moisture and dry weather, among the several seasons. The clearing of lands opens them to the sun, their moisture is exhaled, they are more heated in summer, but more cold in winter near the surface; the temperature becomes unsteady, and the seasons irregular.[106]

This argument was made more complex by the fact that the Earth was emerging from a cool period known as the Little Ice Age at the time of these discussions. Does the clearing of land and conversion to agriculture make a region climatically more demanding, particularly the colder winters (Webster)? Or does the climate become more moderate, as a beneficent Creator rewards those who tame the land and control the wildness of nature with a warmer climate (Jefferson)?[107]

In assessing the change in winters, Webster noted that the "warmer winter" arguments, made by such luminaries as Abbé Du Bos, Buffon, Hume, Gibbon, Jefferson, and others, often drew from Bible-based evaluations of the past climates of Palestine and Judea. To refute these arguments, he cited the rich discussion of weather phenomena, particularly in the Book of Job and elsewhere in the Bible, concluding, "But the most positive evidence which can possibly exist to prove that the climate of Palestine has not suffered any increase of heat, for more than three thousand years, is the production of certain fruits in the days of David, which will not thrive in any but mild, warm countries; as pomegranates, olives and figs."[108] These latter plants were the product of the Secondary Products Revolution, which we have just discussed. Two centuries after Webster's presentations, computer models of North American climate change predicted that a relatively deforested America would differ significantly in its climate from a forested, precolonial version of the nation. These findings supported Webster's contention that "frost followed the plow"—earlier onsets of cold weather in the fall across the middle of a cleared America.[109]

Webster's observations of land-cover change and its coupling with regional climate change is not a one-off occurrence. Such early observations of the consequences of land alteration came from several other scholars and historical figures. Some of these observations were produced by comparisons between regions in the Age of Exploration. For example, Christopher Columbus wrote to his son that he thought the forests of the West Indies caused more rainfall there than observed in the deforested Azores—an observation supported by computer-based climate model studies almost five centuries later.[110] Other observations of the feedback between regional land cover and regional climate, like those of Webster, were fueled by the rapidity of land change often from the advent of the growth of human populations and urban markets. Antoine-César Becquerel in the mid-1800s famously answered his own question, "How do they [the forests] modify the temperature of the country?" with the conclusion that the presence of forests and other vegetation affected the climate. Becquerel's conjecture and an equivalent one from Humboldt at about the same time have subsequently been borne out by modern studies.[111] Just as scholars of the past have wondered how the change in the cover of the land might affect climate, climate-vegetation feedbacks remain a central topic for today's researchers attempting to appraise the actions of our modern society and its technological capability to alter the land. That our scale of alteration has increased from the regional scale to the global scale heightens the importance of understanding how our planet functions as a dynamic whole.

CONCLUDING COMMENTS

Beginning with the whirlwind question of domestication, this chapter has discussed the effects of plant and animal domestication on human societies and the lands they occupy. People have a remarkable power to alter landscapes, and that power is increasing. Landscape manipulation is part of our mode of survival from well before the discovery of agriculture. People alter landscapes and, as was noted immediately above, regional climate. Current technology has provided the means to change the face of the Earth. Indeed, if one looks from an airplane window, one sees the hand of humanity upon the landscape. Does the fire-stick farming of the Martu, which seems to be applied in a sustainable fashion, have any resemblance to our own (un)sustainable use of the landscape? At best, we have a lot more to learn. Some of what we need to know is adumbrated in our answers to other of the Joban whirlwind questions, questions involving the understanding of how the planet functions in a physical and biological sense.

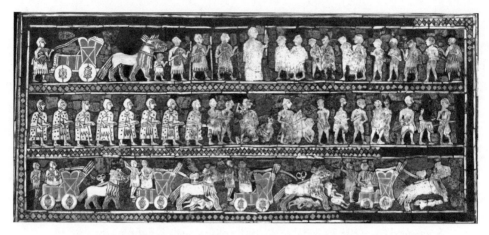

The Royal Standard of Ur. From the royal cemetery of the Babylonian city of Ur located south of what is now Bagdad. Dating from about 2500 BCE, it was discovered by Sir Leonard Woolley, who took it to be some sort of battle standard carried on top of a pole. The image shows the Standard of Ur as reconstructed and preserved in the British Museum. Onager tails' top parts have short hairs, with longer hairs creating a tuft toward the end of the tail, somewhat like the tail of a lion. Many of the horselike animals pulling four-wheeled chariots have an onager's tail and ears that are proportionally larger than horses' ears. *Source*: Used with permission of the British Museum.

4

Freeing the Onager

Feral and Introduced Animals

Who has let the wild ass go free? Who has loosed the bonds of the swift ass, to which I have given the steppe for its home, the salt land for its dwelling place? It scorns the tumult of the city; it does not hear the shouts of the driver.

—Job 39:5–7 (NRSV)

What is the animal whose bonds have been loosed in these verses? When one initially reads these, one immediately thinks of wild donkeys in the hills that have somehow eluded domestication—more or less an equine parallel to the oxen in the field and the aurochs in the wild, which were discussed in the last chapter. However, donkeys are derived from an African animal, the African wild ass (*Equus africanus*),[1] which is native to Ethiopia, Eritrea, and Somalia.[2] While there are prehistoric records of *E. africanus* from Arabia and elsewhere in the Fertile Crescent,[3] these are not the animals being described in the Book of Job. The Joban wild ass is an animal called the onager (*Equus hemionus*).[4] Its preferences for dry grasslands (mountain steppe, steppe, semidesert, and desert plains)[5] are well characterized by the habitats described in Job 39:6 ("the steppe for its home, the salt land for its dwelling place . . ."). Onagers also have a direct etymological connection to the Job text. The word onager actually means "wild ass," derived from the ancient Greek ὄναγρος, from ὄνος (ass) combined with ἄγριος (wild).

THE ONAGER

Onagers somewhat resemble donkeys, but they are larger.[6] They are horselike in many of their features. Indeed, the *hemionus* species name in its scientific name is from the Latin (*hemi* or half, *onus* or ass), meaning "half ass." This refers to the animal's appearance, which is halfway between that of donkeys and horses. They are not hybrids between donkeys and horses. A donkey-horse hybrid is called a mule, with a horse as its mother. A hinny (or jennet) is born from a donkey mother. Onagers are a separate species. Their color varies seasonally: reddish-brown in summer, yellowish-brown in winter months. A black stripe with a white border extends down the middle of the back. The tail of the onager has short hairs at its base and long hair on the lower section of the tail, a distinctive difference from the flowing tails of horses. Onagers are between horses and donkeys in height and weight, somewhat closer to the size of a donkey. Their ears are intermediate between the large donkey's ears and the smaller horse's ears.

Who tamed the onager? Was it domesticated? These questions are a matter of some debate at this time. A 1930s excavation of the Sumerian site at Tell Asmar (third millennium BCE)[7] in Iraq uncovered large numbers of onager bones.[8] This led to the speculation that the bones were from domesticated animals being raised for their meat.[9] From Çatal Hüyük, a pair of wall paintings each of a man holding a tamed onager dates to the sixth millennium BCE.[10] There are also a range of other onager-related objects: clay tablets from the city of Susa in the Elam civilization in southwestern Iran (dated between 3250–3000 BCE), with four onagers in a row with a group of people; a later clay tablet from Elam (3000–2800 BCE) decorated with several onager heads; a painting from Tell Halaf in Syria showing onager-drawn chariots, from 3000–2800 BCE.[11] These have been taken as indications of an early taming and potentially an early domestication of onagers.[12]

THE ONAGER AS A DRAFT ANIMAL

One particularly striking onager-related artifact is the Standard of Ur, an inlayed hollow wooden box with a background of lapis lazuli, with shell and red limestone inlayed details, dating from 2500 BCE. The box ends are trapezoidal, and the sides are twenty by forty-eight cen-timeters. One of the side panels shows a battle scene. This "war" panel illustrates the Sumerian spearmen, other infantry, and four-wheeled chariots trampling the enemy. The other side's panel shows a peace-ful scene, perhaps after the subsequent victory. In all, seven pairs of equids appear on the standard.[13] It was found by Sir Leonard Woolley in the royal cemetery of the Babylonian city of Ur, which was located south of what is now Baghdad.[14] Woolley was of the opinion that the object was a battle standard carried on a pole (hence the name); others speculate that it might be part of the sound box of some sort of musical instrument. It now resides in the British Museum.

The Standard of Ur has been the focus of debates on whether the onager was actually domesticated or not. Was the onager used as a draft animal? Among the pairs of equids on the panels, there are some animals, which clearly display the distinctive onager tails, draw-ing chariots. Then why haven't onagers been found with the wear marks of bits on their teeth, as is the case with ancient horses? The onagers on the standard seem to be harnessed with a strap around the jaws and another behind the ear, but without bridles.[15] This method of control also was used with horses, and the nostrils of horses were sometimes slit to keep the nose straps from hampering their airflow.[16] When unharnessed, the onagers illustrated on the standard have rings in their noses.[17]

The Horse and the Onager

It is generally agreed that with the advent of the horse, which was stronger and easier to handle, the widespread use of the onager as draft animal ceased. The horse (*Equus callabus*) as a ridden animal seems to have originated in southern Ukraine and Kazakhstan starting about six thousand years ago.[18] Horse bones increase in frequency in archeological sites in this region at this time.

In the middle of the fourth millennium BCE, a horse-oriented people in what is called the Botai culture from northern Kazakhstan in the Eurasian steppe possibly had domesticated horses. There is a pattern of wear on a few of the animals' teeth that imply the use of a bit, an indicator that these were ridden animals. This is not definitive for a domesticated horse as either domesticated or tamed animals could have been ridden. Additional evidence for domestication includes measurements of Botai horse leg bones (metacarpals), which demonstrate they resembled Bronze Age domestic horses and not Paleolithic wild horses from the same region.[19] Also, the pottery from the sites contains chemical evidence of products derived from mare's milk. Taken as a whole, these are strong indications of the presence of domesticated horses in the Botai culture.[20] Artistic evidence of domesticated horses only dates from the end of the third millennium BCE, so this pushes back the evidence in dated texts and illustrations.

The key to the domestication of the horse may have been a genetic change predisposing some horses to breed in captivity. The domesticated horse seems to have derived from many mothers but from only one father. Genetically, the Y-chromosomes found in a wide collection of male horses indicate that the horse was derived from a single stallion or a limited number of related stallions.[21] The mitochondrial DNA of horses indicates a large number of matriarchal lines.[22] The implication is that a chance genetic change produced a stallion that was tolerant of the conditions involved with breeding in captivity. This gave humans an ability to control horse breeding and is consistent with the low number of horse "founding fathers." During the continuing process of domestication, tamed or wild mares from different regions were crossed to sires from this line of stallions.[23] Thus the modern horse has many founding mares. This pattern of propagation might have allowed the rapid expansion of horse populations throughout Europe around 2000 BCE.[24]

SETTING THE ONAGER FREE

If the onager was domesticated, as the Standard of Ur and other artifacts indicate, then "Who has loosed the bonds" of the domesticated onager? This puzzle may be revealed in translations of the ancient writings of the Sumerians. In 1840, the royal library of the Assyrian King Ashurnanipal (688–627 BCE), consisting of twenty thousand clay tablets written in cuneiform script, were found at Nineveh, on the eastern bank of the Tigris River opposite modern Mosul, Iraq. This great library and other cuneiform writings have been the focus of translators ever since.[25] It has been found that there are several words used for different equids, including terms for donkeys, onagers, horses, and hybrids between onagers and donkeys.[26] One tablet from 2050 BCE provides a three-year account, which records the ownership of thirty-seven horses, 360 onagers, 727 onager-donkey hybrids, and 2,204 donkeys.[27] From the texts, these hybrid onager-donkeys (using male onagers as sires) were large, powerful animals. They were nearly horse sized and were yoked in teams to draw chariots.

In mules and other equid hybrids, the appearance of the heads and tails of the offspring generally reflect the male.[28] By this logic, the onagers on the Standard of Ur could have been these onager-donkey hybrids, which would have had smaller ears than the donkeys but also onager-like tails. To maintain these valuable hybrids, tamed onagers from the wild would be regularly captured as foals, tamed, and maintained to breed with donkeys. When the horse became available as a domesticated species, tamed male onagers as breeders to donkeys were no longer needed. From around 2000 BCE in Sumerian writings, the horse is mentioned increasingly more frequently, and the mention of onagers disappears:

"Presumably once the horse, which could not only breed hybrids but was also in itself a better animal and trainable, had appeared on the scene, there was no longer any call for keeping the onagers to do nothing but eat, drink, and enjoy the pleasures of their harems."[29]

The hybridization of onagers and donkeys appears to be an intensive practice of the ancient Sumer empire in the western arm of the Fertile Crescent in the Tigris and Euphrates rivers' lower drainage region. The practice of the hybridization of onagers and donkeys was occasionally reported in the eighteenth and nineteenth centuries CE,[30] but large-scale hybrid production using onagers ceased millennia earlier with the advent of the domesticated horse. So, in answer to this chapter's whirlwind question, the onager was freed by humans. It was "set loose," or, more accurately, it no longer continued to be tamed for reasons of economics—it was easier and more effective to use horses and mules as draft animals to produce agricultural products.[31]

The opening actors in this narrative, the wild horses, wild donkeys, and onagers, have not fared well in the passage of history. The wild species and source of the domestic horse, the wild horse (*E. ferus*), is now represented by the 325 free-ranging, reintroduced, and native-born Przewalski's horses (*Equus ferus przewalskii*) in Mongolia. This horse subspecies, which is not likely to be the horse subspecies that produced the domesticated horse (*E. callabus*), became extinct in the wild in 1969. All living Przewalski's horses descended from thirteen or fourteen individuals assembled from zoos and propagated in a captive breeding program starting in 1992.[32] A second subspecies of the wild horse, the tarpan (*E. ferus ferus*, also known as the Eurasian wild horse), became extinct relatively recently.[33] The last tarpan died in captivity in Russia in 1909.[34] The African wild ass, the progenitor of the domestic donkey, is a critically endangered species.[35] The onager is an endangered species that is much restricted in its former range and is decreasing in numbers.

The sad situation of these remarkable wild equids, which were the source stock that contributed so greatly to the development of human civilization, is part of a current pattern of species endangerment and extinction. This loss of biotic diversity springs from a range of human pressures including change in the land cover of the Earth.

HUMANS AS A KEYSTONE SPECIES

The previous chapter discussed anthropogenic landscapes. Humans and their technology, domesticated animals, and machines have and continue to change landscapes remarkably. They are also a powerful force in the evolution of plants and animals and on the assemblage of novel ecological systems.

The domestication and release of animals along with the transportation of plant and animal species has changed the terrestrial landscapes of the world. The development of agriculture and the power of draft animals to perform the heavy work of pulling plows and moving large stones, trees, and eventually wheeled carts gave humans the capacity to alter their surrounding ecosystems greatly. Modern technology has amplified this trend. From prehistoric times into modern times, humans have progressed from being keystone species at the local to the regional to the continental to the planetary level.

What is a keystone species? Basically, such a species controls the structure and function of the ecosystems in which they occur. The name refers to the keystone in an arch. Remove the keystone, and the arch tumbles under stress; remove the keystone species, and the ecosystem collapses with change.

The original keystone-species concept was produced from a study on the role of a starfish (*Pisaster ochraceus*) on the rocky intertidal zone in the U.S. Pacific Northwest coast.[36] The rocky intertidal is found on rocks along a seashore in the band between high and low tides. The rocks there are covered with various barnacles and mussels, which are the prey of the predatory starfish. When the starfish were experimentally removed from areas of the rocky intertidal, white acorn barnacles (*Balanus glandula*) initially increased to take over 60 to 80 percent of the space. Nine months later, these barnacles were crowded out by small, rapidly growing California mussels

(*Mytilus californianus*) and goose-neck barnacles (*Mitella polymerus*) along with some rock barnacles (*Semibalanus cariosus*). Eventually in areas with the starfish removed, mussels became dominant across the area, with occasional patches of goose-neck barnacles.[37] The predator starfish was deemed a keystone species because its presence strongly controlled the ecosystem on the intertidal rock surfaces. Its presence increased the overall diversity of species in this case, and its removal decreased species richness. Other cases demonstrate the opposite diversity pattern.

In other ecosystems, herbivores serve as keystone species. African elephants (*Loxodonta africana*) can be remarkably destructive animals with regard to the forest and shrub cover in savanna ecosystems.[38] Variation in elephant densities significantly affects the mixture of grass and woody plants in savanna vegetation and thus the fuel loads for wildfires.[39] Changes in wildfires and in woody cover both can alter the nature of the savanna ecosystem.

Plants can also function as keystone species. In the central wheat belt of Western Australia, an assemblage of nonmigratory birds called honeyeaters (in the family Meliphagidae) feed on flower nectar and simultaneously pollinate an array of shrubs and small trees in the Proteaceae family. Honeyeaters are attractive, mostly small birds colored in combinations of yellow, black, green, and red; the proteaceous shrubs display flowers, mostly in the red, orange, and yellow spectra that typify a bird-pollenated species. The different flowers are sized and shaped to accommodate honeyeaters of appropriate sizes and with particular bill lengths. The plants flower at different times, a phenomenon associated with timing reproductions so as not to compete with other flowers for the services of an appropriately sized honeyeater. However, there is a short time of year when only one shrub species, the orange banksia (*Banksia prionotes*), blooms and sustains all of the honeyeaters. Eliminate the keystone orange banksia, and the honeyeaters, as well as all of the plants they pollinate, could well disappear from the region.

In the last chapter, Neolithic people, Polynesians, specifically the Mangaiains, functioned as keystone species capable of dramatic modifications of the islands of the Pacific. Hunting/gathering people using fire as a landscape-management tool also place humanity in the keystone species role. The capability of technological human society to modify landscapes, regions, and the planet is easily demonstrable. More subtle is the effect of the mere presence of low densities of hunting/gathering people on the ecosystems they inhabit.

One can easily imagine positive feedback systems in which the seeds of fruits that are discarded after being carried by their dispersers (or that pass through the digestive tracts of their dispersers unharmed) would evolve to forms that are increasingly preferable by their disperser. The coadaptations between animal dispersers and plants have evolved a remarkable array of adaptations.[40] One would expect this to be no less the case for human dispersal systems as well.

In a process referred to as domiculture,[41] species adapt to human "homes." They evolve mechanisms to promote their dispersal to human camp sites, latrines, and waste dumps. They develop life histories that increase the likelihood of their survival once there. Domiculture was initially used to describe the observable effects of Australian aboriginal people living in hunting-and-gathering societies on the Cape York Peninsula on the surrounding flora.[42] Basically, the transport of the best edible fruits back to temporary base camps selects for larger, tastier fruits on plants near these camp sites.

The domiculture of a species provides a potential starting point for eventual domestication of some plants. Dating the age of archaeological sites requires samples of formerly living material. Frequently and depending on the location, these often are the cherry pits, pawpaw seeds, olive pits, and melon or other seeds that have been dropped, spit out, or excreted by the inhabitants of the sites. Fruits and vegetables with large fleshy edible parts that encourage the dispersal of hard, digestion-resistant seeds fill the produce sections of modern and traditional markets. Many of these may have begun their path to becoming domesticated cultivars through domiculture in response to humans as a keystone species.

Negative Change in Human-Dominated Ecosystems

In the previous chapter, human cultures were shown to alter landscapes, particularly with the aid of domesticated animals. These changes can be positive, as in the case of domiculture, but the alteration of ecosystems by humans induces changes in the species composition of natural systems and promotes behavioral and natural change. Species in natural ecosystems can respond behaviorally or evolutionarily to the human keystone species in less benign ways. The same can be said of the plants, animals, and microbes in agricultural systems. These include adaptations that produce weeds, pests, and diseases. These represent the darker side of domiculture. This dark side may be worsened by the release of domesticated species and by the introduction of species from one place into another.

Weeds

The earliest weeds to have been dated are from 3200 BCE. These are the seeds of the cockle (*Lolium temulentum*), found mixed with cereal grains from two archaeological sites located near the Dead Sea.[43] Cockle grows in the same zones as wheat. It so resembles wheat that it has the alternate common name of false wheat. It mimics wheat through most of its life cycle. Only when the plant develops seeds does its wheat mimicry break down. Its seed head is lighter than that of wheat, and it stands erect in mature wheat fields; wheat heads droop from their weight.

There were likely weeds well before the cockle's seeds from 3200 BCE. Weeds are also likely to exist for some time to come. Fields prepared for planting food crops represent a high-value location for other species of plants as well. Such sites are selected as locations where plants will grow vigorously—combinations of deeper soils with more nutrient elements, well watered, good drainage, and easily tilled. Weeds are sometimes characterized as plants in the wrong place. From the weed's perspective, agricultural land is clearly the right place. The agricultural sites represent desirable prime locations for what have been call "ruderal species" of plants.[44] Ruderals have good dispersal abilities, grow rapidly under favorable conditions, and are relatively short lived. In more natural ecosystems, the ruderal life history allows plants to colonize and pioneer the initial sites created following a disturbance. In agricultural systems, these same attributes produce a predilection to colonize agricultural fields.

Weeds may not be edible at all; they may even poisonous. Because they compete with crops, they lower the yield from the harvest. Thus, agronomists have been locked in battle with weeds since agriculture was invented. In this battle, weeds have evolved two tactics, both under the heading of mimicry. The first tactic is created by natural selection, which produces plants that resemble the crop plants. This makes it difficult for humans to separate the weed plants from the crop plants. In the second tactic, weed seeds mimic the crop seeds. This allows the weed seeds to "hide" among the crop seeds and be planted in the next crop cycle.

Evolution in seed mimicry in weeds counters human mechanical strategies to clean the weed seeds out of the next season's crop seeds. Sift the seeds for size, and the selection favors weed seeds the same size as the crop; sort the seeds for color, and selection for seed colors produces matching colors in the weed seeds; winnow the seeds to sort for density, and the weed seeds match the densities of the crop. These evolutionary changes can develop quickly and can modify the plants considerably. There are some weeds, the aforementioned cockle and the field bineweed (*Convolvolus arvense*), for example, which now are found only in fields.

Other weeds have evolved other complex strategies for self-perpetuation in agricultural fields. An initial adjustment in weeds often tunes the time of the maturation of the weed seeds to match the crop harvest interval. The wild oat (*Avena sterilis*) has a germination pattern such that only about half of a given year's production of seeds germinates in the following year. The rest remain dormant to produce for future germinations in later years.[45] Some weeds in wheat fields (wild rye,

Secale cereal; and hooded canary grass, *Phalaris paradoxa*) have seed spikes that shed their lower part, which sprinkles seeds onto the soil. The seeds of the upper part of the seed spike remain. These seeds are threshed with the wheat and then show up next the year among the planted wheat seeds.[46] Herbicides are a chemical counteroffensive in the struggle between humans and weeds, but herbicide-resistant weeds are appearing in several areas.[47]

Pest Animals

Agricultural fields full of edible, high-energy plants and granaries loaded with seeds represent a strong incentive for animal species to alter their behavior or for natural selection to favor adaptations that allow access to these resources. Legions of species of birds, mammals, and insects and other arthropods were either preadapted or have rapidly adapted to take advantage of the stores of materials in human-dominated landscapes. Other creatures have taken advantage of the shelter and the moderation of the environmental extremes provided by human constructions. The "built environment" has its own menagerie, often of less than pleasant animals.[48] The agricultural landscape is no less full of noxious creatures. As is the case with weeds, some animals have undergone significant behavioral change or, in other cases, considerable evolutionary change.

A single pest animal removal service in Atlanta, Georgia, listed their services as removal of bats, beavers, carpenter bees, coyotes, dead animals, feral cats, flying squirrel, foxes, gray squirrels, groundhogs, honeybees, hornets, mice, moles, opossums, pigeons, raccoons, rats, skunks, snakes, squirrels, yellow jackets, wasps, and woodpeckers. This diverse assemblage of pests is not unique to Atlanta; this firm could be franchised in major cities all over the world. The length of this list is surprising, but the ecological roles of these animals also command one's attention. Some of the creatures, such as the house mouse (*Mus musculus*) and Norway rat (*Rattus norvegicus*), have cohabited with humans for millennia and have evolved to fit the human environment. Originating in Asia,[49] they are essentially found everywhere there are people, with the exception of Antarctica, the Arctic latitudes, and a few remote islands. Their close proximity to humans has led to domesticated varieties, many of which are widely used as laboratory animals. Other widespread creatures are formerly domesticated animals reverted to a wild state: pigeons (*Columbia livia*) and feral cats (*Felis catus*). Still others on the list are essentially local wild animals: they open garbage bags in search of food, eat food left out for pets (or eat the pets themselves), munch on garden plants, chew up parts of houses, dig up the lawn, find nest sites to shelter their young, and generally adapt to a human-dominated world. This adaptation to humans is an ongoing process underway across the Earth.

The control of pest animals, particularly the insects that destroy crops, has prompted a chemical war of long duration. Indeed, chemical warfare research for World War I and II catalyzed a new stage in this ongoing war against pests.[50]

Pesticides and insecticides have been in use for a long time. Pliny the Elder suggested, "a remedy against ants, which are by no means the slightest plague in a garden that is not kept well watered, to stop up the mouths of their holes with sea-slime or ashes."[51] Among a lengthy collection of similar stratagems, he also recommended the use of salt and ashes to prevent worms in figs and beaten larkspur to kill vermin.[52] Prior to 1800, pyrethrum made from the ground flowers of chrysanthemum was used as an "insect powder." Pyrethrum was harvested by hand and thus was too expensive to use broadly in agriculture.[53] Paris green, an arsenic powder, was used as an insecticide in the 1860s and 1870s in the United States, and lead arsenate was used at the turn of the twentieth century to control a great range of insects.[54] World War I brought a marriage between science and the military. In the United States, the Chemical Warfare Service stimulated research on war gases as insecticides.[55]

In 1874, O. Zeidler synthesized the chemical, dichlorodiphenyl trichloroethane, DDT.[56] DDT was discovered to be an effective insecticide in 1939 by P. H. Müller, who received the 1948 Nobel Prize in Physiology and Medicine for his efforts. DDT came into widespread use in World War II as a weapon against the mosquitoes that transport malaria. After the war, the insecticide was applied in large-scale agriculture against pest insects. In 1962, Rachel Carson published her book *Silent Spring*, which informed the public of the widespread presence of DDT in wildlife.[57]

Silent Spring communicated the potential risk of DDT and other similar insecticides called chlorinated hydrocarbons to the public. Within a few years, birds such as the peregrine falcon, brown pelican, and bald eagle were found to be laying thin-shelled eggs, an apparent consequence of the concentration of DDT up through the food webs that supported them. Decreases in these species were being observed and reported by birdwatchers by the later 1960s.[58] There was an obvious connection to human public health, in that the human agricultural food chain was also delivering DDT in food to people. DDT was banned in Germany and the United States in 1973 and by the United Kingdom in 1984. It was outlawed by the Stockholm Convention in 2004.

The Nobel Prize in Medicine for Müller in 1948 stemmed from DDT's efficacy against malaria. Fifty-six years later, its ban derived from its potential hazard as a widely dispersed chemical. A "principle of intended consequences,"[59] which dictates that positive changes can produce other, sometimes negative responses in inter-active ecological-environmental systems, creates difficult planetary decisions. If there is a moral to the DDT story, it would be that it is hard to do but one thing when modifying the environment. This moral also pertains to other cases involving introduced exotic species, the alteration of important environmental variables, changing land cover, and the other planetary issues raised by the Joban whirl-wind questions.

Disease

The transport of old diseases to new places and the emergence of novel diseases are also entwined to the human role as a planetary keystone species. New ecosystems and the movement of species of plants, animals, and microbes to new locations all form part of the human imprint on ecosystems. Zoonoses are diseases transmissable from vertebrate animals to humans. More than 75 percent of human diseases are known to have a link to wildlife and domestic animals and are zoonoses. One of an obviously large number of examples, Lyme disease (*Borrelia burgdorferi*), which was initially reported from Lyme, Connecticut, has spread across the United States. There are now cases in Europe and Asia. The disease is spread to humans by "hard ticks" of the genus *Ixodes*. Lyme disease bacteria have reservoirs in deer as well as mice, chipmunks, and other small rodents. Suburbanization and increased human interactions with ticks from wildlife have increased the transmission of the disease to people.

As organisms adjust to human-created ecological systems, the possibility of new diseases crossing over from wild disease reservoirs is increasingly likely. Diseases can be classified by the agent (infectious viruses, bacteria, fungi, and parasites) and the number of hosts in the infectious cycle.[60] Two-factor complexes consist of agents transmitted directly from person to person (one agent, one host), such as poliovirus. Three-factor complexes involve transmission through a vector or invertebrate intermediate host, such as a snail, mosquito, or tick or other arthropod. Dengue and malaria, for instance, are transmitted from person to mosquito to person. Four-factor complexes involve transmission between a nonhuman vertebrate and an arthropod, with humans usually being accidental hosts. Eastern and western encephalitis viruses, for example, are maintained in a reservoir of mosquitoes and birds, with transmission to humans as a spillover.

When people change ecological conditions, they may create breeding sites for disease vectors and vertebrate hosts. For example, malaria has become epidemic in the western Amazon region of Brazil, where the population has grown tenfold since 1970. The immigrants are involved in gold mining and forestry and have settled in areas undergoing rapid deforestation.[61] *Anopheles darlingi* mosquitoes breed in the standing water of open-cast mining sites and forest clearings. Initially, people from malaria-endemic regions of Brazil arrived already infected to seed the area. As a second example, the triatomine bugs ("kissing bugs"), which can transmit Chagas' disease, live in the mud walls of homes in Brazil and Argentina.[62] A change in construction methods could eliminate the bug's hiding places. Spraying homes with insecticides with residual activity in some cases also has controlled the insect.

Viruses, parasites, fungi, and bacteria have evolved from their ancestral forms along with the ancestral forms of their animal reservoirs. As new species have evolved, so have new versions of the microbes that inhabit them. The high diversity of plants and animals in the tropics is accompanied by high species diversity of their associated microorganisms. Greater host diversity implies greater diversity of disease agents.

Human incursions into undeveloped regions serve to increase the exposure of humans to new, so-called emerging diseases. Some of these are simultaneously terrible and lethal, such as the viral disease Ebola hemorrhagic fever. In 1976, Flemish nuns in a missionary hospital in the Ebola River Valley in the Republic of the Congo first discovered this disease. It has had periodic and deadly outbreaks in the region since that time. Fruit bats may be a host for the disease.[63] Focality is a term that describes the degree of localization of a disease. The relatively sudden appearance of a deadly human disease that clearly seems to have a reservoir in some rainforest animals, such as Ebola, represents a disease with high focality. Particularly in the tropics, the often local distribution of the reservoir species creates localized diseases with high focality. This implies that there are potentially numerous and as yet undescribed disease agents that are infecting wild vertebrates and vectors in tropical forests and have the potential to cause disease in people.

Hopefully, our remarkable progress in medical science will insulate us from emerging diseases, at least to some degree. The whirlwind question, "Who has set the wild ass free?" foreshadows the modern era: humans are setting all manner of things free, often in new locations. Next, let us consider the consequences of these releases.

Humans have been keystone species for millennia and are consolidating their role at larger and larger spatial scales. As the planetary keystone species, humans currently are engaged in an ad hoc program of Earth's alteration. Because the planet is so strongly modified by human action, the term "Anthropocene" has been proposed to describe our current state, as if it were a new geological epoch.[64] Several aspects of the changes that humans have wrought are expressed on the chemistry of the planet and particularly the chemistry of the atmosphere. These changes are the topics of chapters that follow. Earth, in the Anthropocene epoch, is undergoing profound ecological rearrangement as well. Human-altered ecosystems change under the influence of the human keystone species and in the human release of plants and animals to new parts of the planet. One pronounced biological effect of the Anthropocene has been the dispersal of introduced species across the planet, with an associated spreading of feral versions of our domesticated animal portfolio.

INTRODUCED SPECIES

An old-fashioned term from the formative era of ecology is chorology, the science of the geographic spread of organisms and the study of the rules of distribution of living organisms across the Earth's surface.[65] Chorology has revealed changes in the positions of continents over geological time and the geographic origins of different plants and animals. The riddle of the underlying causes of species distribution has promoted speculation and theory building in ecology for the past two hundred years. One early chorological theory was produced by the great Swedish biologist Carl von Linné ("Linnaeus"), the same person who invented and developed an important system of describing species. The modern basis for naming and relating species today derives from the Linnaean classification system.

The classification of species using a genus/species binomial name was developed by Linnaeus. The first edition of his book, *Systema Naturae*, was a short pamphlet produced in the Netherlands, where he served as a medical doctor, in 1735.[66] It continued to be issued in new and large editions throughout his life. His use of the parts of flowers as a basis to classify plants caused rancorous debate. Johann Georg Siegesbeck criticized and eventually ridiculed him for being too focused on sexual matters (flowers and flower parts) in his classification.[67] The ensuing smear campaign originating from Siegesbeck left the issue sufficiently touchy that the Pope forbade Linnaeus's works from the Vatican.[68]

Linnaeus's important tenth edition,[69] which included a classification with scientific (binomial) names of animals, was produced in two volumes in 1758 and 1759. He had listed binomial names of plants in his earlier book, *Species Plantarum* (1753).[70] His classification of humans, which grouped them with the apes, also touched off a firestorm.[71] The issue was religious: if man was created in God's image, it was blasphemy to associate their appearance with that of simians.[72] Darwin's *Origin of the Species* prompted the same response a century later. For all of this religious tumult, Linnaeus himself was a strongly religious person. He was drawn to natural theology, a popular philosophy in the eighteenth century. Its premise posited that since God created the world, one can understand God's wisdom through studying the natural world, God's creation. Considering the Book of Job's whirlwind questions would be an obvious pursuit in natural theology.

Chorologically, Linnaeus thought that unchanging species had spread across the Earth from the landing point of the biblical Noah's ark (Genesis 8:4) in the mountains of the ancient proto-Armenian kingdom of Ararat (860 BCE–590 BCE). Ararat was also known as Urartu.[73] The inhabitants of Noah's ark would have dispersed from there to generate the floras and faunas of the regions of the world. Now, if this were true, he reasoned, species experiencing similar environmental conditions but in different locations would have originated from Noah's landing point and should be the same. Further, the diversity of species of a given sort should be lower the farther one moved from the Ararat dispersal point.

Linnaeus's Ararat-centered chorology wilted in the face of scientific exploration, which simply showed this not to be the case. By 1762, Comte Georges Louis Leclerc de Buffon observed that the large, ecologically similar mammals of the tropical parts of the Old and New Worlds are quite different taxonomically.[74] Alexander von Humboldt expanded Buffon's observation that unrelated species are found in separated but environmentally similar locations (sometimes called Buffon's Law) to flowering plants, birds, reptiles, insects, and spiders.[75]

Modern chorologists might have had a more difficult time in discerning the geographical patterns of flowering plants than Buffon or Humboldt. Locations with a long history of trade and transport with other regions have large numbers of alien species that have been transported from elsewhere: 36 percent of the vascular plant species in New York State are introduced alien species, 28 percent in Ontario, 11 percent in Poland, 16 percent in Finland, 33 percent in Norway.[76] Islands both nearby and remote have even higher numbers: the British Isles have 43 percent alien plant species, the Canary Islands 35 percent, Bermuda 65 percent, New Zealand 40 percent, the Galapagos 30 percent, Hawaii 44 percent.[77] Today, disproving Linnaeus's Ararat theory of plant and animal distributions by reading the current patterns of species would be made more complex by the "haze" of alien species in unexpected locations all over the world.

INTRODUCED FERAL SPECIES

The planetary increase of introduced plants and animals and particularly the establishment of novel feral animals immediately come to mind when one hears the Joban whirlwind question, "Who has set the wild ass free?" The wild horse and the wild donkey are teetering near the brink of extinction, but the feral versions of their domesticated cousins are prospering. When domesticates escape or are released into the wild, they often are occupying human-altered landscapes. These "semiwild" landscapes, created by the human keystone species, share attributes with the agrarian landscapes in which the formerly domesticated species were husbanded.

Domesticated animals have been selected for several different traits, one of which is elevated reproductive rates. Often domesticated animals have more frequent reproductive intervals than do their wild primogenitors. Creatures that breed once a year in the wild breed might twice a year when domesticated. Species in which the presence of an alpha female suppresses reproduction in subordinate females may lose this trait when domesticated. Species with multiple births may have larger litters. Artificial selection in domestication emphasizes fertility. The high reproductive capacity allows feral domesticated animals to explode into human-altered habitats and to take rapid advantage of novel situations.

Domesticated animals are transported to new places where, if they escape captivity, they may enjoy biological advantages over the native fauna. Australia is an excellent example of this phenomenon. Australia has feral populations of water buffalo, horses, camels, pigs, cats, rabbits, and an introduced menagerie of other creatures. Along with these escapees are other invasive species that have been released either accidentally or purposefully all over the world. For all these animal cases, there are matching and often more numerous plant examples. Two principal factors contribute to this epidemic of feral and invasive species. One is the substantial modification of habitats, worldwide, by human actions. Another is the spread of species from one place to another by humans.

The introduction of the rabbit to the Australian continent is an excellent example of the conjunction of both of these factors: habitat change and human transport leading to the successful invasion of much of a continent by an alien feral animal.[78] Domesticated European rabbits (*Oryctolagus cuniculus*) and European people arrived on the same boats, the First Fleet to Botany Bay, to become Australian residents in January 1788.[79] Subsequently, the rabbit was shipped to several other ports in Australia. It was introduced to the island of Tasmania (in 1820), where it eventually became established as a feral animal.[80]

On Christmas night, 1859, the rabbit's explosive invasion of the Australian mainland started in Melbourne, with the arrival of twenty-four rabbits unloaded off of the Black Ball clipper ship *Lightning*, sailing out of Liverpool.[81] These would have been feral domesticated rabbits: neither was the rabbit native to the British Isles.[82] The rabbits, along with sixty-six partridges and four hares, had been shipped to Thomas

Austin by his brother, James, who was still in England. Thomas may have immediately released some of his rabbits onto his estate. By 1862, Thomas Austin's rabbits, held in a fenced enclosure on his estate, Barwon Park, near the town of Winchelsea, Victoria, numbered in the thousands. The rabbits became so numerous that they soon were chewing the bark off of the trees, and eventually they either escaped or were released.[83] They were damaging neighboring farms by the early 1860s. By the mid-1860s, rabbit hunts and rabbit coursing were popular field sports among the wealthy in several locales where rabbits at been loosed on the land.

Without native rabbit diseases, the rabbit moved rapidly into a climate and habitat that matched its requirements very well. In the decade of the 1860s rabbits spread like a wave over Australia, advancing as much as one hundred kilometers per year—the most rapid spread known for any colonizing mammal anywhere on Earth.[84] The 1860s was a dry decade in Australia. Pastures were degraded. Sheep populations crashed to as much as half their former density. The rabbit population exploded into the very sort of degraded landscape for which it had been bred by its human domesticators. The combination of a small fluctuation in climatic conditions and a human-altered landscape set the conditions for the rabbit to take over Australia.

The rabbit's invasion of Australia is a story that repeats in other places at other times. At the dawn of history, the European rabbit was restricted to the Iberian Peninsula (present-day Spain and Portugal). The species was also found in a small region in northern Africa, but it may have been transported there by the Phoenicians. The animals have been spread across the world, often intentionally, by people. Pliny the Elder reported outbreaks of rabbits that had been introduced to the Balearic Islands off of the Mediterranean coast of Spain: "There is also a species of hare, in Spain, which is called the rabbit; it is extremely prolific, and produces famine in the Balearic islands, by destroying the harvests. . . . It is a well-known fact, that the inhabitants of the Balearic Islands begged of the late Emperor Augustus the aid of a number of soldiers, to prevent the too rapid increase of these animals."[85]

Through time, sailors have transported rabbits and have released them onto islands to provide future provisioning while en route on subsequent voyages. This practice started with the early Phoenicians (1500–300 BCE). Roman sailors brought rabbits to the Balearic Islands around 30 BCE and to Corsica by 230 CE. Subsequent sailors have spread rabbits to over eight hundred islands worldwide.[86]

DABBLING WITH PLANETARY DIVERSITY: THE PURPOSEFUL INTRODUCTION OF SPECIES TO NEW PLACES

Near the city of Ontago on the Southern Island of New Zealand, a stone monument sits on the side of the road. It reads,

> This monument commemorates the centenary of the first libera-tion of Red Deer in Ontago in March 1871. The deer were pre-sented to the Ontago Acclimatisation Society by the 11th Earl of Dahlhousie of Brechen in Scotland and shipped to Port Chalm-ers on the *City of Dunedin* and *The Warrior Queen*. Seven deer were liberated in this area after being shipped to Oamaru in the paddle steamer *Wallace* and transported over Lindis Pass by bull-ock wagon. These deer formed the basis of the world renowned Otango South Westland red deer herd. Erected by the New Zea-land Deerstalkers Association Inc. North Ontago Branch, 1971.

The effort to capture red deer (*Cervis elaphus*) in Scotland, ship them halfway around the world on two boats (the *City of Dunedin* and the *Warrior Queen*), then load them onto a paddle steamer and then into ox carts to haul them to the outskirts of Ontago for release seems like a remarkable amount of work and expense. One would expect such an endeavor to be attempted (and succeed) but once. That would be incorrect. Starting in 1861, red deer, mostly from Scotland, were intro-duced at different locations in New Zealand more than 220 times![87]

Included among the brace of mammals (thirty-three species), birds (forty-four species), and other species introduced to New Zealand from elsewhere were a remarkable number of species of deer from all over the world.[88] Hunting may have been the sport of nobles in England, but it was the pastime of the commoner in New Zealand.[89] This egalitarian spirit may have compelled the eventual introduction of the red deer as well as sika deer (*Cervis nippon*), Javan rusa deer (*C. timorensis*), sambar deer (*C. unicolor*), wapiti or Rocky Mountain elk (*C. elaphus nelsoni*), fallow deer (*Dama dama*), white-tailed deer (*Odocoileus virginianus*), chamois (*Rupicapra rupicapra*), and the Himalayan thar (*Hemitragus jemlahicus*). The diversity of deer in New Zealand appears to be affecting the ground vegetation from their grazing, but the long-term effects are far from clear.[90]

Many of the species released in New Zealand and elsewhere were the product of "acclimatization societies," founded in several nations, many of them European colonies. Inspired by the writing of the French naturalist Saint-Hilaire,[91] acclimatization societies spread worldwide after being first founded as the Societe Zoologique d'Acclimatation in Paris. Their objective was to establish species for pleasure and food in locations around the world. Some of the rationales of these releases seem surreal from a modern viewpoint. For example, the line from Shakespeare's 1597 *Henry IV*, "Nay, I'll have a starling shall be taught to speak nothing but 'Mortimer,'" inspired the American Acclimatization Society in New York, who were intent on introducing all the species mentioned in Shakespeare's works to the New World, to release European starlings (*Sturnus vulgaris*) in Central Park in New York City.[92] Starlings have subsequently become a pest animal. Some of the acclimatization societies' other projects included the introduction of house sparrows (*Passer domesticus*) around the world, the already mentioned importation of various game animals to Australia (rabbits) and New Zealand (deer), Australian possums to New Zealand, and European carp and trout to river drainages worldwide.[93]

Should the release of red deer and a menagerie of other creatures to New Zealand seem the product of eccentric mores from other era in a faraway place, today "game ranches" in the American West are being stocked with a brace of exotic game animals for fee-paying trophy hunters. One ranch in Texas selected at random as an example has fifteen species of exotic large mammal species from elsewhere in North America, Europe, Asia, and Africa. There is a branch facility of the same operation in England that offers eight species of deer for the hunter. These are commercial ventures that house valuable animals. The cost to shoot the largest of the African antelopes, the eland (*Taurotragus oryx*), is, at the time of this writing, $4,750. The owners of these animals are unlikely to want them to wander off of their property. Nevertheless, escape is always possible. These established populations of introduced animals are not radically different from the efforts of the acclimatization societies of over a century ago.

THE MAGNITUDE OF THE DIVERSITY OF INTRODUCED ALIEN SPECIES

The intentional introductions of species are accompanied by nonintentional commercial transportation of species in agricultural produce and soils and as hitchhiking seeds. There are also poorly conceived releases of exotic pets, such as those that have allowed the establishment of Burmese pythons (*Python molurus bivittatus*) in the Florida Everglades. We are awash with introduced alien species. In 1958, Charles Elton, a population ecologist at Oxford, wrote about the new planetary mixture of species,

> The whole matter goes far wider than any technological discussion of pest control, though many of the examples are taken from applied ecology. The real thing is that we are living in a period in earth's history when the mingling of thousands of kinds of organisms from different parts of the world is setting up terrific dislocations in nature. We are seeing huge changes in the natural population balance of the world.[94]

How much of a novel mixture of species is involved in the modern transfer of species? In terms of novel species tallied by continents (excluding Antarctica), there are on average a minimum of 8,350 species of plants found on each of the six continents that have come from someplace else. These are minimum estimates because they tally the species that appear to be established and have been listed. There may be other species that have not been documented. Other minimum estimates of alien species are 155 species of introduced mammals on average across the six inhabited continents, thirty-four reptiles and amphibians, and thirty-nine species of birds.[95] These do not include the thousands of invertebrates or the tens of thousands of microorganisms that have been displaced to new continents.

The magnitude of this biological expatriation is staggering, but darker implications hide in the details.[96] In North America, it is thought that at least twenty thousand exotic microorganisms have established themselves from other continents. How might an inspector of agricultural products know that one of these was *Cryphonectria parasitica*, or chestnut blight, a plant disease that eliminated the American chestnut (*Castanea dentata*) as a canopy species from the eastern North American forest, where it once was a dominant species? European rabbits had been present as an escaped domesticated animal since the first colonists arrived in 1788. How one could expect that a small shipment of rabbits from England seventy-one years later would touch off the European rabbit invasion of Australia? Who would have guessed that the mongoose (*Herpestes javanicus*) brought to Fiji to control rats would turn out to be a voracious predator of birds, small mammals, reptiles, and sea turtle eggs—especially after all its good press in Kipling's *Rikki-Tikki-Tavi*? The potential risk in the large wave of invasive species challenges research scientists to provide practical solutions, a difficult task, to say the least. The basic issues boil down to a few questions: Which species or groups of species are most likely to invade? How rapidly can they invade? What makes an ecosystem vulnerable to invasion? How can we control or eliminate invasive species?[97] These are not easy questions to answer.

Introduced Diseases

One of the more worrisome aspects of the current high rate of bio-logical exchange is in the possible transfer of novel diseases from one place to another. We have past historical examples of the potential effects of novel disease on human populations. The initial contacts with the indigenous peoples of the New World featured horrific levels of death associated with Old World diseases. Thomas Hariot, exploring on the behalf of Sir Walter Raleigh, traveled through the region now known as the Outer Banks of North Carolina in 1585. His book reporting his travels included invaluable illustrations of the indigenous inhabitants and their villages. In one striking section of this narrative, Hariot noted,

> There was no town where . . . we leaving it unpunished or not revenged (because we fought by all means possible to win them by gentleness), but that within a few days after our departure from every such town, the people began to die very fast, and many in a short space, in some towns about twenty, in some forty, in some sixty and in one six score, which in truth was very many in respect to their numbers. This happened in no place that we could learn but where we had been. . . . The disease was so strange, that they neither knew what it was or how to cure it, the like by report of the oldest men in the country never happened before, time out of mind—a thing specially observed by us, also by the natural inhabitants themselves.[98]

It is chilling to imagine a small band of British soldiers and explorers, all young and healthy, visiting villages in a newly discovered land and leaving death in their wake. But it clearly illustrates the virility of novel disease on human populations.

Even worse were the earlier Spanish accounts of outbreaks of small-pox. Smallpox was imported from Spain to Hispaniola in the Carib-bean in 1507. It exterminated entire tribes, but then the disease died out. The African slave trade reintroduced the disease to the Caribbean as well as to Mexico and Brazil.[99] An outbreak in Hispaniola originating with African slaves working in the mines wiped out about one-third of the native inhabitants. The outbreak spread to Cuba in 1518 and then to Puerto Rico in 1519. It killed over half of the people on both of these islands.[100] Narváez landed on the Mexican coast near modern Vera Cruz in 1520 with an African slave with smallpox on board. The result was described by a Spanish friar in 1525:

> At the time that Captain Pánfilo de Narváez landed in this country, there was in one of his ships a negro stricken with smallpox, a disease which had never been seen here. At this time New Spain was extremely full of people, and when the smallpox began to attack the Indians it became so great a pestilence among them throughout the land that in most provinces more than half the population died; in others the proportion was little less. For as the Indians did not know the remedy for the disease and were very much in the habit of bathing frequently, whether well or ill, and continued to do so even when suffering from smallpox, they died in heaps, like bedbugs.[101]

The lethality of Old World diseases such as smallpox was also seen in other diseases that had crossed the Atlantic Ocean with European explorers, whose ancestors had been in long contact with domesticated animals and crossed-over diseases. These would include measles (a viral disease originating from the cattle disease rinderpest) and influenza (a diversity of virus strains from humans and several domesticated animals including fowl and swine).[102]

The novel-contact disease remains a problem for relatively isolated people who have had minimal contact with the outside world. Hopefully, modern medicine will insulate most of us from the plagues of the past, but this is clearly not a time to relax our collective vigilance in the field of disease control. The spread of plant and animal diseases is another matter. The diversity of potentially infectious plant and animal diseases, as well as the spread of a growing assortment of insect pests, presents a major problem that has only been magnified by the increased globalization of trade.[103] The volumes of things that are moved intercontinentally to feed, clothe, and equip the world's societies are simultaneously providing potential vectors for all sorts of organisms.[104]

ECOSYSTEMS AND ANTHRO-ECOSYSTEMS
ON A HUMAN-DOMINATED PLANET

Ecologists study relatively pristine ecosystems to understand the consequences of many complicated ecological processes that interact to produce observable patterns. Mature, relatively undisturbed ecological systems provide an essential baseline to understand how small environmental changes might alter these natural systems. In a procedural sense, astronomers and ecologists approach their objects of study in a similar manner. Ecosystems and solar systems are not easy experimental objects. The modus operandi is to observe the systems, compare them, and understand them in terms of underlying processes that cause what is observed.

Comparison of large-scale patterns of diversity across different continents yields a regular relationship between the size of a continent and its species diversity. Larger continents have more species. What if the increased continental exchanges of species made the Earth's land diversity function as if it was a large interconnected continent? If the current relationship between continent size and diversity holds, there should be a sizable reduction in biodiversity on a connected Earth. For mammals, this land area–to-species relationship implies that a thoroughly connected world might have about half as many mammal species as are now found on the separated land areas of our planet.[105]

We live in an era in which truly undisturbed, mature ecosystems are becoming rare.[106] The studies of natural systems that are not greatly modified by humans provide necessary insight on ecosystem function. Early ecological research focused on such systems, and they are no less valuable now. However, an additional understanding of human-modified ecosystems and their functioning commands study because the planet continues to be altered by human action. We know from the fossil and geological record that past ecosystems with different mixtures of species and different environmental conditions coalesce, persist, and then eventually change over time. The instances in the geological past in which floras and faunas mingled after the formation of land bridges often featured the extinction of many species. For example, with the formation of the isthmus of Panama around three million years ago, the remarkably diverse marsupial mammal fauna of South America collapsed and was replaced by more advanced placental mammals from North America. Thus, we have reason to believe that the species we have loosed across the Earth will also change the planet.

The question remains: "What will these new ecosystems be like?" One would like to be optimistic. Along a roadside, introduced trees and shrubs colonize wasteland, substituting greenery for bare earth. We are drawn to gardens and arboreta loaded with exotic plants from all over the world, and we take pleasure in this human-created biological diversity. In New York City, peregrine falcons swoop down to catch pigeons from the canyons between skyscrapers and feed their young in nests on the high building ledges. Pigeons are a domesticated cliff-dwelling rock dove. The peregrines are rebuilding their populations, which had been decimated by DDT use. The entire scene is an urban version of falcons hunting their prey on the face of a Scottish cliff—a scenario featuring a truly wild event that most people would never hope to see.

While there are positives, at least from some points of view in all of this, there are also significant negatives. Many of these stem from feedback loops between the exotics and the ecosystems they inhabit. Fire-tolerant alien plants prosper under fires and create additional fuel for more frequent or hotter fires. Introduced fish eliminate natural fisheries that support coastal towns. Inedible or even poisonous weeds invade pastures and prosper when the native grasses are overgrazed by cattle or native animals. Eucalyptus trees introduced in the subtropics to supply fuel wood for cooking or heating plunge their roots into the ground water and reduce the already scant water supply. Housecats functioning as predators and subsidized by their owners reduce bird populations in suburban areas. The tiger mosquito (*Aedes albopictus*), an extremely invasive species worldwide and imported to the United States in cargo from southern Asia, can vector several diseases, including West Nile virus, yellow fever, St. Louis encephalitis, dengue fever, and Chikungunya fever.[107] Once one realizes that some of these alien species are a problem, it can be virtually impossible to remove them—it is hard to get the genie back in the bottle.

CONCLUDING COMMENTS

The onager as depicted in Job is an analogue to the *unicorn* translation of the *re'em* in the King James translation of Job 39:9 discussed in the previous chapter. The onager (the wild ass or swift ass) and the *re'em* (when translated as unicorn, as in the *King James Bible*) represented a creature of untamable wildness that would remain so despite human efforts. The onager was an unusual case in that the "domesticated" form was a hybrid. When it was replaced by the horse, the hybrid was no longer created through a process of onager taming and interbreeding with donkeys. The domesticated hybrid disappeared and only the wild form remained. It was "set free" in a rather special sense. Unfortunately, many other creatures that we have domesticated, tamed, or captured do not have such internally programmed self-destruct buttons when set free. Instead, many of them have come to fill niches in other locations and, in particular, in places influenced by the human keystone species. The global implications of this topic will be taken up again in chapter 8.

Thus, the whirlwind question of "Who has set the wild ass free?" has hundreds of similar questions that can be asked all around the world. In New York City, "Who has set the starling free?" In suburban London, "Who has set the cat free?" In Meso-America, "Who has set smallpox free?" In Australia, "Who has set the rabbit free?" In Fiji, "Who has set the mongoose free?" For these and myriad analogous questions, the answer is often, "We have." Out of intent or by accident, casually or with great planning, for plants or animals (or microbes), for organisms large or small—we have set them free on the surface of a planet that we have modified but do not fully understand.

Glacier, ice, and ocean from Alaska's Inner Passage. *Source*: Photograph by the author.

5

Bounding the Seas, Freezing the Face of the Deep

When the Sea Is Loosed from Its Bonds

Or who shut in the sea with doors when it burst out from the womb?—when I made the clouds its garment, and thick darkness its swaddling band, and prescribed bounds for it, and set bars and doors, and said, "Thus far shall you come, and no farther, and here shall your proud waves be stopped"?

—Job 38:8–11 (NSRV)

From whose womb did the ice come forth, and who has given birth to the hoarfrost of heaven? The waters become hard like stone, and the face of the deep is frozen.

—Job 38:29–30 (NSRV)

As was pointed out in chapter 2, the "whirlwind speech" creation narrative is different from other biblical creation stories (Psalms 74:13–14, 89:10–14; Isaiah 51:9–10) and from other creation accounts in the region. In the Babylonian *Enūma Elish,* a lightning-welding warrior god named Marduk splits the body of the primordial sea goddess, Tiamat, in half and thus forms the world. As in the Joban account, Marduk constrains the seas within an enforced boundary demarcating the sea and the land.[1] While the boundary aspects are similar, the *Enūma Elish* account differs in others. It is violent and combative. This is the case for myths of a divine warrior conquering the sea (or sea monsters), which is seen as a primordial force and emblem for chaos in many Near Eastern texts.[2]

By contrast in Job, God functions in the whirlwind speech as midwife to the sea's birth. He is the sea's collaborator and not its opponent.[3] The sea bursts from the womb and is dressed in clouds. It is swaddled in darkness to calm its thrashing. The sea is birthed, calmed, and put into its proper place in the cosmos. This stands in sharp contrast to the divine warrior dominating sea monsters in other regional mythologies.[4] In the latter part of the whirlwind speech in Job, God sports with the fearsome forces of his creation—the Leviathan, a fire-spitting, heavily armored sea dragon, which God can play with as if it were a bird, and the Behemoth, a gigantic, amphibious beast that only God can approach. In dealing with the sea, the Leviathan, and the Behemoth, God is at ease with powerful entities that He has created and that He nurtures and enjoys.

Standing on the shore of an ocean, the waters seem to be at odds with an unseen opponent. The waves wash in and recede only to come back again and again in a regular rhythm. Watch the waves long enough, and one realizes that it is a rising tide. The sea is winning its incremental warfare to gain the land. But come back a few hours later, and the tide has turned: the sea is losing to the land. It is calming, even spiritual, to watch. One knows that the tide is driven by the positional change of the Sun and Moon with respect to Earth, but the sea seems alive and pulsating. Return to the same beach after the storm of the century or the tsunami of a millennium, and one understands the fierceness of the sea crashing against its bounds.

THE TIDES

The most obvious pulsations of the seas, its heartbeats, are the tides. The drivers behind these oceanic rhythms are the Moon and, to a lesser extent, the Sun. Most of the world's coasts, including much of the North and South American Atlantic coasts, experience two cycles of tides daily, with tidal of peaks and troughs of roughly the same size for both. In some locations (for example, Cuba, New Guinea, and much of the Pacific coasts of North and South America), one of these tides is more extreme than the other. Finally, there are coasts, such as those of the western Gulf of Mexico, of Western Australia, and of the southern end of the South China Sea, that have but one tidal cycle daily. These variations are attributable to complex consequences of the shape of the ocean floor (bathymetry) and the geometry of the shoreline.

These same bathymetric considerations produce extreme tides in some locations. Bodies of water have frequencies of variation that relate to depth, size, and shape. Imagine holding a large bowl of half full of water. If you wiggle the bowl to and fro, there is a frequency at which the period of "slosh" of the water in the bowl matches frequency of your moving of the bowl—the "waves" will increase until the water spills over the edge and onto the floor. Similarly, when the frequencies that the tides push and pull a large basin of water (such as the roughly 12.5-hour periodicity in the tide from the gravity of the Moon or the twelve-hour component caused by the influence of the Sun) match the period of the natural "slosh" of the water body, tides are amplified. This same concept, starting in the Victorian era, was used to construct tide-prediction machines, which will be discussed later in this chapter. The record-holding location for tides has long been considered to be the Bay of Fundy, between New Brunswick and Nova Scotia. The Bay of Fundy's tides may actually be tied with those of Ungava Bay, a large bay off of the Hudson Strait and surrounded by northern Quebec. Both have high tides of around seventeen meters.[5] Ungava Bay is covered with ice most of the year; the Bay of Fundy is ice free.

The connection among tides, Moon, and Sun has been known since ancient times. Pytheus of Massalia sailed to northwestern Europe and circumnavigated Great Britain in 325 BCE. Along with documenting polar ice and the midnight Sun that does not set in the far north on the summer solstice, Pytheus also reported that the Moon caused the tides. This is apparently the first written record noting this interaction.[6] The Babylonian astronomer Seleucus of Seleucia in the second century BCE linked the variations of the tides with the Sun and Moon as part of his demonstration that the Earth rotated on its own axis as it orbited around the Sun.[7] Seleucus understood the tides were attributable to the attraction of the Moon, and he also realized that a longer-term variation in the height of the tides depended on the Moon's position relative to the Sun.[8]

While Pytheus and Seleucus may represent early writings on the tides and celestial movements, the understanding of such relationships goes back even further into history. This will be discussed in chapter 6 as a consideration of another of the whirlwind questions, "Do you understand the ordinances of the heavens" (Job 38:33). The cycles that are seen in nature are driven by cycles, such as the orbital cycles of the Moon and Sun or the spin of the Earth on its own orbit, which do not multiply and divide into one another equally. This makes reconciling a calendar based on solar days, lunar months, and solar years quite complex. One gets a glimpse of this complexity when considering the tides and their regular variations.

The relative position of the Moon and Sun are indicated by how much of the face of the Moon is illuminated. Along coasts in which there are two tides a day, the differences in the height of high and low tides are at a maximum when the moon is full and again when the moon is new. These are the two times when the Sun, Moon, and Earth lie on a straight line. Such an occurrence happens about once every two weeks and is known as syzygy,[9] which would be a killer Scrabble word (if Scrabble had more than two *Y*'s). The tides at syzygy are larger, typically about 20 percent greater than normal, because the tidal effect of the Moon and Sun reinforce the other. Such maxima are called spring tides, in the sense of spring meaning "to move or jump suddenly or rapidly upwards."[10] Spring tides do not refer to the spring season. When there is a first- or last-quarter moon, the Moon's and Sun's tidal effect partially cancel the other, and the difference between high and low tides is at a minimum. This condition is called a neap tide. The meaning of "neap" is not certain, but it originates in Old English.[11]

The regular orbital variation in tides also has longer-term patterns attributable to different combinations of the conditions that affect the orbits of the Sun, Moon, and Earth. The trifecta of orbital combinations, called a perigean spring tide, occurs when a lunar perigee (the closest approach of the Moon's elliptical orbit to the Earth) occurs when Earth, Moon, and Sun are in syzygy. Under these conditions one gets much greater tidal range, with higher highs and lower lows than average. When the Moon's orbit is furthest from the Earth (apogee) on a first- or last-quarter moon, there is an apogean neap tide with a greatly reduced tidal range.

PERIGEAN SPRING TIDES AND APOGEAN NEAP TIDES

The perigean spring tide and its antithesis, the apogean neap tide, appear both in literature and in history.[12] As an early literary example, "The Franklin's Tale" in Chaucer's *Canterbury Tales* is thought to be related to a perigean spring tide that occurred in December 1340. Geoffrey Chaucer had a keen interest in astronomy, which is particularly evident in his works after about 1385.[13] In the Franklin's Tale, Averagus seeks knightly honor and leaves his wife, Dorigen, alone near the town of Penmarc'h in modern Brittany. Dorigen has two principal concerns, the premonition that her beloved husband's ship will wreck on the black rocks of the Brittany coasts as he returns and the unwanted attention of a would-be suitor, Aurelius. Hoping to discharge Aurelius's attentions, she tells him,

Looke what day that endelong Britayne	On whatever day that from end to end of Brittany
Ye remoeve alle the rokkes, stoon by stoon,	You remove all the rocks, stone by stone,
That they ne lette ship ne boot to goon—	So that they do not prevent ship nor boat to go—
I seye, whan ye han maad the coost so clene	I say, when you have made the coast so clean
Of rokkes that ther nys no stoon ysene,	Of rocks that there is no stone seen,

Thanne wol I love yow best of any man;	Then will I love you best of any man;
Have heer my trouthe, in al that evere I kan.	Have here my pledged word, in all that is in my power.[14]

Aurelius hires a magician, the rocks are covered by an exceptionally high tide, and the plot thickens.

In military history, the tides and phases of the moon (and their interactions) have long been a part of military tactics, particularly amphibious landings. In September 13, 1759, the British general Wolfe calculated the complex tidal delays of the St. Lawrence Estuary and timed his attack on the French forces on the Plains of Abraham.[15] Wolfe's forces paused awaiting favorable tides on a dark night and altered the history of Canada and North America.

In the run-up to the American Revolution, a colonial ship's pilot steered the British ship the HMS *Cancaeux*, which was en route to reinforce Fort William and Mary near Portsmouth, New Hampshire, behind a shoal on a spring high tide. The *Cancaeux* was trapped for days, giving Paul Revere ample opportunity to spread the "Portsmouth Alarm,"[16] which warned the colonists to raid the undermanned fort for its gunpowder before the *Cancaeux*'s reinforcements arrived. The Boston Tea Party occurred on the low of a perigean spring tide, which explains, among other things, why the tea dumped into the Boston Harbor persisted there into the next day and why the water was so low that "three of the vessels lay aground."[17]

More recently in World War II, in the early hours of October 14, 1939, the German submarine *U-47* used a perigean spring high tide to slip into the British Navy's anchorage in the Scapa Flow in the Orkney Islands by an unexpected route. The U-boat torpedoed the battleship HMS *Royal Oak* with a loss of life of 833 men and boys—120 of those who died were "boy sailors" between fourteen and eighteen years of age. The sinking of the HMS *Royal Oak* was the British Navy's largest such loss of boy sailors either before or since.[18] The U.S. Marines' amphibious assault on Tarawa in the World War II Battle of the Pacific was launched on an apogean neap tide. The landing crafts hung on the reefs, making the eventual victory more costly.[19]

TIDE-PREDICTION MACHINES AND THE NORMANDY INVASION

In World War II, Hitler's planned 1940s invasion of England (*Unternehmen Seelöwe*, or Operation Sea Lion) was assumed by the British to occur on high tide to minimize the width of open beach to be crossed. On Churchill's request, the British Admiralty computed the most likely time for such an invasion as "when high water occurs near dawn, with no Moon" and computed such times for beaches near eight British ports. These calculations were developed on mechanical devices that computed the different frequencies that affected tidal change.[20]

The first such device was a machine of dozens of gears and pulleys designed to mimic the various Sun-Moon-Earth interactions. The device inked the tide changes for a given location on a rolling sheet of paper. The first such device considered eight different periodicities of tides and was developed by Lord Kelvin in 1872.[21] In 1882, William Ferrel developed a machine computing nineteen tidal components for the U.S. Coast Guard and Geodetic Survey.[22] Edward Roberts designed a forty-component device based on Lord Kelvin's design for the Bidston Observatory's Liverpool Tidal Institute in 1906,[23] and a second U.S. machine with thirty-seven tidal components was designed by Rollin Harris and built in 1912.[24] These huge brass machines were applied to compute the tides over the course of a year for reference stations all over the world.

Having failed to wipe out the Royal Air Force, Hitler cancelled Operation Sea Lion. The British Admiralty ceased to publish tide-prediction tables from the machines so as to not aid the German war effort—probably for naught. In 1915–1916, the Germans had constructed their own tide-predicting machine, which they used in World War I. A second German machine, developed by Heinrich Rauschelbach and built between 1935 and 1939, is now on display in the Deutsches Museum in Munich. Computing sixty-two tidal components, this was the largest such device ever built: 7.5 meters long, two meters wide, and a weight of about seven metric tons. This second German machine was used in the World War II U-boat campaigns.[25]

A most remarkable application of the British devices was in computing the tides for the Normandy invasion for the Allied Forces' Operation Overlord.[26] An overhauled version of the original 1872 Kelvin machine and the 1906 Roberts machine were housed in the Bidston Tidal Observatory in Liverpool. The devices were maintained in two separate buildings so that one bomb would not destroy them both. Six young women under the supervision of Arthur Thompson Doodson, the world's leading authority on tidal predictions,[27] worked long shifts, seven days a week, using these devices to compute the Admiralty tide tables.[28] In October 1943, William Farquharson, the Admiralty's superintendent of tides, requested Doodson to compute the hourly tides for Position Z (the codename for Normandy Beach) for April through July 1944. Farquharson had only incomplete tide data from some nearby beaches along with some recent data from midget submarines and small boats on the currents and tides. It is not clear how he managed to provide Doodson with the calibration settings for the machines.[29]

General Rommel, anticipating an Allied invasion that he felt would occur at high tide, had filled the possible French invasion beaches with underwater obstacles that would be underwater at midtide. These were spotted by reconnaissance airplanes. There were so many such obstacles that there was one "on every two or three yards of the front."[30] The invasion would need to be initiated on a low tide, so engineers in the advance forces could destroy these obstacles. On Normandy Beach, with six-meter tides, the tide rises at over one meter per hour from low tide, so the engineers destroying beach barriers would have to arrive on time and work very fast. There were other constraints as well. Bruce Parker notes:

For secrecy, Allied forces had to cross the English Channel in darkness. But naval artillery needed about an hour of daylight to bombard the coast before the landings. Therefore, low tide had to coincide with first light, with the landings to begin one hour after. Airborne drops had to take place the night before, because the paratroopers had to land in darkness. But they also needed to see their targets, so there had to be a late-rising Moon. Only three days in June 1944 met all those requirements for "D-Day," the invasion date: 5, 6, and 7 June.[31]

D-Day was delayed from the planned date of June 5 by bad weather. Eisenhower decided to go on June 6 based on a forecast of a thirty-six-hour clearing in the weather.[32] Rommel, believing that the Allies would invade on a high tide and noting the foul weather, assumed June 6 would not be the day and was in Germany celebrating his wife's birthday when the invasion began.[33]

The explanation of tidal phenomena seems to require a very phys-ics- and mathematics-oriented discourse. Considering Chaucer as a literary example of an unusual perigean spring tide or contem-plating the historical role of tides as constraints on military tactics hopefully makes this topic more approachable. The great brass tide-prediction machines, which could grind out predictions of tides for any given location, operated by combining periodic compo-nents of tides (Earth, Moon, Sun, their orbits and combinations of these orbits) with calibrations appropriate to the coastline and the bathometry of the location, are a tangible realization of the predic-tion of tides by the frequency components of the driving factors. The calculations that Doodson and his assistants churned out over days to schedule the D-Day invasion can be done in seconds on a modern digital computer.

One might ask how long these cycles upon cycles of tidal com-ponents can be. On January 4, 1912, a perigean spring tide (under a full moon) occurred within a few minutes of the Earth being at its closest to the sun in its orbit. The result was a lunar distance from the Earth of 356,375 kilometers, the closest in 1,400 years.[34] But there are even longer cycles in the orbital interactions that affect the climate.

PAST SEA LEVELS

If one considers the bounds of the sea (or the more mundane term, "mean sea level") over longer time periods, such as the Pleistocene epoch, the bounds of the sea seem much less proscribed by the "bars and doors" of Job 38:10. If one takes the long view over one hundred million years and considers the earlier stands of the seas, the relative height of the seas was 170 meters (between eighty-five and 270 meters) higher than those today.[35] These long-term variations in the height of the seas are strongly a product of the size of the sea basins, which can change in their dimensions under several geological actions.[36]

In geologically more recent times, say over the past 2.5 million years, the sea level has varied thanks to the formation and melting of the continental glaciers of the glacial Ice Ages. Twenty thousand years ago, the oceans stood about 130 meters lower than they do today. Reduce the ocean's height by 130 meters, and the U.S. Atlantic coast lies two to three hundred kilometers seaward from the current coastline. England connects to Europe and Ireland. The Thames and the Rhine flow together into the Dover Strait of the English Channel. Australia and New Guinea merge, as do the North and South Islands of New Zealand. Japan joins the Asian mainland; so do Java, Sumatra, Bali, and Borneo.

For the past 2.7 million years of the Pleistocene epoch, a remarkable amount of water (roughly fifty million cubic kilometers) has moved back and forth between the oceans to glaciers on the land as the climate has cooled and warmed.[37] This is unusual over the long run of geological time, which certainly has witnessed other ice ages of considerable duration. Alternating glaciations represent a minority of the total geological history. Freezing large volumes of ice on the land in vast continental glaciers evokes the Joban question, "From whose womb did the ice come forth, and who has given birth to the hoarfrost of heaven? The waters become hard like stone, and the face of the deep is frozen" (Job 38:29–30). One of the significant effects of ice stages during ice ages has been dramatic change in the sea level over the past 2.7 million years.

Geologically, we are in a time of the Earth when glaciers have regularly expanded and contracted. Cyclical changes in glaciation/ deglaciation during the Pleistocene epoch seem driven by variations in the orbit of the Earth and the angle of its axis. These cyclical changes are called the Milankovitch cycles after the Serbian mathematician, Milutin Milanković, who worked on the theory during his internment as a prisoner of war in World War I. Milankovitch cycles cause systematic variations in the incoming solar radiation from the Sun. For example, there is less incoming radiation when the Earth is furthest from the Sun in its orbit and more when it is closer. There are periodicities of around 400,000, 100,000, 40,000, and 26,000 years, which can be determined from interpreting long sediment cores extracted from the ocean floor. There are also variations in solar activity that can affect global heating and cooling.[38]

The wobble in the Earth's axis, one of the Milankovitch cycles, will be discussed in chapter 6 as a necessary correction in analyzing ancient monuments and megaliths for their possible use in predicting astronomical events. The 26,000-thousand-year wobble of the axis of the planet affects whether the land masses of the Northern Hemisphere are pointed toward the Sun when the Earth-to-Sun distance is smallest. At the other extreme of this cycle, the water-dominated Southern Hemisphere faces the sun at the minimum distance (or some position between these two extremes). Other Milankovitch components involve the cycle of the shape of the Earth's orbit between a circular or elliptical orbit, the cycle of change in the inclination of the Earth's orbit, the cyclical change in the tilt of the Earth's axis, and cycles of change in the location in the Earth's orbit where the spring and fall equinoxes occur.

These cycles interacting with other cycles reminds one of the periodic components of the tides. Indeed, calculations of the Milankovitch variations of solar variation at the top of the Earth's atmosphere could have been performed on the big brass machines that Doodson and his assistants used to calculate the tides for the Normandy invasion. Predicting the consequences of these changes on incoming solar radiation to the Earth's ocean, land, ice, and atmosphere to understand the climate dynamics is much more complex. The modeling of these simultaneous interactions strains the capability of the most advanced modern computers.

It appears that the amount of summer solar radiation on the Northern Hemisphere's polar regions, where ice sheets and continental glaciers have formed in the past (around 65°N), correlate well with formation of continental glaciers. Incoming solar radiation at high latitudes in the south could also have similar consequences, but with different timing.[39] There is certainly a tremendous amount of ice stored in Antarctica today.

What made the current ice age, during which we originated as a species, and the interglacial, in which we now live? It seems the Milankovitch variations in solar input are important in understanding the waxing and waning of the Pleistocene's crop of glaciers, but these cycles have presumably been in operation over all the Earth's history. To form vast ice sheets on the land requires relatively warm winters and the transport of moisture to the high latitudes. The summers need to be relatively cooler so that the ice accumulates and does not melt. What triggered the transition to the icy Pleistocene?

One clue is provided by the ratios of isotopes of oxygen in fossils in the ocean sediments. Foraminifera, or "forams," are a group of protozoa found in the oceans and in brackish water that are armored with shells made of calcium carbonate. Dead planktonic forams rain down to the ocean floor. There their shells are preserved as fossils in the ocean sediment. Remarkably, these fossil shells on the bottom of the ocean can tell us about the amount of ice on the surface of the Earth when the forams were alive. How? Water contains two hydrogen atoms and one oxygen atom. The oxygen atom usually has eight protons and eight neutrons in its nucleus (denoted ^{16}O); a small proportion of the oxygen forming water molecules has two extra neutrons (denoted ^{18}O). Water made of ^{18}O is called heavy water. When water evaporates, the heavy ^{18}O isotope of oxygen tends to be left behind. When there is lots of ice (from snow from water evaporated from the ocean) piled up on the land, there is proportionally more ^{18}O that has been left behind in the oceans.[40] Thus, the ratio of oxygen isotopes in the carbonate "skeletons" of fossil forams in the sediment records the amount of ice on Earth.

Two deep-ocean sediment cores representing about five million years of deposition of material in the deep ocean show the isotope signatures of significant ice deposition starting about 2.7 million years ago. One of these cores is from the Caribbean Sea;[41] the other is from the equatorial Atlantic Ocean.[42] The variations in the abundance of ^{18}O in these and other similar records match the patterns of variation predicted by the Milankovitch theory.[43] At about the same time, melting icebergs calved from glaciers and afloat in the North Atlantic began to sprinkle rocks onto the deep ocean sediments below.[44]

Sometime around three million years ago,[45] the Isthmus of Panama closed and potentially directed the flow of warm Atlantic water northward to supply the moisture for heavy ice-sheet-forming snows.[46] Other theories identify potential conditions for an ice age around three million years ago, including reduced concentrations of carbon dioxide in the atmosphere, producing a "reverse greenhouse effect" of planetary cooling, or complex positive feedback loops in the planet's land, ocean, and atmospheric systems producing a threshold response toward cooling.[47] Changes in the structure of North Pacific waters could increase input of water vapor to augment fall and winter snowfall.[48]

Over the Pleistocene, the Earth has been in an icy, glaciated condition (a "glacial") much longer that it has been in a less icy "interglacial" condition. It is clear that there is a strong interaction between the height of the ocean and climate,[49] but there is also a need for more information to understand how this all works. Interglacials, such as the current condition of the Earth, have been in place for only about 10 percent of the time over the past two million years.[50]

We have more sources of information about changes over the past interglacial/glacial/interglacial cycle, for example, actual samples of prehistoric air, which have been trapped as bubbles in the glacial ice of Greenland and Antarctica for hundreds of thousands of years.[51] There is also more information on the relative height of the sea and land for this more recent timeframe. One difficulty in understanding changes in sea level is that if the land is rising, the sea appears to be falling and vice versa. When millions of tons of ice accrete across subcontinental-scale areas, the crust is pushed downward by the weight. When the ice melts, this crustal compression is released, and the land rises. This "glacial rebound" can be used to interpret the amount of mass that might have been in the former glaciers,[52] but it is also a complication to understanding sea level rise.

FUTURE SEA LEVEL RISE

Understanding changes in sea level of tens and hundreds of meters, which derive from kilometer-thick continental glaciers storing water as ice over tens of thousands of years, is a different problem from understanding changes in sea levels in the range of meters or less over the next century. However, the large-scale, long-term changes inform the more immediate problem: "What will be the height of the seas in the year 2100?"

The Past Is Prologue

We know that the past height of the seas has differed greatly from its height today. And we know from the past that the seas can change relatively rapidly. During the melting and warming ending the last glaciation, "abrupt" sea level changes of as much as five meters per century were observed.[53] These changes were associated with the melting of the great Laurentide ice sheet that sat over eastern Canada. The Laurentide sheet's thickness exceeded over 2,500 meters over vast areas (across Quebec and Labrador; also over much of the southern part of Nunavut).[54]

The Laurentide ice sheet is now gone. A time in the past that is more analogous to the present-day distribution of ice occurred at the end of the last interglacial, about 120,000 years ago. At this time, global mean sea level is estimated to have been four to six meters higher than today.[55] This was the response to a few millennia of elevated temperatures that appear to have significantly melted the Greenland and West Antarctic ice sheets.[56] The seas rose 1.6 meters per century from a 2°C average global temperature increase.[57] This is significant since a 2°C average global warming lies well within the range predicted by the current assessments of human-generated "greenhouse gas" global change.[58] It also agrees with some "high" predictions that have been made for a 1.0±0.5-meter sea level increase by the year 2100.[59] It is a cautionary finding, to say the least.

LOOKING TO THE NEXT CENTURY

The Intergovernmental Panel on Climate Change produced a report in 2007 discussing the current state of our scientific understanding of the possible change in the Earth's climate from human alterations of the air, land, and oceans.[60] The report evaluates the expected sea level rise resulting from a global change in the climate. This is a complex endeavor involving such intricacies as prediction of the use of fossil fuels and other sources of greenhouse gases. To understand future atmospheric emissions requires prediction of the future economy, future population, and future technological innovation. This combination of factors, all of which influence emissions to the atmosphere, sometimes are called the Kaya identity.[61] Different scenarios of patterns of global economic development and governance constrain these intrinsically difficult projections. The *Intergovernmental Panel on Climate Change Report* attempts to wrestle with these uncertainties and other potential uncertainties using a multidisciplinary synthesis. Given the complexity of its task, it is a quite readable document and is available on the Internet.[62]

The report notes that the measured sea level rise between 1961 and 2003 amounted to 1.8 millimeters per year. This rate is 10 percent of the much higher rates seen at the end of the last interglacial.[63] In the last decade of this measurement interval (1993–2003), the rate of rise accelerated to about 3.1 millimeters per year. Sea level rise was predicted to increase to between 0.26 and 0.59 meters by the year 2100,[64] for the "warmest" of the climate change scenarios considered,[65] which, averaged across a collection of different global climate models, produces a range of 2.4°C to 6.4°C global warming by the year 2100. This increase is on the low side of the rate of sea level rise computed for the last interglacial, which saw a similar degree of warming. The projections of sea level rise have been exceeded by observed increases over the relatively short time interval since the report.[66]

Sea level increase also stems from the expansion of water in warming oceans. This is the source for about half the sea level rise seen from 1993 to 2003. The melting of glaciers and the ice caps of mountains account for about a quarter of the increase.[67] The Greenland and Antarctic ice sheets each contribute about one-eighth of the observed sea level rise. By the decade of 2090 to 2099, the projected rate of sea level rise predicted by the Intergovernmental Panel on Climate Change averages to 3.8 mm per year. This prediction derives from several different scenarios of climate change along with applications of different computer models of oceans.[68]

However, since the IPCC report, other evaluations have pushed the expected sea level increase upward to a meter or more.[69] Around 70 percent or slightly more of the increase in sea level over the century is expected to come from the expansion of the warming ocean waters. By end of the century, mountain glaciers and mountain ice caps will largely be gone, and their contribution to sea level rise will become relatively small.[70] Greenland and western Antarctica will increase in their contribution to sea level rise, but the amount is uncertain. Indeed, the uncertainty of sea level increases in the coming century is relatively high, in no small part because the sea level integrates a cascade of uncertainties from other changes: change in global temperature, in the amount of ocean warming, and in the dynamics of change in the Greenland and western Antarctic ice sheets.

What Might a Sea Level Increase Mean?

The millions of people who live adjacent to coasts are the immediate concern when considering the effects of a 0.5-to-1.5-meter increase in sea level.[71] Worldwide, about 10 percent of the world's population, disproportionally from the least developed countries, lives on coastal lowlands (areas ten meters or less above sea level).[72] In the United States, 2.7 million people live below the one-meter-above-high-tide mark.[73] The hazard lies with the change of the combination of tides and high water from storms. In one study, sea level increase by the middle of this century could significantly change the occurrence of storm surges.[74] In some exceptional locations, events now expected to occur once every one hundred years will become annual events. In one-third of the sites studied, a once-in-a-century event could be expected to occur once a decade. The authors summarize the patterns:

> Pacific coast locations are most in danger of seeing their historical extremes frequently surpassed in the coming few decades, followed by the Atlantic. Gulf locations appear in least danger of a rapid shift, despite rapid relative sea level rise, due to the high amplitudes of historical storm extremes, which render the relative effect of sea level rise small.[75]

One can expect these patterns to translate to other locations as well worldwide.

These are some locations that are extremely vulnerable to sea level increase on the order of one meter or less.[76] There are several coastal cities that are sinking from land subsidence and from the sheer weight of the human construction there. Venice and New Orleans are examples of such cities. Of course, the flooding of these cities (for example, New Orleans during Hurricane Katrina) is exacerbated by sea level rise. Half of Europe's coastal wetlands are expected to disappear as a result of sea level rise.[77] The annual cost of protecting Singapore's coast is expected to be between US$0.3 and $5.7 million by 2050 and to triple that by 2100.[78] Studies in Thailand indicate that a loss of land in response to a sea level rise of one meter could decrease gross domestic production by 0.69 percent (approximately US$600 million per year).[79] Alexandria, Rosetta, and Port Said in the Nile Delta could see two million people displaced given a sea level rise of 0.5 meters. This could cost 214,000 jobs and produce a land loss valued at US$35 billion.[80] While the potential effects are widespread, poorer nations and their citizens are particularly vulnerable to many of the consequences of sea level increases.[81] The biological and implications of sea level change are also nontrivial. In the modern world, human action is playing a role in the alteration of the oceans and their boundaries.[82]

CONCLUDING COMMENTS

The dynamics of the oceans: the tides; the change in sea level over time; the interaction among the land, sea, and air; the increasing role of humans in altering the functions of all these complexities—make the Joban questions of bounding the seas and freezing the face of the deep very important ones for our time. In Job's answer to these (and the other) questions of the whirlwind speech, he deferred to God's superior knowledge and power: "I know that you can do all things, and that no purpose of yours can be thwarted. . . . Therefore I have uttered what I did not understand, things too wonderful for me, which I did not know" (Job 42:2–3, in part). It is likely that we know quite a bit more than Job about the planetary function of Earth, particularly since Job professed to be able to answer none of God's questions. If we know more, nevertheless it remains to be seen if we know enough.

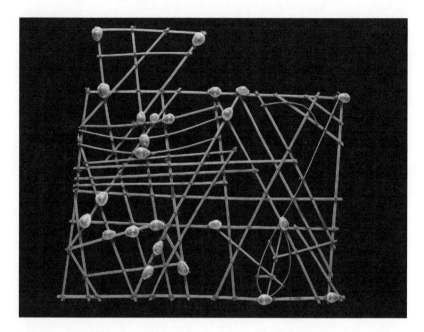

Marshall Island stick navigation chart. These string, stick, and shell constructions are used as models of the patterns produced by waves interacting with islands and atolls to navigate the open ocean. Stick navigation charts are used on land to teach interpretation of the pattern of waves hitting against a boat and allow young navigators learn to read the positions of surrounding islands and atolls when at sea. This knowledge, but not the objects themselves, is carried on voygages. At the same time, knowing the positions of stars and their rising provides a star-based compass.

6

The Ordinances of the Heavens
and Their Rule on Earth

Adaptation and the Cycles of Life

Can you bind the chains of the Pleiades, or loose the cords of Orion? Can you lead forth the Mazzaroth in their season, or can you guide the Bear with its children? Do you know the ordinances of the heavens? Can you establish their rule on the earth?

—Job 38:31–33 (NRSV)

Do you know when the mountain goats give birth? Do you observe the calving of the deer?

—Job 39:1 (NRSV)

Is it by your wisdom that the hawk soars, and spreads its wings toward the south?

—Job 39:26 (NRSV)

Along the shoreline in the Torres Islands,[1] Vanuatu (the archipelago nation formerly known as the New Hebrides), a medium-sized gray sandpiper appears on the beaches in October. It is the *tuwiä*, or, in English, the wandering tattler (*Tringa incanus*), a migratory bird that breeds in the summer along the alpine mountain streams of Siberia, Alaska, and northwestern Canada. In the nonbreeding season, it migrates and eventually circumnavigates the Pacific.[2] It pushes southward down the edge of the Pacific Rim. Then it crosses the Pacific Ocean, arriving at the Hawaiian

Islands in August, Samoa in August, Fiji by late August, and the Torres Islands in October.

When the Melanesian yam farmers of the Torres Islands see the *tuwiä* on their shore, they sing to it, "*Tuwiä, tuwiä, nút mad gor?* [*Tuwiä, tuwiä*, has the Palolo worm come yet?]" The head-bobbing behavior of the birds is taken to mean "Yes," and it signals that the seas will soon swarm with the tasty segments of Palolo sea worms (*Palola viridis*). The Palolo worm is a reef creature about twenty-five centimeters long. Two times a year, on the days of the waning quarter of the moon in October and November, the Palolo sheds its rear segments, which float to the surface and emit sperm and eggs.[3] The seas can be full of these worm segments, which are harvested with baskets woven specially for the occasion. Locally, the worms are considered a delicacy. Palolo feasts bring people together and set the stage for annual planting and harvest times. The *tuwiä* is the harbinger of the Palolo worms and of the agricultural lifecycle. When harvesting the first fruits of the yam crop, the farmer sings, "May the *mena*[4] of the *tuwiä* give the people much food." *Mena* (or, more generally, *mana* in several Polynesian languages across Oceania) is a term that on the Torres Islands means "capacity" and refers to the capability to evoke transformative forces inherent in the living world.[5]

This example from remote islands located between the Coral Sea and the South Pacific may seem jarringly unfamiliar to the Western observer, but it has direct similarities to the manner by which familiar religious celebrations such as Easter, Passover, or Ramadan are determined. The Palolo worm and the *tuwiä* combine the two aspects of the Joban whirlwind questions cited for this chapter: knowing the ordinances of the heavens and knowing the cycles of nature.

Neolithic cultures depend upon their agricultural productivity for survival. Knowing when to plant crops is and was essential to their success and sustenance. If one knew that the year had about 365 days, then one could simply count the days to know when to plant. This would provide an approximate solar-based calendar. Unfortunately, difficult challenges hide in understanding the "ordinances of the heavens." The year is not *exactly* 365 days long. The time it takes for the Earth to revolve around the Sun, based on a fixed reference such as a star (a sidereal year), is 365 days, six hours, nine minutes, and 9.76 seconds. If the Earth's orbit-based year is measured as the time between equinoxes or solstices (the tropical year), the slight shift in the Earth's orbit, called precession, produces an average year of about 365 days, five hours, forty-eight minutes, and forty-five seconds. A heliacal year is the interval of time between the rising of a star. If the star is the star Sirius, the brightest star in the sky, the year is called a Sothic year. These are also affected by precession. There are Gaussian years, Besselian years, Draconic years . . . While this may enrich the reader's ability to solve really hard crossword puzzles, one should stop here.

Recognizing that there are many ways, all with slight variations, to calculate a year's duration, we normally develop calendars using a year of exactly 365.25 days. This is the Julian year. This is also the year measurement used by astronomers for calculating the light-year (the distance that light travels in one year). If one does not add a leap year with an extra day each fourth year, the day-counting Julian calendar would slowly drift forward and lose accuracy.

This is all quite complicated. Not knowing the exact number and fraction of days in a year, for a straightforward way, one used by early peoples, to tabulate a year's passage, one could count moon months. How many times has there been a full moon or the rise of a new moon since the last planting season? Moon-based calendars would appeal to fertility-oriented early agricultural people, particularly given the synchronization of human female-reproductive cycles to moon phases. A moon month, or lunation, is roughly 29.5 days in length. The solar year does not have an even number of moon months. Twelve moon months are 354.37 days. The ordinances of heaven based on moon calendars provide a calendar that drifts over solar years.

The Islamic calendar is a lunar calendar. Each month starts with the sighting of the first crescent of a new moon. The moon-based year in the Islamic calendar, as well as other lunar calendars, is a bit less than eleven days shorter than the solar year. For this reason, Islamic holy days, such as Ramadan, migrate throughout the seasons over time.

The calendar is corrected for this holy-day migration in the Jewish calendar, a lunar calendar that has twelve twenty-eight-day months. To correct for the calendric drift, a thirteenth month (Adar II) is included in the calendar. Over the course of a nineteen-year cycle, this "extra" month occurs in the third, sixth, eight, eleventh, fourteenth, seventeenth, and nineteenth years. Passover occurs from the fifteenth to the twenty-first of the month of Nisan—which is the month right after the "extra" month of Adar II.

Similarly, the Borana people of northern Kenya and southern Ethiopia calculate a lunar-month calendar to time agricultural activities and religious events.[6] The Ayantu, who are experts on the observation of the sky, regulate the Borana calendar by adding an extra month about once every three years. The Ayantu make their calendric alteration based on the rise of different clusters of stars (constellations) above the horizon in relation to the rising of the moon.[7]

Easter is also a "moveable feast" based on modifying moon calendars to reset their drift against the solar calendar. The result is called a lunisolar calendar. In Western churches, Easter is observed on the first Sunday after the full moon that occurs on or after the spring equinox, which is fixed as March 21.[8]

Boranas, Jews, and Christians are, in these examples, all using the astronomical observations of the sun, stars, and moon to correct a lunar calendar so that it remains close to an annual calendar or to constrain important dates from drifting through the year. Just as the arrival of the spring equinox resets a moon calendar to pin the date of Easter as occurring sometime between the dates of March 22 to April 25, the arrival of the wandering tattler and the spawning of the Palolo worms reset the lunar harvest calendar in the Torres Islands.

THE ORDINANCES OF THE HEAVENS AND
THEIR RULE ON EARTH

The Neolithic world bristles with monumental architecture, some of which are thought to be used to observe and commemorate the ordinances of the heavens. The collection of striking and ancient architecture from the earliest times of the invention of agriculture and even earlier are found in disparate locations: Mayan ruins in Central America, Incan high-altitude cities in the Andes, stone circles in Europe, and prehistoric temples from Mediterranean islands.[9] Machu Picchu, Chichen Itza, the Ring of Brodgar, and other similar places are simultaneously mystical and compelling. One's imagination races to understand the almost tactile impact of these places. Perhaps it is the magic that seems to reside there that inspires the development of the various pseudoscientific explanations of their intent.

ARCHAEOASTRONOMY

The term "archaeoastronomy," the interdisciplinary study of astronomy and anthropology, was first used in a paper by Elizabeth Chesley Baity in 1973.[10] The scale of her review paper is epic. It has 867 references to previous work, comments from nineteen other researchers (of the fifty to whom she sent advance copies of the paper), and her responses and/or rebuttals to these comments. Sponsored by the International Union of Astronomers, archaeoastronomers attended an initial successful conference at Oxford in 1982 and since that time have had several subsequent "Oxford" conferences at different locations around the world.[11] The Ninth "Oxford" International Symposium on Archaeoastronomy met in January 2011 in Lima, Peru. There is also a scholarly journal, *Archaeoastronomy*. Needless to say, scientifically rigorous archaeoastronomers are plagued by popularizations that range from highly speculative assertions to pure lunacy.

One of the more challenging aspects of interpreting the alignment of megaliths and monuments is that in the deeper past the apparent locations of rising and setting stars and other celestial events were different from today. The Earth wobbles on its axis over very long intervals, on the order of 26,000 years. This wobble is caused by interactions in the gravitational pull of the Sun and the Moon on the Earth's bulge at the equator. Any star directly over the northern axis is a pole star, which currently is the star Polaris in the constellation Ursa Minor (the Little Dipper). Polaris would have also been the pole star 26,000 years ago. Halfway between these two times, 13,000 years ago, the star at the imaginary point in the sky above the North Pole would have been Vega, in the constellation Lyra. Vega would have been a great pole star: it is the second brightest star in the northern sky. For archaeologists, what this means is that one does not simply stroll into an archaeoastronomical site and start siting lines to star-related positions. One must know the time the feature was constructed and then compute the correction for the wobble or, as it is technically known, "precession."

Recently, the International Council on Monuments and Sites (ICOMOS) and the International Astronomical Union compiled a list of significant heritage sites (and objects from them) for astronomy and archaeoastronomy worldwide.[12] A sampling of these locations and artifacts provides at least a glimmer of the sorts of astronomical knowledge acquired by Paleolithic and Neolithic people:

1. *The Decorated Plate of the Geißenklösterle*—a small rectangular plate (38 mm × 14 mm × 4.5 mm) made of mammoth ivory and found in a collapsed cave near the village of Weiler, Baden-Würtenburg, Germany. It dates from the Aurignacian period, probably 32,000 to 35,000 years ago. On the side of the plate is a carved human (or human-feline) figure with upraised arms. This figure appears to be a representation of the group of stars that form the constellation Orion. The other side and the four edges of the plate contain a series of notches for day counting. There are eighty-eight counting notches, equal to three lunar months. Thirty-three thousand years ago, Betelgeuse, the bright red

star in Orion, would have disappeared from view about fourteen days before the spring equinox. Three lunar months later, it would rise back into the night sky. If conception occurred at a time close to the reappearance of Betelgeuse, birth would take place after the severe half-year-long winter of a glaciated ice-age planet. Also, there would be enough time for sufficient nutrition of the baby before the beginning of the next winter. The plate is a counting system linking the moon, the rise of Betelgeuse, and human pregnancy.[13]

2. The Thaïs bone—a piece of bone, a bovine rib, excavated from Thaïs cave (Saint Nazaire en Royans, France) in 1968–1969. The bone dates from 12000 BP and is engraved on both sides. On one side there are seven long lines scratched into the bone, with perpendicular cross-hatchings across these lines. The Thaïs bone documents the existence of a calendar system based on the phases of the moon with a seasonal factor provided by observing the solar solstices.[14]

3. *Stonehenge World Heritage Site*—an extended set of monuments (Stonehenge, Avebury, Durrington Walls, etc.) along with over seven hundred archaeological features located in the county of Wiltshire, in the United Kingdom. Construction at Stonehenge appears to have initially begun with the digging of three pits with timber posts around 8000 BCE. Around 3000 BCE, a ditch and bank were dug, and "bluestones" from distances as much as 240 kilometers away were added, as were "sarsens," more-local stones weighing up to forty metric tons. All of the main monuments in the larger area and Stonehenge itself have astronomical features.[15] Stonehenge aligns with the solar solstices, as do several of the other monuments in the vicinity. Archaeological evidence on the timing of feasts and the solstice-aligned avenue at nearby Durrington Walls implies interpretations involving seasonal processions and rituals.

4. *Chichen Itza, Yucatan, Mexico*—a regional civic and religious-ceremonial center as early as the sixth century CE. Its peak was in the Mayan Late Classic and the Terminal Classic periods, roughly the tenth and eleventh centuries CE. A civil war caused a decline after 1221 CE. The site is filled with ornately decorated large monuments with

astronomical capabilities. Examples of archaeoastronomical features include El Castillo, or the Temple of Kukulcan, which displays light and shadow at the solar equinox; El Caracol, with several astronomical features such as solar solstice sunset alignments, the northernmost setting position of Venus, sunset at the equinoxes, sunsets on April 29 and August 13 and others; The Great Ball Court or Temples of the Jaguar, which points to the sunsets on April 29 and August 13; Las Monjas, decorated with images of the Mayan zodiac; the Temple of Venus, which has eight solar years (or five Venus cycles) represented by decorative icons; and many others.

Along with these are many other sites. Ko'a Holo Moana on the island of Hawai'i overlooks the ocean from a cliff. It is a navigation temple, called a *heiau* in Hawaiian. It consists of an assemblage of large standing stones, which are directional markers providing the headings to distant islands. These are part of the navigational skillset developed in Micronesia, Melanesia, and Polynesia and an example of a Neolithic star-navigation system, the topic of the next section.

POLYNESIAN KNOWLEDGE OF THE ORDINANCES OF THE HEAVENS: STAR MAPS AND LAND FINDING

The people of the South Pacific have developed a remarkable capability to explore, colonize, and exploit the islands scattered across the vast Pacific Ocean. This is done using star compasses based on the rising and setting of stars along with keen interpretation of observations of their surrounding environment. Polynesians are remarkably capable navigators. Tupaia was a Polynesian navigator taken on board the HMS *Endeavour* during James Cook's initial exploration of the Pacific.[16] Tupaia was originally from the island of Ra'iatea in the Society Islands and joined the *Endeavour* as a guide near Tahiti. His recruitment is noted in Joseph Banks's journal of the expedition on July 12, 1769,

This morn Tupia [Tupaia] came on board, he had renewed his resolves of going with us to England, a circumstance which gives me much satisfaction. He is certainly a most proper man, well born, chief Tahowa or priest of this Island, consequently skilled in the mysteries of their religion; but what makes him more than anything else desirable is his experience in the navigation of these people and knowledge of the Islands in these seas; he has told us the names of above 70, the most of which he has himself been at.[17]

Tupaia, according to journals of both Cook and Banks, knew the location of the ship at sea and of future landfalls far better than the Europeans. In December 1770, just eighteen months after he joined the voyage of discovery, Tupaia died in Batavia (modern Jakarta, on the island of Java). He left two copies of a remarkable chart of the Pacific Islands centered on Tahiti and showing many islands strewn over a 4,200-kilometer span of the Pacific Ocean.[18] His chart is not a conventional map. It illustrates headings to Pacific Islands based on a star compass and the distance in terms of time sailing.[19] Also shown are positions of shipwrecked European vessels.[20] One chart was owned by Richard Pickersgill, an officer who had sailed both with Cook on the *Endeavour* and with Captain Samuel Wallis, the discoverer of Tahiti in 1767. It is now lost. The one surviving copy, once the property of Sir Joseph Banks, resides in the British Library.

Polynesian navigation involves dead reckoning (estimation of how far a vessel has traveled at sea) and the use of stars as a compass. Since the stars and their constellations rise and set at the same location when one is on the Equator,[21] one can determine the heading from one location to another from the stars. Studies on the Caroline Islands found that indigenous navigators and their "star compasses" involve the navigator memorizing the relative rising and setting positions of about fifteen stars.[22] At any time or place, the navigator observes the stars that are near the horizon and then "imagines" the rest that cannot be seen below the horizon.

These navigation methods are augmented by the ability to detect islands that are over the horizon by observations of the flight paths of birds and from patterns of sea waves. A device designed to read the complexities of wave patterns, a Micronesian example of a navigation stick chart from the Marshall Islands (Ratak Chain, Wotje Atoll), is illustrated at the beginning of this chapter. This particular example was collected in 1960. Several of these charts are shown in different ethnographic papers over the past hundred years,[23] for example, the 1899 *Smithsonian Report*.[24] Navigation stick charts augment the teaching of how one interprets the positions of distant islands using complex variations in the regular patterns of waves. They use the "swell," the rhythmic waves produced by storms and winds that can travel long distances over the oceans.[25] Go to a beach: one can watch swells; they generate the rhythm of the waves. When the swells interact with islands and shallow water they can bend and travel in different directions. Marshallese navigators observe these patterns by observing and feeling the ocean's waves slapping against the hulls of their boats. To detect the wave patterns generated by distant islands, they lie against the hulls in their dugouts and feel the small vibrations.[26] The Marshallese likely had navigational skill beyond that of their first European contacts.

Similarly, there is a certain irony that Tupaia's navigational capabilities seemed on a par or better than those of Cook, particularly given that the mission of the HMS *Endeavour* on this voyage was to make astronomical observations intended to improve the ability of European navigators to determine longitude.[27] Their navigation skills gave the Polynesians the remarkable capability to travel and populate the vastest region of any peoples on Earth.

BIOLOGICAL AND ECOLOGICAL TIMING

The roles of the wandering tattler and the Palolo worms in setting the Torres Island yam-planting calendar falls into the general category of auguries, observable events or indications that portend future events.

The use of auguries is very old, dating back well before history. By the time of Pliny the Elder in first century CE, the Romans had compiled significant lists of omens and auguries for all manner of predictions. Priests, thought to be wise in these matters, interpreted auguries as part of personal and national decision-making processes. Speaking of these, Pliny wrote,

> There are a thousand other facts of this kind and the same Nature has also bestowed upon many animals as well, the faculty of observing the heavens, and of presaging the winds, rains, and tempests, each in its own peculiar way. It would be an endless labour to enumerate them all; just as much as it would be to point out the relation of each to man. For, in fact, they warn us of danger, not only by their fibres and their entrails, to which a large portion of mankind attach the greatest faith, but by other kinds of warnings as well. When a building is about to fall down, all the mice desert it before-hand, and the spiders with their webs are the first to drop. Divination from birds has been made a science among the Romans, and the college of its priests is looked upon as peculiarly sacred. In Thrace, when all parts are covered with ice, the foxes are consulted, an animal which, in other respects, is baneful from its craftiness. It has been observed, that this animal applies its ear to the ice, for the purpose of testing its thickness; hence it is, that the inhabitants will never cross frozen rivers and lakes until the foxes have passed over them and returned.[28]

These auguries provide and have provided humans with clues to predicting the natural world. They often are thought to work in mysterious ways. Some are born by a confluence of coincidence and superstition. It is certainly not difficult to imagine that the observation of the heavens for indications of time and season in conjunction with the interpretation of auguries could produce the practice of astrology at some time deep in the past. The Mazzaroth (מַזָּרוֹת) mentioned in Job 38:32 likely refers to some sort of system of zodiacal signs. This remains speculative, for "Mazzaroth" is a hapax legomenon, a word only found once in a body of work—in this case, in the Bible.

Plants and animals integrate cues in the environment to time their biological activities with the variations in the seasons. Such timing is omnipresent in living organisms. Internal biological systems have evolved to maintain appropriate times for the different phases of plant and animal reproduction. Interpreting observations of the natural world as auguries and adages have a deep history—and prehistory—in human culture. Some of these, such as the example of using the wandering tattler to reset a lunar calendar, are very effective. Others, such as using the hairiness of wooly caterpillars to predict the harshness of winter, are probably less reliable.

Some auguries work in one location but not at all in another. The American robin (*Turdus migratorius*) is a year-round resident in the American South but is the harbinger of spring in the northern United States and Canada.[29] A century ago in Quebec, the first robin was thought to bring good crops and good luck on whomever's farm it nested.[30] Perhaps an indication of the desire to see the end of winter here in Virginia, one often hears, "I've seen a robin, so spring must be coming soon." One might think that the perceived value of auguries would depend upon their predictive capability, but evaluating how good an augury is in its predictions may be a more complex task than it might appear at first. For robin watchers in the South, if one senses the lengthening of days, knows the date, and is ready to see the end of winter, then one notices the "first" robin whether they have been around, unnoticed, all winter long or not.

In some quixotic cases, auguries may "work" by not being predictive at all. The use of bird auguries has been described for several different tribes in Borneo and especially for the Kantu' people of West Kalimantan.[31] The Kantu' cultivate dry-land rice and other food crops in the cleared forest patches, or swiddens, used in cut-and-burn agriculture. Traditionally, the Kantu' live in longhouses of ten to twenty families. Seven species of birds serve as auguries to determine what crops to plant and when and where to plant them.[32] These omen-birds' songs, locations, and behaviors all instruct individual farmers on their future agricultural decisions.

Individual farmers differ in their orthodoxy to the auguries; "stronger" omens were more likely to be followed than "weaker" ones. The complexity of the interpretations is less important here than to recognize that the omen-bird auguries add an element of randomness into what each farmer does with his swidden planting. Why might such randomness be important? The region is influenced by the El Niño climatic cycles, which manifest as floods in some years and strong droughts in others.[33] If all the farmers planted what worked best last year (or over the last few years), everyone would go down together with crop failures in the periodic bad year.

In this example, using the bird-omen auguries is not predicting anything; instead it is adding a random "bet-hedging" strategy to the local agricultural cropping system. Significantly, planting using bird-omens is not the most productive agricultural strategy in a given year, but it has significant implications for the long-term sustainability of the farming practice. Similar use of auguries to promote a random component and diversity in survival strategies include such diverse cultures as the Naskapi Indians of Labrador[34] and swidden agriculturalists on the Philippine Island of Luzon.[35]

Other auguries are the products of complex chains of interactions that are reliable in their predictions. The arrival of the wandering tattler, the spawning of the Palolo worm, and the best time to plant yams in the Torres Islands all are driven by natural seasonal cycles. The *mena* of the wandering tattler (and its power as an augury) certainly is mysterious. Simultaneously, it is the product of keen observation and the consequence of several interrelated underlying causes.

BIOLOGICAL CYCLES AND PHENOLOGY

Phenology is the systematic observation of the timing, particularly the annual cycles, of natural changes. The dates of the first frost, the return of migrating birds, the blooming of flowers, and the bud burst of leaves all are topics of phenology, as is the understanding of the causes of these patterns. Phenology is a strongly observation-based branch of science. There are phenological networks and societies in many locations. Joining in these data-collection efforts is an excellent opportunity for interested citizens to participate in coordinated environmental research.

Phenology has deep roots in several cultures, and there are ancient traditions of compiling dates of significant natural events. For example, dates of cherry blooming (and cherry blossom festivals) in China and Japan have been recorded for well over a millennium. Modern phenology as a scientific discipline originated with English naturalist-gardeners in the eighteenth century. From his estate Stratton Strawless in Norfolk, Robert Marsham published his record of phenological events (thrush sings, snowdrops flower, swallows appear, wood anemone flower, etc.) for each year from 1736 to 1787.[36] Gilbert White from Selbourne, Hampshire, and William Markwick from Battle, Sussex, observed similar seasonal events for over four hundred plant and animal species between 1768 and 1793. These data, reported as the earliest and latest dates for events, appeared in an augmented edition of White's *Natural History and Antiquities of Selborne*.[37] *Natural History*[38] has been continuously in print since its initial publication in 1789 and may be the fourth-most printed book in the English language.[39]

Plant phenology is largely regulated by light and temperature. The potential complexity of the environmental control of plant phenology can be remarkable. For example, to germinate seeds of the flowering dogwood (*Cornus florida*), collect the seeds and then place them in a warm moist environment for sixty days. If rushed for time, put them in sulfuric acid for three or four hours or scratch the seed coat with a file to simulate the conditions of being digested by a bird, the normal way these seeds are dispersed in nature. Then, put them in a cool place—a refrigerator will work—for sixty to 120 days. Finally, cover the seeds with about a centimeter of topsoil, water them, and hope for the best.[40] The process is a bit reminiscent of the witch's brew in Shakespeare's *King Lear*, not quite as complex as *Macbeth*'s "fillet of a fenny snake, in the caldron boil and bake; eye of newt, and toe of frog, wool of bat, and tongue of dog," but complex, nonetheless.

Different species of plants or varieties of the same species in different locations typically have such complicated formulae to unlock and time the key phases of the life of the plant. In wild plants, the selection pressure pushes evolution to "tune" the plant's biochemistry to match the environment correctly. This produces local adaptations. Horticulturalists, agronomists, and foresters intensively research the phenological responses of the species so that they can plant varieties to match regions.

The phenological responses of plants often depend on external environmental signals to set the timing of a plant's growth and development patterns. Moving species from one location to another has produced myriad examples of potentially useful plants failing to prosper when their phenological life history was mistimed to the environment of a new location. Climatic change can produce an analogous mismatch between plant timing and environmental timing. Indeed, climate change has the potential to disrupt ecological processes at several different levels.

In the dogwood seed example, the seeds needed to be warmed for a period of time. Indeed, many plant phenological events need exposure to some cumulative amount of warm conditions, sometimes calculated by number of hours above a base temperature or the summation of temperatures above a reference temperature for each day (called growing degree-days). Also like the dogwood, other plants need to be exposed to a summation of hours or degree-days below a certain temperature (chilling degree-days) before their seeds will germinate or flowers will bloom.

Plants also have the ability to register biochemically the changes in quality, quantity, and duration of light, which provides information on changes in the local environment and in the seasons. For detecting changes in light, plants have evolved a series of highly specialized chemical receptors. These chemicals include the red and far-red light-absorbing phytochromes and the blue/ultraviolet light-absorbing cryptochromes and phototropins. These biochemical molecules "sense" light, but the incorporation of this information into the overall biochemistry and timing of the plant is complex.[41] Exposure to progressively longer days induces many species to bloom in the spring. Other plant species are the opposite case. They bloom on shortening days in the fall. Different species use these environmental sensors in one way for one process (flowering on lengthening days in the spring) and in another way for a different process (preparing to drop leaves induced by shortening days in the fall). Manipulating temperatures and day lengths allow horticulturalists to produce blooming orchids, lilies, and other plants year round or to induce tropical poinsettia plants to flower during the holiday season.

PHENOLOGY AND GLOBAL CLIMATE CHANGE

Robert Marsham's phenological observations from his estate in Norfolk, England, the first such records published in English,[42] began in 1736 and were continued by five generations of his relatives on the estate until 1947.[43] Another significant long record, but of a different sort, was the fifty-year one collected by Charles Keeling, who began measuring the levels of carbon dioxide in the atmosphere atop the volcano Mauna Loa in Hawaii in 1958.[44] Keeling's data showed increasing levels of carbon dioxide in the atmosphere. These changes were associated with the burning of fossil fuels and the associated atmospheric release of carbon dioxide. Increases in carbon dioxide and other "greenhouse gases" in the atmosphere strongly imply an increase in global temperature. It is ironic that the Marsham family's two-hundred-plus-year record of the coming of spring at Stratton Strawless ended before the record of atmospheric carbon dioxide at Mauna Loa began. While the Marshams' record may have ended, the interest in phenology by early naturalists inspired others to make similar observations. These observations have implications for the magnitude of effects that might come from human-induced climate change.[45] Indeed, they seem to be signaling climate change in the form of a warming over the over the past several decades.

A phenological network across Europe (twenty-one European countries) has produced a data set of over 125,000 observations of 541 plant species and nineteen animal species. Recent analysis (for the interval of 1971 to 2000) has shown that spring is coming earlier to Europe, about 2.3 days earlier each decade.[46] Seventy-eight percent of all leafing, flowering, and fruiting records have advanced to earlier times in the season. The primary driving factor appears to be temperature. Indeed, in Europe the pattern of change for temperature measurements indicative of spring have advanced 2.2 days each decade—a strong agreement with an advance of spring seen in phenological observations that we have just discussed.[47] These temperature measurements all across the Northern Hemisphere indicate that its average "coming of spring" has advanced about 1.3 days each decade.[48] Other similar temperature findings are an increase in growing-season length, averaging seven days longer since the 1960s[49] and 1.3 days per decade advance in the last spring freeze in the United States.[50] There is evidence from British plant phenology (based on 385 plant species) that the rate of spring becoming earlier is accelerating.[51]

A different measurement from satellites determines the change in the greenness of the planet. Using daily records of surface conditions obtained from weather satellites, the "greening up" of the vegetation over the Northern Hemisphere became earlier over the decade of 1981 to 1991.[52] The greening of spring came earlier over this decade, in total about eight days earlier, and in patterns consistent with computer models of the climate changes expected from greenhouse gas planetary warming.[53] Of course, global warming is an expected consequence of doubling the atmosphere's greenhouse gas content. This was initially predicted in calculations on the physics of the atmosphere made over one hundred years ago.[54] The magnitude of these estimates (around a $2.5°C$ increase in average global temperature with a doubling of the amount of carbon dioxide in the atmosphere) has been surprisingly consistent ever since—even with a progression of increasingly sophisticated climate-system models, as scientists have progressed from hand calculations to modern high-speed computers.[55] This will be the topic for the following chapter.

Three different ways of looking at the Earth's climate—through collating elaborate phenological data, through processing temperature data, and through inspection of satellite-based observations—all point to an earlier spring in the recent decades. Presumably, increased temperatures drive this change. The ecological responses are seen both at the level of individual species (in the phenological record) and at the level of the responses of the entire ecosystems over a large area (in the satellite data). Given that three different perspectives all point to similar change, it is both wise and prudent to understand the cause of such patterns well.

KNOWING WHEN THE MOUNTAIN GOATS GIVE BIRTH

One of the whirlwind questions quoted at the head of this chapter asked if Job knew when certain animals give birth. The length of life, number of offspring, and the number of times organisms reproduce under different conditions can vary significantly among species as well as within different varieties of the same species.[56] To survive, it is essential that the life stages of different plants and animals are coordinated with their environment—the genes of individuals prone to have young during times of resource shortage are soon selected out of a population.[57] This evolutionary selection produces the timing of the births of deer and mountain goats mentioned in the whirlwind question.

How do plants and animals maintain their biological timing? They have "clocks" that can carry timing. These clocks can also be reset or adjusted by external factors in the environment. Many invertebrates have annual spawnings that are synchronized by lunar cycles.[58] The striking regularity of the Palolo worm's biological cycle is something of an extreme example. The Palolo worm's clock is precise to the day within the year for its spawning behavior, even if clouds hide the moon. The animal and its relatives carry an internal clock that tabulates lunar time.[59] Most animals do not keep a lunar clock,[60] but bacteria, fungi, algae, plants, and animals display an internal daily periodicity close to a day, the *circadian* rhythm.[61] The details of these vary across organisms, but they essentially work by having internal biochemical "oscillators" that keep track of the time of day. One or more of these chemical oscillators, called "pacemakers," can set their timing by daily variations in the environment, typically the daily cycle of light and dark. These pacemakers then interact with the other oscillators to produce a daily clock.

There is also another internal cycle of about a year, the *circannual* rhythm, which is found in a wide variety of animals and plants.[62] In mammals, there are two designs for the biochemical mechanisms that provide the individual with an accounting of the passing of years. Some mammals have what can be thought of as a "timer" that is reset by the seasonal change. Once reset, it has a long interval before it can be reset again. When these interval timers are between being periodically reset by environmental conditions, they provide an internal calendar to the animal. If the animals are placed in constant conditions, then the "timer"-type species run through a series of states: nonreproductive, physiologically able to reproduce, shedding fur, growing new fur, etc. When the timer for these events runs out after six months or so, the individual ceases to change until the timer is reset. Other species have an internal timer that operates more like a clock. These clocks can be reset to the time of year, but when the individuals are placed in constant conditions for long periods of time, their clocks and patterns keep the schedule for years. Like the circadian rhythms, the two types of calendars, timers and clocks, are products of biochemical networks and their internal interactions. It is interesting that there is great variability in how species maintain timing to match the changes in the seasons. Both the clock and the timer circannual systems have evolved independently many times in different species.[63] Timers and clocks seem to be the two of many possible ways to anticipate change in the season.

Migratory birds perform remarkable feats using their clocks in conjunction with a surprising capability to travel over large distances. They provide an example of a particularly sophisticated set of internal capabilities.

MIGRATION: WHEN BIRDS TURN THEIR WINGS TOWARD THE SOUTH

The phenomenon of bird migration represents a notable example of synchronizing a species' biological cycles and the periodicities in the environment. The whirlwind question asks by what wisdom a bird flies south (Job 39:26). This issue is alluded to in other parts of the Bible as well. For example in Jeremiah (8:7), "Even the stork in the heavens knows its times: and the turtle dove, swallow and crane observe the time of their coming; but my people do not know the ordinances of the Lord."

Migration in birds reveals one of the most complex integrations of biological and environmental information to be seen in any organism. Probably among the most remarkable of these avian feats, the wandering tattler sends itself on one of the most arduous migrations of any animal. Departing Siberia and Alaska after the breeding season, it crosses the equator and migrates down the east side of the Pacific, then crosses the great ocean to return home to its nesting grounds by flying up the west side of the Pacific. Not every creature follows migration pathways as challenging as those of the tattler. In birds and other animals with simpler patterns of migration, nomadic movements allow the animals to utilize areas that sometimes would be unusable or even hostile to survival. In either case, the animals must be able to return home.

The classification of animals as migratory or nonmigratory is really a simplification of a range of behaviors and adaptations. Some species do not migrate at all but reside in a given region for all of their lives. Other species appear to be quite opportunistic and erratic in their migrations. Still others are highly programmed and perform improbable feats, like flying across vast expanses of open ocean water or returning precisely to the location where they were born from wintering grounds thousands of miles away. Movement over great distances involves risks. Large-scale migration costs energy and also produces physiological stress. Thus, one would expect migratory behavior to evolve in cases in which the benefits minus the costs of being a migrant are greater than the benefits minus the costs from remaining in place.[64]

Probably the most straightforward form of migration is the irruptive movements seen in species that are periodically stressed by shortage of food or environmental stresses. Goshawks (*Accipiter gentilis*) and snowy owls (*Nyctea scandiaca*) feed on small mammals and birds in the far northern boreal forest and tundra. When there are negative shortages in prey, these striking predators invade regions in North America and Eurasia that are thousands of kilometers further south, often to the great excitement of naturalist birdwatchers. Similarly, seed failures in the cone-bearing trees that make up the dark, coniferous boreal forest spill flocks of evening grosbeaks (*Coccothraustes vespertinus*) and red crossbills (*Loxia curvirostra*) south to search for food and to decorate the bird feeders of suburban homes.

OBLIGATE MIGRANTS

The opposite extreme is found in obligate migrants—species that breed in one region during a favorable period and then migrate to far-away wintering grounds when conditions become unfavorable. Obligate migrants are adapted to take advantage of regions with temporary high productivity where there is a pronounced and predictable seasonal variation in environmental conditions. We have already spoken of the wandering tattler. Another obligate migrant is the tiny blackpoll warbler (*Dendroica striata*).[65] These birds breed across the boreal forest of North America to as far west as Alaska. From as much as 2,400 kilometers away for the birds on the Alaskan breeding grounds, blackpoll warblers gather in staging areas in New England, perhaps to as far south as Virginia. Fueling up on insects and storing fat, they form small flocks and make an eighty- to ninety-hour, open-ocean flight of 3,500 kilometers across the open Atlantic Ocean to their wintering grounds in South America. This is a truly amazing feat, and it is accomplished by a tiny warbler that normally cannot land at sea. One of the reasons we know that they are out there is that they occasionally do land on ships in the open ocean.[66]

Between the species of either irruptive or obligate migrants are the partial migrants—species composed of some individuals that migrate and others that do not. In some cases, these partial migrant populations differ genetically with respect to traits that involve migration. In other instances, partial migrants have some segments of the populations (such as younger birds or behaviorally subordinate individuals) that are migratory while others are not. In this latter case, an individual can be migratory at one time in its life and nonmigratory at another.[67]

FINDING THE WAY BACK HOME

Irruptive, partial, or obligate migrants all need to be able to return "home" to the breeding grounds once conditions have become more favorable there. This is an intrinsically difficult challenge. As well as a sense of location and an ability to navigate to a particular location, the animals need to know when to return. Most people even when armed with maps and compasses would be challenged to get themselves home over several thousand miles of ocean. It is hard to comprehend how small animals with little brains can accomplish such feats. Their toolkit includes clocks, compasses, maps, and evolutionary programming:

Clocks

Imagine a small songbird has migrated to the Amazon basin for the winter. There, sitting on the Equator, all of the days are the same length (twelve hours), regardless of the time of year. After six months of identically timed days in a tropical environment with no strong seasonal change, the bird's physiology changes. It becomes restless, does not sleep, and soon flies north to its breeding ground located in northern Canada.[68] If it leaves a fortnight too early, the bird arrives at the breeding areas in a time of inclement weather and could die. Should it leave a fortnight too late, other birds of its species will have acquired the best mates and territories. Clearly, evolution would favor the individuals with the correct solution to this timing problem.

Birds, like the other organisms we discussed in the section above, have two clocks built into their physiology. One of these clocks, the *circannual* rhythm, is more or less like an annual calendar. The other, the *circadian* rhythm, is more of a daily clock. Much as the Palolo worm and the other examples in the section above, the daily cycle appears to be reset by timing cues from the environment such as the light/dark cycle of the day or daily temperature fluctuations.[69] The circannual calendar involves the regular progression of internal physiological cycles that are associated with breeding, molting, seasonal changes in plumage, and gains and losses of weight. This circannual cycle is maintained even when caged birds are placed in constant conditions with twelve-hour days and twelve-hour nights. Programmed into the cycles of obligate migrants is a condition called *Zugenruhe*, a German term for restlessness. In *Zugenruhe*, birds become more active at night than day. This is the condition associated with readiness for migration in obligate migrants.

Compasses

If you were boxed in a closed crate and then stuck into a windowless van and driven around the countryside, you likely could not immediately point the correct direction home when discharged in a remote and unfamiliar forest clearing. Pigeons can.[70] The sport of homing pigeons is based on returns from just such distant release points under just such conditions. When released, homing pigeons usually fly immediately in the direction of their home roost. There seem to be two aspects in accomplishing this remarkable feat: a "map" to determine position and a "compass" to move in the direction indicated by the map.[71] Birds appear to have four different compasses.

The first such compass is the one most familiar to humans. If one knows the time (which for birds would involve utilizing the circadian clock mentioned above), then the position of the sun in the sky indicates direction. The sun is in the east in the morning and in the west in the afternoon. Homing pigeons apparently can use a "sun compass" to find their way back home. However, many of the obligate migrants fly by night, so they must be equipped with other navigational tools as well.

The second compass is the ability to read stars. At night, caged birds in a state of migratory unrest or *Zugenruhe* will hop on the side of their cages in the direction that they would move if they were free and making a migratory flight. When these birds are placed in a planetarium, experiments on bird navigation can be conducted under the planetarium's artificial starry skies, which can be manipulated by changing the projector. In the planetarium with changed skies, birds change their direction of hopping, as if they are star navigating. It appears that birds learn the stars of the night sky in the first year of their lives and are able to use the stars as a second compass.

The surprising third compass that birds (as well as several other animals) appear to have is the ability to sense magnetic fields. The lines of force in the Earth's magnetic field are vertical over the magnetic poles and become flatter as one moves away from the magnetic poles. A bird uses its sense of magnetic fields and determines the *dip angle* or verticality of these lines of force and thus is provided a sense of direction.[72] Migratory birds in *Zugenruhe* held in the same sorts of experimental cages used in the planetarium experiments can be placed in controllable magnetic fields using Helmholtz coils. When the magnetic field surrounding the birds is altered, so is the direction of hopping movements in the cages.[73]

Birds have a fourth compass, a sense that humans also have but rarely utilize. This is the ability to sense polarized light. If you look at a lamp through a polarized sheet of plastic (or even polarized sunglasses), a yellowish hourglass figure appears. It rotates when you rotate the plastic sheet. This figure, called "Haidinger's brush," points in the direction of the polarization of the light.[74] It can also be seen looking at a white area on a LCD flat-panel computer screens. Haidinger's brush is relatively easy to observe under the correct conditions, but our brains usually edit this phenomenon out of our perception. It is not clear what birds see, but they can detect the pattern of the polarized light that is created by the atmosphere acting as a weak polarizing filter on sunlight. Thus, birds know the position of the sun from patches of sky even when the sun is obscured from view. Similarly, insects can detect polarized light and use this information as a sun compass.[75]

Maps

Geese and cranes dependably return to the same stopover points during migration in different years. This indicates a rather precise ability to navigate from year to year. Armed with their multiple compasses, birds have the information to sense direction under a range of conditions: daytime or night, cloudy or clear. An accurate clock and the knowledge to read the stars, tell north from south, detect the dip of the Earth's magnetic fields, and determine latitude are certainly sufficient information for navigation. However, the calculations are complex, and the clocks and senses would need to be remarkably accurate to allow birds to migrate with the exactitude that some birds are known to have. Without a superaccurate clock but armed with a compass (or multiple backup compasses), one still needs a map to navigate back to home base. We know less about how this aspect of avian migration works.

Much of the work to understand the map used by birds has been done with homing pigeons. That homing pigeons can return to their lofts even when fitted with frosted contact-lenses implies that senses in addition to sight are involved.[76] Birds are now known to have small organs in their heads that contain magnetite crystals—this provides a magnetic sense.[77] Pigeons appear to use the magnetic fields and magnetic anomalies (places where, for example, large iron ore deposits disturb the pattern of the Earth's magnetic fields) as maps. There is evidence that pigeons and other birds may use a sense of smell for certain odors blowing in winds from different directions to find their way.[78] It has also been proposed that birds can use the inaudible, extremely low frequency sound called "infrasound" that is generated by the Earth's surface as a kind of map based on sound reflections.[79] It seems plausible that long-distance migrants could use their sense of direction and clocks to travel in appropriate directions at appropriate times and to then use magnetic, chemical, or infrasound clues to find their homes. The direction taken by birds on their first migration appears to be inherited genetically.[80]

THE DEVELOPMENT AND LOSS OF THE ABILITY TO MIGRATE

The migratory phenomena can develop (or be lost) from a population in a relatively short period of time, even in obligate migrants.[81] Ten thousand years ago, the breeding areas that blackpoll warblers now target for their spring return were covered by continental glaciers or equally inhospitable polar deserts. Thus, the complicated behaviors of long-distance migration must have developed in this obligate migrant during this period of time. Ten thousand years is relatively short in the scope of evolutionary changes, but development and loss of migratory behavior has occurred in short enough time periods as to be observed by human observers—in time spans measured in decades rather than in millennia.

The nonmigratory house finches (*Carpodacus cassinii*) introduced to Long Island, New York, in the 1940s have now spread across the upper half of the eastern United States. The species is a common bird-feeder visitor in suburbs of the east and has become a partial migrant. Rufous hummingbirds (*Selasphorus rufus*) from the Pacific Northwest of the United States and Canada that normally migrate to Mexico for the winter now have learned to migrate by the hundreds to backyard hummingbird feeders in the southern states of the Gulf Coast. Other migratory species when introduced to new situations have become nonmigratory in equally short periods of time.[82]

CONCLUDING COMMENTS

In a time when some see religious faith at odds with science, it is worthwhile to remember that the scholarly observation and interpretation of the natural world were significant positive attributes in biblical sages.[83] For example, King Solomon, the epitome of a royal sage, was characterized as a man of great wisdom and insight into the workings of the Earth:

> He would speak of trees, from the cedar that is in the Lebanon to the hyssop that grows in the wall; he would speak of animals, and birds, and reptiles, and fish. People came from all the nations to hear the wisdom of Solomon; they came from all the kings of the earth who had heard of his wisdom.
>
> (1 Kings 4:33–34)

One sees this in the account supposedly from King Solomon himself (more likely a Jewish writer in Alexandria, Egypt, between 100 BCE and 100 CE)—the Wisdom of Solomon:

> For it is he who gave me unerring knowledge of what exists, to know the structure of the world and the activity of the elements; the beginning and end and middle of times, the alternations of the solstices and the changes of the seasons, the cycles of the year and the constellations of the stars, the natures of animals and the tempers of wild animals, the powers of spirits and the thoughts of human beings, the varieties of plants and the virtues of roots; I learned both what is secret and what is manifest, for wisdom, the fashioner of all things, taught me.
>
> (Wisdom of Solomon 17:17–22)[84]

Wisdom and knowing are virtues are shared by religion and the sciences, perhaps each with its own objectives, but shared nonetheless.

The calendric aspects of the Joban "ordinances of the heavens" were made more difficult for humans to understand because none of the relationships synchronized—the solar year does not have an even number of solar days, the solar year also does not have an even number of lunar months, and the lunar month does not have an even number of solar days. Perversely, all the obvious natural timekeepers do not match up. Clocks and calendars had to be reset by correlating daily sunrises with the positions of stars and with the lunar months, somehow, so that Neolithic cultures could time their planting and harvest cycles. Knowing the ordinances of the heavens was essential knowledge for survival. Perhaps we are beyond all of that now, but the essential need to time our life cycles and to integrate the ordinances of the heavens remains a problem for the biota that share our planet. These plants and animals have evolved the capacity to anticipate the daily cycle of the sun and to time the coming of seasons. This time sense is essential to survival for them, just as it was for our Neolithic ancestors.

The Palolo worm in the narrative that initiated this chapter was an example of a creature whose breeding behavior was strongly entrained by lunar cycles. In a world with changed conditions, particularly different climate conditions, it is possible that this exact timing could produce a breeding event during a time of unfavorable conditions. Indeed, this breakdown of environmental conditions and time of years is one of a suite of changed conditions (increasing acidity of the ocean, pollution, changes in exploited populations) that could alter the world's tropical reefs in the future.[85]

Including the effects of such changes on the reef systems are the planetary consequences of changes that would confuse the clocks and timers of the biota of the planet. We know that these systems have a degree of flexibility because we can move species from one location to another and they can in some cases adjust their physiology to the new situation. We know that birds can develop (or lose) the complex behavior and internal time sense to migrate. We know that natural selection can eliminate from the gene pool those individuals with inappropriate phenological responses. But these responses all require time. Which species can adapt rapidly enough and which ones cannot? We do not really know this part of the answer to the Joban whirlwind questions of "Is it by your wisdom that the hawk soars, and spreads its wings toward the south?" (Job 39:26) and "Do you know when the mountain goats give birth?" (Job 39:1).

Zephyr, the Greek god of the west wind. Zephyr was the messenger of spring and was dominant over plants and flowers, particularly as his Roman equivalent, Favonius. *Source*: Photograph by the author from the Villa da Schio, Costazzo di Longare, Venito, Italy.

7

The Dwelling of the Light and the Paths to Its Home

Winds, Ocean Currents, and the Global Energy Balance

Have you entered into the springs of the sea, or walked in the recesses of the deep?

—Job 38:16 (NRSV)

What is the way to the place where the light is distributed or where the east wind is scattered upon the earth?

—Job 38:24 (NRSV)

God's interrogation of Job highlighted the divine control of the weather as a direct challenge: "Can you lift up your voice to the clouds, so that a flood of waters may cover you? Can you send forth lightnings . . . ?" (Job 38:34–35) and "Who has the wisdom to number the clouds? Or who can tilt the waterskins of the heavens . . . ?" (Job 38:37–38). That we now seem prepared to apply our current knowledge about these questions fuels the discussion in chapter 10. The Joban questions for this chapter also concern the weather and the mechanisms that make it. The path of light to its home and the springs of the deep are taken to refer to the systematic patterns of winds and ocean currents that circulate the heat from the sun's radiation ("the light") about the planet. A place to start is with some of the more regular winds.

There are places where the winds have names.[1] Zephyr, shown in the illustration opening this chapter, was the Greek god of the west wind, the gentle wind that signaled the coming of spring. There was a panoply of Greek and Roman wind gods, each with a different direction and a different personality. There are other, less divine, named winds as well. A sirocco is a scorching, dry, dust-filled November wind that blows out of the Sahara with speeds as high as one hundred kilometers per hour.[2] A chinook is a dry wind that has been warmed by compression as wind moves downslope over the Front Range of the Rocky Mountains and onto the Canadian plains and the U.S. Great Plains. Chinooks can raise winter temperatures from well below freezing to warm enough to melt as much as a foot of snow in a day. A föhn is a similar wind that blows downslope from the Alps.[3] The Fremantle Doctor is a cool afternoon wind off of the Indian Ocean that provides relief from the blazing hot summers of Perth, Western Australia.[4] The monsoon is a complete reversal in the pattern of winds that brings the rainy season to areas near the equator and even at higher latitudes.[5] A calima is a dry summer wind from the southeast that blows Saharan dust across the Canary Islands. Ultimately, transported Saharan dust fertilizes the Amazon with phosphate, an often limiting plant nutrient.[6] Nor'easters are fierce winter storms that batter beaches and houses along the Atlantic coast of Canada and the upper U.S. Atlantic coast.

These winds gain their names from their regularities—the similarity of their patterns and when they reoccur generates a seeming familiarity. We know them by their look, smell, and feel; by the timing of their arrivals and departures; by the changes they wreak. The processes that cause the regular features of winds arise from underlying fundamental physics of the way heat is distributed.

FÖHNS AND CHINOOKS: PHYSICAL PROCESSES BEHIND THE PATTERN

Föhns and chinooks are straightforward examples of how a unity in underlying physical processes creates similar wind patterns in mountain winds in the lee of the Alps in Europe and the Rocky Mountains of the western United States. When a wind blows upslope over a mountain or mountain range, the air cools from expansion, because the air pressure lowers as height increases. Depending on the amount of water in the air, this cooling eventually causes clouds to form and precipitation to fall on the mountain's side. When water vapor condenses into liquid droplets, it releases the heat that was originally needed to evaporate the liquid water that supplied the gaseous water vapor to the air. Moist air lifted up a mountain cools at a rate of 5°C per thousand meters of rise. When this same parcel of air crests the mountaintop and then moves down the other side of the mountain, the now-dry air warms by compression from the increasing air pressure. Its temperature increases 10°C for every thousand meters of decrease in altitude. By the time the air reaches the leeward base of the mountain, it is warmer—sometimes considerably warmer if a tall mountain range is involved.

How much can a chinook change the local temperature? The most extreme twenty-four-hour temperature difference recorded for the United States (a change of 57°C) was produced by a chinook wind in Loma, Montana.[7] On January 23, 1943, a chinook in Spearfish, South Dakota, generated a 27°C change in temperature in two minutes, a record that still stands.[8] Föhns, the chinooks' European cousins, help warm central Europe and provide it with a milder climate than would otherwise be expected there.

TROPICAL CYCLONES, HURRICANES, AND TYPHOONS

Hurricanes, typhoons, and tropical cyclones are large, remarkably powerful, organized tropical storms. These storms are called hurricanes when they occur in the Atlantic and Eastern Pacific,[9] typhoons in the northwestern Pacific,[10] severe tropical cyclones in the southwestern Pacific Ocean and southeastern Indian Ocean,[11] and severe cyclonic storms in the northern Indian Ocean. In all cases, they feature maximum sustained winds greater than thirty-three meters per second (roughly seventy-four miles per hour), although the U.S. definition of a maximum sustained wind differs from that used by most other countries.[12] For convenience, they will be referred to as hurricanes in this section unless the more geographically specified term is appropriate.

"Hurricane" derives from the Carib/Taino/Arawak storm god Juracán, itself from the Mayan god Hunrakan, a creator storm god who blew his breath across the chaotic waters and created dry land. The meteorologist and the British East India Company's president of the Marine Courts of Inquiry at Calcutta, Henry Piddington, coined the word "cyclone" in 1844, with, "I am not altogether averse to new names . . . we might perhaps for this last class, of circular or highly curved winds, adopt the term 'cyclone' from the Greek Κυκλως (which signifies amongst other things the coil of a snake)."[13] Snaky etymology is found in another name for these storms, typhoon, which may originate with the Greek monster, Typhon,[14] a horrifically large creature with the lower half of its body as a gigantic serpentine coil, a human upper half, and serpents affixed as fingers on each hand.[15]

The fierceness and devastation of tropical cyclones, hurricanes, and typhoons are etched on the memories of their victims and sometimes on human history. Their impact inspires personification. *Kamikaze* (Japanese for "divine wind") refers to two major typhoons in 1274 and 1281, which destroyed the Mongol Kublai Khan's invasion fleets intent on the subjugation of Japan.[16] Kamikaze subsequently designated the Japanese suicide-attack airplanes used over six centuries later in World War II. For several centuries in the West Indies, hurricanes were named according to the saint's day when it arrived. Thus a particular hurricane could have different name when it arrived in another location on a different day. Terrible hurricanes striking the United States often were designated by place and year: the Galveston Hurricane of 1900 or the Great New England Hurricane of 1938.

The first modern meteorologist to name tropical cyclones was the eccentric Clement Lindley Wragge, who in the 1880s and 1890s worked as a government meteorologist for Queensland in Brisbane.[17] Wragge connected a system of weather stations by telegraph (including an undersea cable to New Caledonia in 1893), which allowed him to study and track typhoons. He took a perverse delight in naming Australian typhoons after politicians as well as applying names from Polynesian mythology and European history.

During World War II meteorologists in the U.S. armed forces in the western Pacific took up the informal tradition of naming tropical storms after their wives and girlfriends. An early feminine naming of tropical cyclones appears in George R. Stewart's 1941 novel *Storm*, which featured a Pacific storm named Maria as its principal character. *Storm* inspired Alan Jay Lerner and Frederick Loewe to write the song "They Call the Wind Maria" for their 1951 musical *Paint Your Wagon*. In 1953, *Storm* and the previous informal wartime convention provided the impetus for officially using women's names for Atlantic tropical cyclones. The practice ended in the late 1970s, in no small part because the naming of agents of death and destruction after women seemed inappropriate to the changing times. Both male and female names are now used alternately.

The Hurricane as a Working Mechanism

A typical hurricane will have a nearly circular "eye" typically thirty-five to fifty kilometers in diameter located near the center of the storm. The eye is an open column of relatively calm, warm air. This core of warm air is what most clearly separates tropical cyclones from other spiral storms.[18] The eye of the storm is surrounded by an eyewall, which contains the strongest winds in the system. Winds in the eyewall spiral upward.

The condensation of water in the rising eyewall vortex produces remarkable volumes of rain. The rising and spiraling eyewall air vents out the top of the eyewall, roughly eighteen kilometers high, and forms a layer of high-altitude cirrus clouds.[19] Some of this air also moves downward into the cyclone's eye and makes it more cloud free. As the winds spiral away from the top of the eye, they deflect to the right in a clockwise direction. At the frigid, high-altitude temperatures of around $-70°C$ ($-94°F$), the air cools, becomes denser, and drops back to the ocean surface. Below the cirrus cloud cap, great curved arms of thunderstorms called rainbands spiral back into the hurricane's center eye and are deflected to the right, in a counterclockwise rotation. These are warmed and hydrated by warm ocean surface water, which provides the energy that fuels the hurricane system.

From a fixed location with a tropical cyclone passing overhead, the bands of torrential rainstorms build and diminish and then build again. Each rain is more intense and longer in duration than the previous one. If the eye happens to pass directly overhead, this increasingly violent progression of rainstorms reaches a crescendo, followed by the eerie calm of the eye's passage. This is a dangerous time. If one leaves shelter into the calm to inspect the situation, the subsequent arrival of the other side of the eyewall can bring tragedy.[20]

Hurricanes are well-defined and stable storm systems, at least as long as they are over warm ocean water. Their spiraling morphology is the consequence of several straightforward agents of motion—rising warm air, falling cold air, and air movement from high to low pressure. In addition to these, there are other forces that give hurricanes their striking organization. How do hurricanes function as working mechanisms?

The Components of the Hurricane

It is initially difficult to think of a hurricane as a machine, probably because one tends to associate machines with solid physical devices— levers and screws for simple machines, cranes or V-6 gasoline engines for more complex ones. The mechanistic workings of fluid mechanisms become more apparent in vortices: whirlpools in liquid fluids and whirlwinds and tornados in gaseous fluids seem more tangible and visible. In the context of the present biblical text, God spoke to Job from a whirlwind (Job 38:1), and other parts of the Joban text identify whirlwinds as powerful forces not to be reckoned with by mortals. Regarding the fate of the unrighteous, "Terrors overtake them like a flood; in the night a whirlwind carries them off" (Job 27:20, NRSV). Regarding the power of God:

God thunders wondrously with his voice; he does great things that
we cannot comprehend. For to the snow he says, "Fall on the earth";
and the shower of rain, his heavy shower of rain, serves as a sign on
everyone's hand, so that all whom he has made may know it. Then
the animals go into their lairs and remain in their dens. From its
chamber comes the whirlwind, and cold from the scattering winds.

(Job 37:5–9, NSRV)

Whirlwinds and other vortices such as hurricanes are phenomena of
surprising appearance and power organized by physical forces.

Gases are accelerated from areas of high pressure to those of low pressure. This motive force is called the pressure-gradient force. The larger the
difference in pressures over a given distance, the greater the wind speeds.
If this was all there was to consider, this movement from high-pressure
regions to low-pressure ones would transfer air to lower the highs and
heighten the lows. One might expect high and low pressure differences
to disappear over time. However, there are other forces that act on a horizontally moving parcel of air (as well as on vertically moving air parcels).

First, on a rotating Earth, the Coriolis force deflects the winds away
from their straight-line, pressure-gradient-force-driven path of motion
from high to low pressure. With a boundary at the equator, the inertia
of moving air with respect to the planet's rotating surface causes it to
deflect to the right in the Northern Hemisphere or to the left in the
Southern Hemisphere. Thus, air moving from the periphery toward
the center of a tropical cyclone spirals counterclockwise toward the
low-pressure center in the Northern Hemisphere and clockwise in
the Southern Hemisphere. The Coriolis effect was noted early, if not
first, in 1651 by Giovanni Battista Riccioli, a Jesuit priest and astronomer, and his assistant, fellow Jesuit Francesco Maria Grimaldi.[21] Their
research was an attempt at a correction in targeting cannonballs, which
they reported to veer slightly to the right in the Northern Hemisphere.

Second, as the spiraling air approaches the eye of the hurricane, it speeds
up from the conservation of angular momentum (mentioned in chapter
2 in regard to the formation of solar systems). As the spinning tightens

around the hurricane's eye, centrifugal force becomes increasingly important. Centrifugal force is the force that allows one to whirl while holding a pail of water without spilling it. This outward force relative to the center becomes larger as the velocity increases or as the radius of the spin decreases.[22] As the hurricane's winds spiral faster and tighter around the eye, the centrifugal force becomes increasingly larger and helps reinforce the boundary between the eye's calm and the high-velocity winds of the eyewall.

The Hurricane as a Heat Engine

Kerry Emanuel points out that hurricanes are what are termed heat engines.[23] Heat engines ideally are driven by heat differences in four stages:

1. Gases are heated and rise to reduce the pressure on the gas. This step is called isothermal expansion. In a conventional engine, this expansion might be the heat generated by exploding gasoline in an automobile engine; in a hurricane, it is the heat and moisture transferred to the inwardly spiraling wind from the warm ocean water.

2. Removed from the heat source, the air continues to rise. The air rises without the addition of heat from the outside but is warmed from the condensation of water, as was the case in the rising air generating a chinook. This adiabatic expansion (expansion without the addition of heat) occurs in the upward spiral in the wall around the hurricane's eye and in the towering storms that form in the spiral arms of the hurricane.

3. At the top of the hurricane's eye, the air spirals outward, and it is cooled in the cold upper atmosphere. The air drops with a nearly constant temperature; heat is radiated into space to balance the heat gained from compression. This approximates a condition called isothermal compression.

4. The cold air continues to fall and is compressed by adiabatic compression. This air would be on the outer part of the hurricane and would become warmer and wetter as it spirals back in toward the hurricane's eye to complete the cycle.

This spiraling atmospheric machine is what the French scientist Nicolas Léonard Sadi Carnot described in 1824 as a heat engine.[24] Carnot saw the motions of winds and ocean currents as products of heat engines: "To heat also are due the vast movements which take place on earth. It causes the agitations of the atmosphere, the ascension of clouds, the fall of rain and of meteors, the currents of water which channel the surface of the globe, and of which man has thus far employed but a small portion."[25] The "fuel" for the hurricane heat engine comes from warm ocean water. The greater the difference in temperature between this warm surface and the cold of the upper atmosphere, the more work it can do. Hurricanes work to move air, and they are remarkably efficient and powerful engines for doing so. If it could be harnessed, the power generated by a typical Atlantic hurricane would light thirty billion standard hundred-watt light bulbs. A Pacific supertyphoon can generate ten times more power.[26]

Hurricanes, once formed, are great atmospheric machines driven by heat and moisture. In general, they require ocean water temperatures of 26.5°C (80°F) to a depth of fifty meters, high humidity, and unstable air. Winds tend to disrupt cyclone formation, and they tend to form more than five degrees of latitude away from the equator. They also require a preexisting disturbed weather system. Hurricanes "feed" on the heat from warm ocean water. They strengthen over warmer waters and weaken over cooler ones. The increase in surface friction when they pass over land disrupts their organization.

Record Hurricanes

Hurricanes can be tracked, particularly nowadays using satellites, but their future movements are simultaneously predictable and capricious. These are charismatic storms that, because of the danger they represent to life and property, light up the radio and television with weather reportage. News of the storms is essential for the survival of the people in their paths.

In the United States, ever since Dan Rather stood on the seawall at Galveston, Texas, in September 1962 to report live on the landfall of Hurricane Carla and was catapulted to national fame as a television journalist, it is hard to keep aspiring TV newscasters off the beach when a hurricane is coming in. Rather also introduced the first radar image of a hurricane to television audiences. On the radar screen, Hurricane Carla was about 640 kilometers wide as it approached the Texas coast. The public realization of the magnitude of Hurricane Carla (the largest hurricane by combination of intensity and size to make landfall in the United States) spurred an evacuation of 500,000 people from the Texas coast and may have saved thousands of lives.[27] The Galveston Hurricane of 1900, which was thought to be of comparable intensity to Carla, killed at least 8,000 people and possibly as many as 12,000.[28]

The Galveston Hurricane of 1900 was the deadliest natural catastrophe to befall the United States, but it pales in lethality to other storms. Of the six storms with mortalities of over 100,000 people, one of these, a typhoon that hit Japan in 1923, produced a death toll of 250,000, many of whom perished in a simultaneous earthquake and the resultant widespread fires.[29] Another typhoon landed in southern China in 1881 with an estimated loss of 300,000 lives. The rest and also the majority of the hundred-thousand-plus killer storms occurred in India or Bangladesh. The Bay of Bengal has high astronomical tides and a coastline that funnels water onto low, flat terrain. The worst storm of them all appears to have been the 1979 Bhola cyclone, which landed on the Ganges Delta region of Bangladesh and killed more than 300,000 people, with 200,000 recorded burials and another 50,000 to 100,000 people missing.[30] This was a moderate-strength cyclone, but it came in on a high tide and brought a twenty-foot storm surge along with it.

Hurricanes illustrate the forces involved with the intensity and directions of winds. The innermost part of a tropical cyclone has very low air pressure. The lowest pressure measured for an Atlantic tropical cyclone was generated by Hurricane Wilma, which in 2005 registered a central low pressure of 882 millibars, about 13 percent lower pressure than the average sea-level pressure and approximately equal to the atmospheric pressure at the height of 1,100 meters (close to the height of Denver, Colorado). The world record of 870 millibars is held by Typhon Tip

(August 1979), a monster storm that at its peak was 2,220 kilometers wide. It eventually made landfall on October 19 in southern Japan.[31]

THE HADLEY CIRCULATION: THE PATHS TO HOME

A hurricane is a very tangible example of a complex structure arising from fundamental interactions involving the pressure-gradient force, centrifugal force, Coriolis force, and friction. There are other structures organized by these forces at different scales in time and space. The rotation of winds around areas of low pressure demonstrates the same interactions. These interactions are also illustrated by the large-scale winds of the global circulation of the atmosphere. Likely, the earliest discovery of these large-scale circulations is the trade winds, which are found on either side of the Equator.

The Polynesian navigators knew of and understood how to use the trade winds in their colonization of the Pacific around 1000 CE.[32] Their likely ancestors, the Lapita people of the western Pacific, were making significant voyages over two thousand years before, around 1350 BCE.[33] It should come as little surprise that the Polynesians, who designate directions on their island homes as "windward" or "leeward," would intrinsically appreciate the trade winds as a part of the fabric of their lives.

Europeans, notably the Portuguese, utilized the trade winds starting in the early part of the fifteenth century to propel their rapid exploration of the oceans. The word "trade" describing these winds derives from a Middle Low German word, *trade* (*trâ*), meaning a track, a course, a path. The trades were sea paths made of winds. The importance of the trade winds to British merchant fleets later inspired the alternate meaning of trade as "commerce." These beneficent winds blazed an ocean path to exploration, discovery (see chapter 8), and commerce.

It is one thing to know that these winds are there; it is another, deeper thing to know their cause. The story of this understanding is a mixture of lost discovery and incorrect attribution, of good science lost in the scientific literature, of opportunity lost.

Understanding the cause of the trade winds attracted the attention of the renowned British astronomer Edmond Halley, who developed the theory, now known to be incorrect, that the tropics followed the daily cycle of the Sun's heating, which moved from east to west, creating a flow of denser (cooler) air toward the west-moving area of maximal heating.[34] In his conjecture, the winds chase the heat of the day as it lifts air and races ever westward around the planet.

Forty-nine years later in 1735, George Hadley explained that air would be heated at the Equator and would rise to be replaced by air drawn in from areas to the north and south, creating a flow of air from both sides toward the equator.[35] Further,

> air, as it moves from the Tropics towards the Equator, having a less Velocity than the Parts of the Earth it arrived at, will have a relative Motion contrary to that of the diurnal Motion of the Earth in those Parts, which being combined with the Motion towards the Equator, a N.E. wind be produced on this Side of the Equator, and S.E. on the other.[36]

Sadly, Coriolis's essay providing the mathematics and physics behind the Coriolis force, the component that generated Hadley's eastward component for the trade winds, was not to be published until a full century later.[37] Hadley's insightful work required a deeper understanding that simply was not available. He was brilliant but before his time.

Hadley's work quickly became scientific "old news" and was lost almost as soon as it was published. George Hadley was a relative scientific unknown who was often confused with his more famous brother John Hadley, the inventor of the octant (a device for calculating an observer's latitude from astronomical observations and a valuable navigation tool) and the developer of methods to make parabolic mirrors for reflecting telescopes.

Shortly before George Hadley's paper, Halley's earlier theory on the trade winds was published verbatim in a popular encyclopedia.[38] By dark coincidence, Hadley's and Halley's names are similar enough that they were often confused, usually in Halley's favor. Subsequent scientists developing theories similar to Hadley's simply did not cite him. It would take a full century for George Hadley to be recognized for his explanation of the trade winds. Along the way, Hadley's concept was reinvented without attribution. In 1793, it was discovered by John Dalton, the renowned English scientist best known for his contributions to physics, chemistry, and ultimately to the basis of modern atomic theory. Dalton eventually learned of the existence of Hadley's much earlier work.

A modified version of Hadley's concept was put forth, again without attribution, in 1837 by Heinrich W. Dove, a stereotypical Prussian meteorologist who ruled his field with a strong and sometimes dictatorial hand.[39] A letter posted on September 5, 1837, to Richard Taylor, the editor of the *Philosophical Magazine* read:

Notice Relative to the Theory of Winds

By John Dalton, D.C.L., F.R.S.
To Richard Taylor, Esq
Manchester, Sept 5th 1837

Dear Friend

I published a theory of the Trade Winds, &c, as Mr Dove has published,—it was forty-four years ago, as may be seen in my meteorology, 1793 and 1834.[40] It was first published by G. Hadley, Esq, in 1735, as I afterwards learnt. It is astonishing to find how the true theory should have stood out so long.

—John Dalton

To the chagrin of Heinrich Dove, this letter, which was published in the *Philosophical Magazine*, also appeared soon thereafter in the *Annalen der Physik*.[41] This ultimately produced a thorough, posthumous, and long-lasting association of George Hadley with meteorological explanations of the trade winds. In particular, the circulation in the tropics at the global scale is conceptualized as a paired system of recirculating winds called the Hadley cells.

THE HADLEY CELLS

The trade winds are deflected by the Coriolis force of the rotating Earth and come together toward the Equator. Where they converge, the air moves upward. The warm, moist tropical air cools as it rises, condenses to form tall convective clouds, and produces the copious rainfalls of the equatorial region. The air that has been transported upward to high altitudes (roughly fifteen kilometers) moves toward the poles, cools as it radiates its heat into space, and eventually sinks downward far from whence it originally rose. The sinking dry air heats from compression as it sinks and produces a zone of intense aridity near 30°N latitude and a mirrored dry zone at 30°S latitude. These are the latitudes of the Earth's great deserts, the Sahara, the Namib, the Great Australian Desert, among others. These arid locations on land associated with sinking air have their analogues on the seas. These are areas of extensive calms called the Horse Latitudes by tall-ship sailors and the Subtropical High by meteorologists.

The rising air near the Equator creates a zone of low pressure; the zones of sinking air (at 30°N and S) become zones of high pressure. Thus there is a flow of air from high pressure to low. This creates the pressure-gradient force, which, coupled with the Coriolis force, drives the trade winds. These paired equator-centered circulation systems are the Hadley cells, posthumously named for George Hadley.

The zone of converging air from the trade winds located near the equator is called the doldrums. At sea, navigating the doldrums left sailing ships still in the water, sometimes for long and potentially deadly intervals as ship's supplies were exhausted:

> Day after day, day after day,
> We stuck, nor breath nor motion;
> As idle as a painted ship

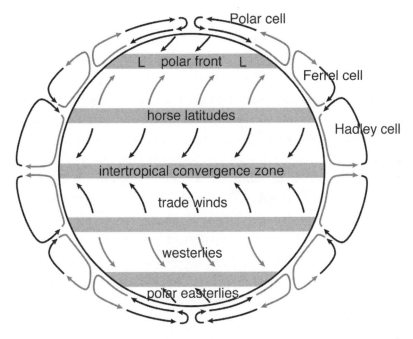

Global wind belts. Driven by circulation cells, winds move air in bands around the world. For example, air near the Equator is heated and rises in the at the Intertropical Convergence Zone (ITCZ). Surrounding air moves into this lower pressure area to replace this rising air. The uplift of warm, humid air at the ITCZ produces towering thunderstorms. This air then moves toward the poles, cools, and sinks as dry air in the region called the horse latitudes, the arid zone where many of the planet's large deserts are found. This entire circulation system is called a Hadley cell, which pushes winds toward the Equator. These winds are deflected by the Coriolis force to become the trade winds. Circulations from other cells produce the prevaling westerlies.

Upon a painted ocean.
Water, water, every where,
And all the boards did shrink;
Water, water, every where,
Nor any drop to drink.[42]

With their life-sapping calms punctuated by strong squalls and thunderstorms, the doldrums were oceanic traps for sailing ships.

They are located where the winds of the two Hadley cells converge. This zone is generally referred to as the intertropical convergence zone or the ITCZ. The ITCZ covers the oceans and tropical lands with a band of clouds. These result from the convective thunderstorms associated with convergence and rising air that are the average condition of the ITCZ; this is easily seen in satellite images of the Earth. It moves with the seasons. The ITCZ is displaced north in the Northern Hemisphere's summer and south in the Southern Hemisphere's summer. It moves northward for half a year (until the height of the Northern Hemisphere summer) and crosses the equator. It then reverses to move south for another half a year. This cycle repeats annually.

Africa straddles the Equator and lies to thirty-five degrees of latitude on either side. It shows clearly the effects of the Hadley circulation on land systems at long- and short-time scales. Africa has extensive deserts and arid zones in the horse latitudes and rain-green savannah and woodland transitioning to equatorial rain forests at the doldrums. This is the large-scale vegetation pattern one would expect for the Hadley circulation climate to develop over a longer period of time. The dominant effect of the Hadley circulation over Africa is the suppression of rainfall. With the exception of the high-rainfall regions in the convergence zones, Africa is a dry continent, with 70 percent of the continent receiving less than fifty centimeters of rain per year.[43]

The striking annual consequence of the Hadley circulation over Africa is the movement of the ITCZ and the movements of rains across the continent. Starting at the height of the Southern Hemisphere summer, the ITCZ is displaced to the south. This position of the ITCZ brings rain to Botswana, in the southern part of Africa. The southernmost part of the nation only gets a taste of the ITCZ rains before the rain belt begins to push northward. The rest of Botswana has rain and thundershowers for a longer time. Then the ITCZ moves further northward, and these regions wait for a dry season until the rain returns from its annual migration. The return takes six months in the wetter northern part of the country and eight or ten months in the dryer, more southern parts—the longer the wait, the longer the dry, and the less the annual rainfall. As the rain belt continues to migrate over Tanzania and Kenya, one of the last remaining great migratory herds of large mammal herbivores migrates across the Serengeti following the greening grass. In some intermediate locations, dry turns to wet as the ITCZ moves over, followed by dry again when it moves completely past, but then there is a second wet season when it passes over again on its return. Near the height of its journey north, it waters the Ethiopian Highlands and feeds the Blue Nile from June to September.

When the ITCZ is well away from the equator in either direction, notably about five hundred kilometers away from the Equator, the Coriolis forces, which are zero at the equator and weak nearby, become stronger. The witches' brew for forming hurricanes and typhoons is assembled (unstable converging air, lots of humidity, warm waters, existing storms, and the Coriolis force) and can begin to cook up tropical cyclones. In satellite images, one can watch the embryology of hurricanes, which are conceived in these nursery grounds, strengthen and steer toward destinations in Florida, or Queensland, or Japan, depending on the location.

THE WALKER CIRCULATION AND EL NIÑO EVENTS

There is a second important large-scale circulation of winds associated with the tropics. The winds from the Hadley cells are moving air from the direction of the poles to create the trade winds. There is a second circulation that moves air around the equatorial zone in directions more or less at right angles to lines of longitude. This circulation is called the Walker circulation, after Sir Gilbert Walker. In 1918, Walker, then director of the Indian Meteorological Department, published his findings regarding longer-term oscillations in weather patterns.[44] The Walker circulation is an interlinked and interactive system of wet and dry regions chained together around the equatorial zone.[45]

For example, in the tropical and subtropical South Pacific, the trade winds interact with ocean temperatures.[46] In the normal circulation pattern, the trade winds push warm water toward the western equatorial Pacific, causing a "pool" of warm water to collect off of

Indonesia and Australia. These warm seas heat the atmosphere, and this warmer humid air lifts to form thunderstorms and considerable precipitation. The lifted air, now high in the atmosphere, flows back to the east to close the circulation cycle by cooling and settling over the warm oceans.[47] The circulation moves heat from east to west across the Pacific. Its strength can be indexed by the difference in sea-level barometric pressure in Darwin, Australia, minus the pressure for Tahiti.[48]

When these values are negative, abnormal conditions occur. The normally cold waters of the Humboldt Current off of Peru are replaced by warmer water. Without the ocean's upwelling nutrient-rich cold waters from the Humboldt Current, the great Peruvian fishery collapses. Local fishermen call the event El Niño ("the little boy," referring to the baby Jesus), because the event normally occurs around the Christmas season. The effect of an El Niño is felt in anomalous weather around the world: heavy rains in East Africa; drought in south-central Africa, southeastern Asia, and northern Australia; and bush fires and dry conditions in the eastern part of Australia and Tasmania, and elsewhere. The opposite condition, La Niña, is a cool phase of the circulation cycle in the Pacific. Africa becomes wetter, as do South Asia and Australia. The periodicities of these episodes are on the order of three to five years.

One of Walker's major accomplishments was to identify through statistical analysis the periodic swings and sways of weather conditions in the equatorial Pacific.[49] He identified three large-scale oscillations of air pressure.[50] The one just described, the Southern Oscillation with periodic El Niño and La Niña conditions, involves the Pacific Ocean, ultimately the Indian Ocean, and beyond. It was the strongest oscillation Walker identified. Of the two others, one was the North Atlantic Oscillation, with swings in the intensity of the difference in the low pressure normally associated with Iceland and the high pressure of the Azores.[51] The other was the North Pacific Oscillation.[52] Gilbert Walker's fundamental idea was that climate variations operating on scales of years to decades arise from interactions between circulations of air and water in the Earth's atmosphere and the oceans. These interactions also connect globally. Such oscillations, feedbacks, connections, and teleconnections will likely be an important topic of research for some time to come.

WINDS AND OCEAN GYRES

Air over the polar regions cools and then warms as it moves toward the Equator. This relatively warmer air rises and cools as it moves back toward the poles. This forms a pair of polar circulations with winds moving in the same directions as the trade winds. Systems of winds called the prevailing westerlies blow in the middle latitudes, between about 30° to about 60°N and S. Separated from the trade winds by the Horse Latitudes, the westerlies blow in the opposite direction of the trades—from the southwest in the Northern Hemisphere and from the northwest in the Southern Hemisphere. The Horse Latitudes' descending cold air forms an extended high-pressure region, the Subtropical High. This descending air moves as part of the Hadley cell toward the tropical low pressure of the ITCZ. It also moves toward the poles, where low pressure is generated by the warming and rising of the cold polar air. This poleward motion turns to the right because of the Coriolis force (and to the left in the Southern Hemisphere). A circulation cell arises from the transport of air poleward from the Horse Latitudes toward the low pressure of rising polar air. This is called the Ferrel cell, after William Ferrel, who theorized its existence in 1856.[53]

Polar cells and the Hadley cells function in a relatively straightforward manner: air moves from a cooler high-pressure area toward a warmer low-pressure area, then rises and recirculates. The Ferrel cell is more complex. It is turns, like a cog between two large wheels, between the Hadley and polar cells. It transports heat toward the pole via its warm air. Its presence is manifested as the passages of great curling eddies of air—the comings and goings of highs and lows, of warm and cold fronts. It is a zone of mixing air, but its prevailing direction is southwesterly in the Northern Hemisphere. From the U.S. East Coast, one sees the effects of the westerlies in the movement of hurricanes, which march from the tropical Atlantic to the Caribbean, eastern seaboard, and Gulf coasts under the steering of the trade winds. They eventually curve back toward Europe when they move far enough north to be steered instead by the westerlies.

In the Southern Hemisphere, the westerlies traverse mostly open water. With little land to slow them down, they become stronger and more organized than their Northern Hemisphere mirror-twins. At around 40°S, the only land of any extent is Tasmania, New Zealand, and the southern tip of South America. The westerlies reach their maximum strength in a zone called the Roaring Forties, after their latitude. "Running the easting down" in the era of sail was the route of clipper ships sailing from England to Australia and New Zealand in the Roaring Forties, a wind expressway. Modern yacht racing uses a similar route. Usually the racers depart the European Atlantic, pick up the Roaring Forties south of the Cape of Good Hope, head west below Australia and New Zealand and swing past Cape Horn back to the European Atlantic coast. The current record time for this voyage is slightly longer than forty-five and one-half days.[54]

OCEAN GYRES AND THE *VOLTA DO MAR*

With the trade winds pushing waves and water to the east and the westerlies doing the same at higher latitudes to the west, the two work together to propel large water circulations in the great ocean basins. The Coriolis force operating on the moving waters also curves the movement and reinforces the direction of movement imparted by the winds. The results are large rotating ocean currents called gyres. There is also movement of water to relatively low-density warmer water from higher-density cold water, a direct analogue to the pressure-gradient force in winds. There are five major ocean gyres: the North Pacific and North Atlantic gyres rotating clockwise in the Northern Hemisphere and the Indian Ocean, South Pacific, and South Atlantic gyres rotating counterclockwise in the Southern Hemisphere. There are also smaller gyres, currents, and countercurrents working in this great clockwork of large ocean gyres.

Needless to say, the early ocean explorers were keenly interested in wind directions and sea currents, as they were a critical part of their voyages. The Portuguese, under the patronage of Prince Henry the Navigator (Infante Henry, duke of Viseu, 1394–1460), made prodigious advances in navigation, notably the technique they called *volta do mar*, or "turn of the sea." The Portuguese were exploring the African coast in search of slaves and converts, which appeared to be interchangeable terms in the mind of Henry.[55] In the process, the *volta do mar* technique was developed. Ships would work their way down the African coast, and to return, counterintuitively, they would sail west, moved by the trade winds and the currents of the North Atlantic Gyre, and then north to catch the westerlies for a fast return to Portugal. The further away from their Portuguese home, the further to the west they needed to go before circling north and then turning east to return. Otherwise, one would have to fight the winds of the trades to get back, a slow and potentially lethal strategy. In the process of working down the West African coast and swinging out on the *volta do mar*, the Portuguese discovered and colonized several remote Atlantic islands: Madeira, the Cape Verde Islands, and the Azores, which all became resupply ports for the return.

The Portuguese discovery of the Cape of Good Hope and the eventual route to the spice supply in India, southeastern Asia, and Indonesia involved a second *volta do mar* that involved heading east almost to Brazil before making the turn to run south of the Cape of Good Hope and heading for India.[56]

The line between heroics and foolhardiness can be drawn very finely; some might say that in some cases there is no line at all. Some remarkable navigators bravely put their faith in the *volta do mar* to navigate vast stretches of unknown ocean. For example, Christopher Columbus on returning from his first voyage to the New World used a *volta do mar* to return to Europe. He sailed north to around 36°N latitude, caught the westerlies, and then headed east, landing in Lisbon on March 4, 1493.

The remarkable Andrés de Urdaneta served on the second complete circumnavigation of the world, the Loaisa expedition in 1525. He became a monk and then sailed again in 1564 on the orders of Philip II of Spain for the conquest and colonization of the Philippines. From there he was sent to find how to get from the Philippines to Mexico across the Pacific.[57] He employed a *volta do mar*, sailing as far north as Japan (38°N) to find the westerlies, pioneering a route from Manila to Acapulco. He arrived on October 8, 1564, with a nearly dead crew.

His discovery, the Urdaneta route, was sailed by the Manila Galleons, two ships that brought Ming porcelain, furniture, silk, carved ivory, and lacquerware from Manila to Acapulco to trade for Mexican silver.[58] The galleons were incredible sailing ships capable of carrying two thousand tons of cargo and one thousand passengers. The trade route lasted until 1815. Some of the remarkable merchandise from the Manila Galleons is on display in the Museo Histórico de Acapulco in Acapulco, Mexico.

The ocean gyres and the eddies, currents, and countercurrents, coupled with the trade winds and prevailing westerlies, gave explorers in the age of sailing ships and exploration the ability to determine efficient routes over the open ocean. The ships initially used by the Portuguese and Spanish explorers were caravels, relatively small ships with lateen sails, which allowed the ships to sail close to the wind and advance (beat) against a headwind using a zig-zag course.[59] Clearly, for a voyage into unknown territory, a ship with the ability to return regardless of wind direction is a wise design choice. For example, Columbus's first exploration to the New World sailed on three vessels: The largest, the *Santa Maria*, was a carrack with both lateen and square-rigged sails; the other two, the *Pinta* and the *Santa Clara* (nicknamed the *Niña*), were lateen-rigged caravels. Once the sea routes were established, fully rigged ships with three or more square-rigged masts plied the ship routes.[60] The fully rigged ships were fast with fair winds but could not sail close to the wind.

THERMOHALINE CIRCULATION

There is one additional aspect of the winds and ocean circulations to discuss. Some of the large-scale ocean circulation is driven by

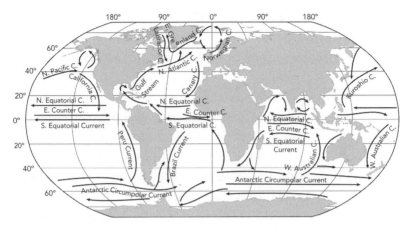

Major ocean gyres. The ocean gyres' are circular patterns in the ocean currents and are driven by the prevailing winds. Note the circulation of currents in a clockwise direction forming a large gyre in the North Atlantic. There are five major oceanic gyres: One in the Indian Ocean, two in the North and South Atlantic, and two in the North and South Pacific. When the currents in the gyres and the prevailing winds move in the same direction, an ocean vessel traveling in that direction had the "fair winds and favoring seas" that were a great advantage for sailing ships. *Source*: Vivien Gornitz, *Rising Seas: Past Present Future* (New York: Columbia University Press, 2013). Used with permission.

differences in the density of sea water. Very cold and very salty seawater, which is produced only in a small part of the ocean surface, sinks and flows as deep ocean currents to return eventually to the surface at a faraway location. The resulting circulation is called the *thermohaline* (temperature-salty) circulation.[61] One way to make really cold and salty sea water requires polar conditions and strong winds both to cool the water further and concentrate the salt by evaporation. These conditions occur in the North Atlantic at two regions, the Nordic Sea and the Labrador Sea. This sinking water forms what is called the North Atlantic Deep Water, which flows south.

Another way to produce cold, salty water is to freeze sea water. Since freezing salt water excludes ice, the sea becomes saltier in areas of extensive ice formation. This occurs in the Weddell Sea off of the coast of Antarctica. In front of the ice shelf, the evaporation by the winds cools and increases salinity to make deep water and to produce the Antarctic Bottom Water.

These flows of dense water are steered by the topography of the ocean basins. The flows of cold, salty waters eventually connect with other surface ocean currents to form a "mechanically-driven fluid engine capable of transporting vast quantities of heat and fresh-water."[62] The joining of the deeper circulation with the surface circulation of water forms a connected circulation of water referred to as the ocean conveyor belt.[63] One loop on this conveyor belt starts with the North Atlantic Deep Water forming in the seas to the south and east of Greenland: the cold, salty, deep water sinks and flows south under the surface warm current the Gulf Stream and then through the basin of the Antarctic Ocean. There one branch of the North Atlantic Deep Water goes into the Indian Ocean and upwells as it becomes fresher and less salty by mixing with the waters of the Indian Ocean. This water is eventually transported into the South Atlantic up the western side of Africa, across the Equator to the west along the Equatorial Current; mixes with the waters forming the Gulf Stream; then moves northward to the seas off of Greenland from whence it came to cool and sink to repeat the cycle. The other loop of the conveyor belt involves the eventual merging of the Antarctic Bottom Water with the second branch of the North Atlantic Deep Water, which eventually mixes and upwells in the Pacific.

The ocean conveyor belt interchanges the waters of the oceans and transports considerable heat toward the polar regions. For example, in the North Atlantic, the Gulf Stream moves warmer water to higher latitudes in the North Atlantic and thus gives Europe a milder climate than one might expect. Palm trees grow in Inverewe Garden along the Highland coast in Scotland at almost 58°N latitude; this is around the same latitude as the frigid Labrador Sea that forms the North Atlantic Deep Water. In the past, events have changed the movement of these ocean currents, with fierce consequences to climatic patterns.

One significant case involves the Gulf Stream and the onset of over a millennium of fiercely cold climate conditions across Europe and elsewhere, called the Younger Dryas.[64] At the end of the last glacial, about 14,000 years ago, the continental glaciers began to melt. From the fossil remains of plant parts, notably plant pollen preserved in lake sediments (see chapters 3 and 8), conditions worldwide were warming— fragmented tropical forests began to coalesce; boreal tree lines were moving north; and a cold, dry biome (the tundra-steppe) began to retreat.

Then, apparently quite rapidly, Europe went into a deep freeze. Vegetation resembling the vegetation of full glacial conditions returned, and glaciers began to build and expand. The event was called the Younger Dryas, after the tundra and high alpine plant, *Dryas octopetala*, a white wildflower indicative of cold conditions and whose pollen is extremely common in lake sediments formed during this time. The event occurred between 12,800 and 11,500 years ago. The Younger Dryas ended abruptly, perhaps in as little as fifty years, and its onset may have been even more rapid.[65]

The Younger Dryas was caused by a shutdown of the ocean conveyor belt by a glacial melting and large influx of freshwater released into the North Atlantic. This stopped the production of cold, salty seawater and, thus, the thermohaline circulation. The Gulf Stream flow slowed, and the transfer of massive amounts of heat to Northern Europe stopped. There are several theories as to what generated this influx of water: a gigantic flush of freshwater from an ice dam holding back a giant lake of melted glacier water finally giving way;[66] a change in climate patterns bringing more precipitation to the North Atlantic and shutting down the making of deep ocean water;[67] the eruption of the volcano that formed the Laacher See near Bonn, Germany;[68] or the impact of a carbonaceous chondritic asteroid or a comet into the North American glacier with heating and a tremendous freshwater release into the North Atlantic.[69]

Regardless of the exact cause, the rapid climate change that seems to have occurred at both ends of the event is a cautionary tale. It appears that alteration of the ocean's circulation and its movement of heat on Earth can have relatively rapid and highly consequential consequences.

LIGHT'S PATHWAY HOME: THE GLOBAL
RADIATION BALANCE

The interconnected motions of wind and water about our planet are fueled by the difference in the heat absorbed by different parts of the Earth's surface. Almost all of this energy is from the Sun (99.97 percent). Other sources of energy to the Earth's surface are obviously small but include energy originating from geothermal heat and radioactive decay leaking to the Earth's surface, tidal energy, and waste heat from fossil fuel consumption and nuclear reactors.[70] The amount of incoming solar energy that would fall on a surface perpendicular to the direction of light is around 1,370 watts per square meter (Wm^{-2}).[71]

However, most of the Earth's surface is not perpendicular to the incoming solar radiation, and half of the Earth is dark at any time. Taking into account nighttime and slanted surface angles, the incoming solar radiation is 342 Wm^{-2}. For the Earth's annual average temperature to remain constant, this incoming radiation must be balanced by a radiation transfer away from Earth. This balance is partially maintained by the 107 Wm^{-2} that is reflected back into space by clouds, aerosol particles, and gases in the atmosphere.[72] An additional 235 Wm^{-2} is emitted by the surface and atmosphere as outgoing long-wave radiation.[73] This long-wave radiation is a function of the temperatures of the surface and atmosphere.

Without the atmospheric greenhouse gases, an Earth of around −19°C would produce enough outgoing radiation to balance the incoming solar radiation (minus reflection). The average surface temperature of Earth is some thirty-three degrees Celsius warmer, at about 14°C.[74] Greenhouse gases absorb outgoing long-wave radiation coming from the planetary surface and heat the atmosphere. This additional absorption of radiation gives us a warmer Earth.[75] This "natural greenhouse effect" is mostly due to two gases—water vapor and CO_2. Clouds also have a similar warming effect from capturing outgoing radiation, but clouds also "whiten" the Earth and reflect more of the incoming solar radiation, which yields a net cooling effect.

The tradeoff between the consequences of more clouds as a positive or negative effect on planetary warming is but one of many of the complexities of the Earth's heat budget that we need to understand better. Without natural greenhouse warming, Earth would be a cold and foreboding planet with a frozen ocean.[76]

From the numerous articles in the news media and the pronouncements of sundry politicians and "talking heads", one might think scientists have only recently discovered the warming consequences of atmospheric greenhouse gases. That "greenhouse warming" has a history in science that goes back almost two hundred years may come as something of a surprise.

The initial discovery that gases in the atmosphere could capture radiation and increase temperatures was found in experimental chambers in an experiment conducted by Horace-Bénédict de Saussure, a Swiss aristocrat probably best known as the founder of mountain climbing as a sport. Based in part on this experiment, Joseph Fourier, a French polymath, theorized in an 1824 paper that the atmosphere, like a greenhouse, "lets through the sun's rays but retains the dark rays from the ground."[77] His "dark rays" would now be called infrared radiation.[78]

An exploration of the effects of CO_2 in the atmosphere was famously developed in 1896 by the Swedish Nobel laureate Svante Arrhenius.[79] On the heels of a divorce, Arrhenius occupied his time through a dark Swedish winter by laboriously solving an estimated ten to one hundred thousand hand calculations on the greenhouse effect of different concentrations of CO_2 in the atmosphere. By the time he was done, he had determined the average change in temperature for every ten degrees of latitude and for four seasons using values of 0.67, 1.5, 2, 2.5, and 3 times the average CO_2 concentrations at the time.[80]

In making these laborious computations, Arrhenius synthesized the experimental and observational physics of his day: measurements produced by Langley on the transmission of heat through the atmosphere, by de Bort on cloudiness for different latitudes, by Ångström on the absorption coefficients of water vapor and CO_2, and by Buchan on mean monthly temperature over the globe.[81] Stephan's law, which states that radiant emission varies as the fourth power of temperature, was used to calculate heat exchanges. Arrhenius estimated the absorption of the surface and of clouds (surface albedo and cloud albedo, respectively) as well as the effects of snow cover feeding back on the radiation budget by decreasing the surface albedo. For all this effort, several significant factors had to be omitted. Neither the horizontal transport of heat (by winds or ocean currents) nor the effects of changes in cloud cover were included in his computations.

Arrhenius's work was part of a long and continuing evolution of our understanding of the climate of the planet. The results of his calculations resemble those derived modern general circulation models (GCMs) that currently are used in assessments of the planetary effects of increases in "greenhouse gases" in the atmosphere. As with the modern GCMs, the warming effects are greatest in the higher latitudes and in the winter. The magnitude of Arrhenius's predictions of 5 to 6°C for doubled CO_2 is not out of line with recent GCM predictions, either.[82] This agreement may be fortuitous in that, as noted above, Arrhenius's calculations do not include important significant changes,[83] notably the horizontal advection of heat, clouds, and vegetation change (to be discussed in chapter 8). The Earth's climate has feedback systems that complicate our evaluation of the consequences of our actions.

For example, increased greenhouse gases could increase the warming and thus cause snow and ice to melt in regions that are otherwise perennially white or to lengthen the ice-free season at other locations. This would make the surface land and water, which are darker than snow and ice, absorb more of the heat from the Sun and promote further warming.[84] This "ice-albedo feedback" could amplify an initial warming caused by increased greenhouse gases in the atmosphere. It is one of the many planetary feedbacks that this chapter has but introduced. Certainly, we have begun to modify the composition of the atmosphere by introducing increased CO_2 from land clearing, cement production, and fossil fuel combustion as well as by producing and releasing other greenhouse gasses.

CONCLUDING COMMENTS

The whirlwind questions that began this chapter are profound and touch upon current issues of exploration and planetary change:

"Have you entered into the springs of the sea, or walked in the recesses of the deep?" (Job 38:16). We have indeed "walked" in the recesses of the deep, and we have seen incredible things there. We have seen lakes of liquid methane, strange organisms that derive the energy that supports them from chemical reactions, organisms of great diversity of forms that live in total darkness and numbing cold. While the creatures of the abyssal depths are remarkable, this chapter has focused on the springs of the sea, the deep ocean currents that transport heat from the warm regions to the colder ones. Changes in these currents can modify the habitability of the planet. They have done so in the past; they could do so in the future.

"What is the way to the place where the light is distributed or where the east wind is scattered upon the earth?" (Job 38:24). The light, the radiant energy that warms our planet, is distributed by the ocean currents of the deep, by the surface currents of the oceans, and by the winds. They do, indeed, drive the east winds and the trade winds and the other large-scale winds that are activated by the circulation of the atmosphere as it moves.

One could only speculate on what Job might make of the scientific answers to these whirlwind questions. The answers are certainly complicated. This chapter has focused on the remarkable feedbacks and interactions of the major Earth systems. The powerful interactions between the atmosphere and oceans to transfer heat from the warmer equatorial zones to the colder polar regions exemplify the interconnectedness of our planet. It is astonishing that we are modifying aspects of our planet in measurable and potentially significant ways, given that we do not fully comprehend our planet's workings.

As people of a technological age, we are accustomed to exercise caution when we modify our own constructions. One likely would not attempt to rebuild an automobile's transmission without consulting the manufacturer's manual. Certainly, such an endeavor would be foolhardy if one were in the middle of the desert and the car was the only way out. The Earth is far more complex than an automobile, and the instruction manual remains a work in progress.

Convergence in form for plants and for vegetation in arid conditions in different regions Top left: A cactus from Mexico, North America. Top right: A spurge (*Euphorbia*) from South Africa. Lower left: Peski Karakum desert, Turkmenistan, Central Asia. Lower right: Kalahari desert, Botswana, Africa. *Source*: First two photos by the author. Latter two photos used with permission of T. M. Smith.

8

Making the Ground Put Forth Grass

The Relationship Between Climate and Vegetation

Who has cut a channel for the torrents of rain, and a way for the thunderbolt, to bring rain on a land where no one lives, on the desert, which is empty of human life, to satisfy the waste and desolate land, and to make the ground put forth grass?

—Job 38:25–27 (NRSV)

In the southern African nation of Botswana, the currency is the *pula*, a Setswana word that simultaneously means rain, blessing, and wealth. This dry nation sees little rainfall, and its Kalahari sands drink what rain there is into its sandy soil. Botswana is a "thirst-land" where water is precious—hence the association between rain and other good things, such as good fortune and money. In arid and semiarid environments, rain is everything. Biblical Uz, the setting for the story of Job, seems a thirst-land as well.

Rain is a commodity not to be wasted in Uz or in the Bible, in general. In Isaiah 35:6–7, God makes rain in the desert to allow the survival of the returning exiles to Zion. However, in Job 38:25–27, the rain falls in the desert, "where no one lives," pointedly emphasizing that the distribution of the water solely for the benefit of humanity is *not* a divine objective in the Book of Job.[1] Indeed, a minimization of humanity's importance in the fabric of nature resonates through the whirlwind speeches,[2] a point noted in chapter 1. From an anthropocentric view, what is the point of making rain fall on a place humans do not use? The whirlwind question reveals that rain is not just for humans or the ecosystems that immediately support them. It also is for nonhumans and their supporting ecosystems. For example, it rains in the desert for the onager from chapter 4, who laughs at humans and makes its home there. Presumably, it rains for all the other creatures as well. The implication of the whirlwind question is that rain nurtures the planet and its biotic diversity but is not solely for humanity's needs.[3]

Deserts are difficult places for many species to survive. For other species, they represent diverse habitats and are filled with remarkably adapted plants and animals. The consideration of how plant and animal species fit into and survive in their environments falls under the general ecological topic of niche theory. The "niche" was formally defined in 1917 by the great California naturalist Joseph Grinnell.[4] It refers to the set of conditions (climate, soil, other associated species, etc.) that determine where a particular species of plant or animal will be found. Grinnell's father, Fordyce Grinnell, was an Indian Agency physician to the Oglala Lakota (Sioux) and their chief Red Cloud (Maȟpíya Lúta, 1822–1909). Joseph Grinnell was born in 1877 near what is now Fort Sill, Oklahoma, and grew up among the Lakota (Sioux) in what is now the Pine Ridge Indian Reservation in South Dakota. His childhood undoubtedly produced a unique ecological understanding and was likely influential in his formalization of the niche.[5] The Lakota, like other people surviving on the land, necessarily developed keen insights on what controls where species, particularly useful ones, are located. Grinnell and his niche theory channel these Lakota concepts to ecologists.

There also can be remarkable similarities between totally unrelated species found in the same environments, as the illustration of desert plants in the front of this chapter shows. The phenomenon is called "convergence" and seems evidence for parallel organization among the species that compose ecosystems in similar environments.[6] Convergence is often seen as evidence for similar niches to be filled in similar but geographically distant environments. The overall look of ecosystems in similar environments/different places can also be strikingly alike. Job, extremely successful at husbanding a diverse array of grazing animals, would have known about deserts, because the patterns of their vegetation can indicate underground water seeps. After all, he owned "seven thousand sheep, three thousand camels, five hundred yoke of oxen, five hundred donkeys" (Job 1:3). To make good decisions as to what to do with such vast herds, one must anticipate the local microclimate from the landscape condition. Knowing how to "read" a landscape is essential to any grazing pastoralist.

Job would not have been surprised by the somewhat unusual occurrence of "rain on a land where no one lives, on the desert, which is empty of human life, to satisfy the waste and desolate land, and to make the ground put forth grass." His herdsman's understanding would have given Job the context to appreciate but not answer God's question. The kernel of this whirlwind question is not the event of rain in the desert; it is, "Why would God waste water that people so desperately need on deserts?" One answer is, "Perhaps God is not as concerned about people (versus other creatures) as Job would like to think."

This chapter focuses primarily on the less theological but most ecological part of the "rain-where-no-one-lives" whirlwind question, namely, of how climate (the rain) is interlaced with the vegetation (the desert). As was discussed in chapter 3, matching plants to their environment was on the minds of the early agriculturalists of the Fertile Crescent, who shifted their crops when climate conditions fluctuated.

In taking an ecological emphasis, the present chapter presents some of our historical and current understanding of the couplings between the atmosphere and terrestrial surface. This is a significant piece of the larger puzzle asked by the whirlwind question, "Have you comprehended the expanse of the earth?" (Job 38:18). This comprehension summarizes the ensemble of topics from other whirlwind questions, which connect sea and atmosphere, atmosphere and land-cover change, and the Earth and the Sun. The equally interesting connections among the natural histories of animals and planetary change were covered in an earlier text of mine.[7]

There is one part of this whirlwind question not treated in this chapter, "Who has cut a channel for the torrents of rain, and a way for the thunderbolt, to bring rain?" because it is the topic of other chapters. People have worked hard to try to learn how to make it rain. Rain gods, rain dances, and other rituals, often attended by sacrifices of different intensities, are part of the mythic structure of many different cultures, and so are the calendric celebrations discussed in chapter 5. Many cultures have developed and designed rituals and festivals both to commemorate and induce the coming of the seasons. Chapter 3 points out that land clearing, initially with the assistance of domesticated animals, appears to modify local and regional climate. This topic will be taken up again in chapter 10, which treats the current possibility of directed human alteration of the Earth intended to modify and control the climate—planetary geoengineering.

Throughout the whirlwind speech, Job is asked to explain what he knows about the workings of nature. Probably wisely, Job realizes that he knows nothing comparable to the wisdom of God. The whirlwind questions touch on how different parts of the Earth's systems function, in the case of vegetation, on their dynamics and on causes of pattern. These are central themes in ecological and environmental science and have been so for quite some time.

At the early edge of history, ancient scholars from non-Jewish traditions were keenly interested in plants, initially for their medicinal properties and later for the factors that control their presence (ancient niche theory). Some of the earliest scientific literature was in the form of botanical herbals, which advised treatment of medical conditions with plants or plant derivatives.[8] These were highly pragmatic treatises. For example, the Egyptian Ebers Papyrus, thought to be from 1500 BCE but possibly copied from texts as old as 3400 BCE,[9] had such cures as, "To kill roundworm [*Ascaris lumbricoides*]: root of pomegranate 5 ro, water 10 ro, remains during the night in the dew, is strained and taken in 1 day."[10] A *ro* was a measure equal to about a tablespoon.[11]

A transition between these ancient medical herbals and a more ecological view of plants and their environment can be seen in work of Theophrastus (372–287 BCE), a student of Aristotle. In the ninth volume of his book *De Causis Plantarum*, also known as *Enquiry Into Plants*, Theophrastus described medical herbs, but he also described their natural habitats and geography.[12] Theophrastus is sometimes called the father of ecology and for no small reason. He experimentally transplanted species to areas outside their natural range to determine if they would grow or flower there. The plant-leaf attributes of deciduousness and evergreenness were observed to vary under different climate conditions, and Theophrastus documented these systematic changes in leaf retention. He also observed that high altitudes and northern latitudes were similar in the pattern of their climate and vegetation.[13] That climate and vegetation are strongly related remains relevant, particularly in today's concern over the potential effects of change in climate on the vegetation of the Earth.[14]

CLIMATE AND VEGETATION

Mountain vistas provide their viewers with a sense of the unity of nature, particularly the constancy of the relationship between climate and vegetation. Looking at adjacent mountain bases, the slopes are covered in forests of the same mixture of species. Move one's eyes up the slope, and all of the mountains transition to a different type of forest at the same altitude. This persists for the position of the change from montane forest to alpine heather and for the elevation of the beginning of snowfields at the mountain's tops. These changes in mountain zones all display a pleasing, ecologically based regularity. Of course, there may be a bit of variation, perhaps thanks to a slope's different steepness or aspect to the Sun, but the scene speaks to a unity of underlying processes working to form a regular, repeatable vegetation pattern.

Theophrastus understood this unity in mountain vistas and in the parallel changes as vegetation changed with climate and location. Theophrastus's teacher, Aristotle (384–322 BCE), felt that the world, assumed to be a sphere, consisted of five zones (in Greek, κλίμα or *klima*; plural *klimata*). There were two frigid *klimata* (arctic and antarctic), a torrid *klima* at the equator, and two temperate *klimata* between the torrid and frigid. This was a simplification of the even-then-ancient practice of dividing the world into seven such climes.[15]

Greek geographers placed the inhabited portion of the spherical Earth into various zones according to the angle of the sun's rays to the celestial axis, the position that today is occupied by the pole star, Polaris, in the constellation Ursa Minor. The word *klima*, meaning angle or slope, refers to this angle, which corresponds to the latitude of the location. The original seven climes devised by the Greeks geographers ranged from 17°N latitude to 48°N. This system of *klimata* later was elaborated by Claudius Ptolemy (a Greek-speaking Roman citizen of Alexandria, Egypt, c. 90–c. 168 CE). In Ptolemy's system, each hemisphere was divided into twenty-four *klimata* or latitudinal bands.[16] Ptolemy used the length of the longest day as his indicator of latitude. Much of what remains of this Greek geographical knowledge fortunately was preserved through the European medieval times thanks to its translation into Arabic.[17] As the word *klimata* passed into Latin, eventually to become the English word "climate," its meaning was broadened to mean a region along with the typical weather associated with it. Thus, in their shared and deep history, climate, mountain slopes, and latitude are conjoined.

By the late eighteenth and early nineteenth centuries, European scientists began a significant synthesis of the relationship between climate and vegetation. In Sweden, Carl von Linné (Linnaeus) developed what was to become the modern basis for naming plants and animals. Linnaeus used a genus/species binomial name for each species in a continuing series of updated and ever more extensive texts starting with plants in 1735.[18] A Czech nobleman, Kaspar Maria Graf von Sternberg, established the massive fossil and mineral collection of the Bohemian National Museum (then called the Gesellschaft des vaterländichen Museums in Böhmen), which was initially housed on the ground floor of the Sternberg Palace.[19] Using the relation between climate and plants,[20] Count Sternberg reasoned that the difference in past fossil plants in different layers of deposits at a location implied that the past climates of Europe had been very different from the present ones.[21] Sternberg also championed the use of highly detailed drawings and lithographs to allow museums to communicate their holdings to other museums, for comparative studies. The Swiss botanist Augustin-Pyramus de Candolle and his fellow naturalist Jean Baptiste Pierre Antoine de Monet, Chevalier de Lamarck, produced the first biogeographical vegetation map in 1805, showing the distributions of plants across France, and commented on the environmental factors controlling plant distributions.[22] Lamarck is better known for his now-rejected theory on the evolution of species, which involved the acquired traits of the parents being inherited by their offspring. For example, in Lamarck's concept, the ancestors of giraffes stretched their necks for high leaves and passed the longer necks that they had acquired through generations to produce the current long-necked giraffe. It is sad that Lamarck is now remembered for his failed evolution theory rather than for his groundbreaking work as a natural historian and geographer, but c'est la vie. This latter work added context to that remarkable era of exploration and discovery.

VOYAGES OF DISCOVERY

The appreciation of the influence of climate on vegetation intensified after the observations of the biologist-explorers on voyages to the remotest parts of the world.[23] The first such voyage of discovery was by the crew of the HMS *Dolphin*, in 1764–1766. Captain John Byron produced a book, *A Voyage Round the World*, documenting his discoveries; this started a practice that persisted with later voyages. In the year she returned, the *Dolphin* sailed again under Captain Samuel Wallis, along with the consort ship, the HMS *Swallow*, under the command of Philip Carteret, and they made a second circumnavigation in 1766–1768. This second expedition reported on explorations of Tahiti, the Solomon Islands, and Papua New Guinea. Other voyages followed in rapid sequence. The HMS *Niger* explored Newfoundland and Labrador in 1766. The *La Boudeuse* and *L'Étoile* made the first French world circumnavigation in 1766–1768. It also carried the first woman to circumnavigate the world, Jeanne Baré,[24] the valet and possible mistress of the ship's botanist, Philibert Commerson. Commerson also named the day-glo pink- and purple-flowered bougainvillea after the chief of the expedition, Louis-Antoine, Comte de Bougainville.

On August 26, 1768, the HMS *Endeavour* departed Plymouth, bound for the island of Madeira and then onto its mission to observe the exact time of the astronomical transit of Venus across the Sun, on June 3, 1769, at Tahiti—this was intended to aid a determination of the longitude of Tahiti and was part of the continuing attempt to perfect a methodology for predicting the longitude of a ship at sea. The *Endeavour* was then to make an exploration of the southern Pacific prior to its return to England. Captained by James Cook, they landed on and claimed the Society Islands, New Zealand, and Australia for Great Britain. They mapped the coast of Australia, ran aground on the Great Barrier Reef, made repairs, and eventually returned to England via Batavia (modern-day Jakarta, Indonesia). The *Endeavour* carried three naturalists. Joseph Banks, the famous British naturalist, headed the group. Banks had hired Daniel Solander, a Swedish student of Linnaeus at Uppsala University, and Herman Spöring, who was Solander's assistant naturalist. Alexander Buchan and Sydney Parkinson (landscape and natural science artists), along with four servants, filled out the Banks entourage.[25] They stopped in Batavia to make more repairs before their return home. Batavia was a pestilent city plagued by malaria and dysentery. It was the death of many of the sailors on the voyages of discovery.[26] Spöring and twenty-nine others from the *Endeavour*, including the remarkable Polynesian navigator Tupaia (see chapter 6), died from diseases there. Of Banks's nine-man group, only four people (Banks, Solander, and two of the servants) survived the voyage.[27] Banks went on to great fame. He became president of the Royal Society and served in that role for forty-one years. He was a great promoter of international science and scientific exploration as well as of the colonization of Australia. As a mentor and promoter, he championed multiple generations of young natural scientists.

The biological discoveries that Banks and his colleagues brought back on the *Endeavour* were scientifically astounding. The continent of Australia had been effectively separated from much of the world for many millions of years and fully isolated when it broke away from Antarctica about fifty million years ago. It had flora and fauna that had been independently evolving since this time of separation. If the voyages of discovery are analogues to the modern era of space exploration, the observations from Australia were equivalent to finding life on the Moon or Mars—novel, surprising, speculation generating, unpredictable, and theory challenging. New information, new theories, and new scientific questions boiled up from attempts to understand the source and function of this alternative continental-scale biota.

The voyages of discovery spanned well over a century, starting in 1764 with the voyage of the *Dolphin* and ending in 1899 with the return of the German steamship *Valdiva* from her mission of charting the deep water on the front of the Antarctic ice shelf.[28] The goals of the explorations transitioned from charting new regions and returning drawings and specimens of their novel flora and fauna to Europe, to the search for the Northwest Passage and surveying polar regions, to the mapping of the world's coastlines and delving the depths of the seas. The role of the exploratory voyage faded, and in its place a system of national maritime study centers with laboratories and dedicated modern research vessels evolved. Nonetheless, the voyages and their reports and biological findings electrified the imagination of natural scientists of their era and continue to do so today.

Two fundamental biological/ecological questions were sharpened by the voyages of discovery. First was the attempt to understand the remarkable diversity of species that were found on these explorations. There were so many more species than anyone had imagined! How did they come into being? Answering this question eventually led to the theory of evolution formulated by Charles Darwin, himself a member of one of the voyages (the HMS *Beagle*, which circumnavigated the world from 1831 to 1836).[29] The co-originator of evolutionary theory, Alfred Russel Wallace, also was an explorer, initially of the Amazon basin and later of the Indonesian archipelago. Formulation of their theories on evolution propelled studies of the nature of genetics and heritability. This was followed by inquiry on the nature of the gene and to modern molecular biology.

The second fundamental question from the voyages asked what caused the regularity in the biological and ecological systems that were found in different environments all over the world, even when the species in different places were unrelated. One facet of this question involves elucidating the basic causes of climate/vegetation relationships. As has already been discussed, this topic has remarkably deep historical roots. The exploratory biologists of the eighteenth and nineteenth centuries were astounded by the array of strange animals, novel plants, and different human cultures they witnessed. Notably, they developed theories on causes of ecological convergence—similar ecological patterns in the plants and animals in distant regions with similar climates.

Alexander von Humboldt stands out as the most conspicuous example. He was a holistic thinker with a unique capacity to record biological and cultural details coupled with a fascination with a diversity of environmental data.

ALEXANDER VON HUMBOLDT (1769–1859)

Friedrich Wilhelm Heinrich Alexander Freiherr von Humboldt was born in Berlin on September 14, 1769—the same year that the HMS *Endeavour* arrived at Tahiti during Captain James Cook's first voyage. His mother, Mary Elizabeth Colomb, was the second wife of Alexander Georg von Humboldt, a major in the Prussian army. His mother was also the wealthy widow of Baron von Holwede. Her estate from this earlier marriage would eventually transfer to her two sons. These funds bankrolled Alexander von Humboldt's career in science.

While growing up on his family's estate near Berlin, he developed a keen interest in natural history. Carl Ludwig Willdenow, one of the earliest and best-known plant geographers, mentored young Alexander and encouraged his interest in plant geography.[30] Humboldt read books from the voyages of discovery, particularly Bougainville's *Le voyage autour du monde, par la frégate La Boudeuse, et la flûte L'Étoile (Journey Around the World . . .*, 1771), which described his circumnavigation, and Georg Adam Forster's *Voyage Round the World* (1777), which chronicled his observations while with Captain James Cook's second voyage to the Pacific.[31] Humboldt was to later meet Georg Forster while a student at Göttingen University. They became close friends.

Humboldt finished his formal education at the Freiberg Mining Academy in 1792 and took up a position with the Prussian Department of Mines as a mine inspector. His work gave him experience with a wide range of surveying instruments and other scientific devices, which served him well during his later explorations.[32] He also appears to have decided upon his course in life. Buried in a footnote of the first paper he produced on primitive plant species found in the Freiburg region, "plant geography traces the connections and relations by which all plants are bound together among themselves, designates in what lands they are found, in what atmospheric conditions they live."[33] This is the question of understanding the climate-vegetation relationship writ large. The illumination of this opinion was to frame the remainder of his life. His father had died when Alexander was nine years old. With the death of his widowed mother in 1796, he resigned his position as mine inspector to follow a career in science and exploration.

He initially planned an expedition to the West Indies, which did not materialize.[34] In 1798, he developed a plan to explore Egypt with the outrageously unconventional fourth earl of Bristol and traveled to Paris to equip himself with scientific instruments.[35] This Egyptian exploration also unraveled. However, while in Paris he met Nicolas

Baudin, who was in the process of organizing another voyage around the world funded by the French government.[36] Humboldt was invited to join the scientific crew of the expedition. One can only imagine Humboldt's delight to be given the opportunity to sail the world himself. One can equally imagine the disappointment when the expedition was cancelled.

During this emotional rollercoaster of developing and dashed expectations, Humboldt met Aimé Jacques Alexandre Bonpland, who was to have been the botanist on the cancelled Baudin expedition.[37] Bonpland and Humboldt joined forces and eventually traveled to Spain. They were granted permission by Carlos IV to explore Spanish America, if they were prepared to pay their own expenses.[38] They departed La Coruña, Spain, aboard the ship *Pizarro*, on June 5, 1799, bound for Cumaná, Venezuela, where they arrived on July 16, 1799.[39] The day he sailed he wrote a friend, Karl Freiesleben, "I shall try to find out how the forces of nature interact upon one another and how the geographic environment influences plant and animal life. In other words, I must find out about the unity of nature."[40] On the way, Humboldt and Bonpland stopped in the Canary Islands, where they climbed the recently active volcano Pico de Teide and descended into the crater. They were to be away from Europe for over five years. Humboldt and Bonpland returned to Bordeaux on August 1, 1804.

In his exploration, Humboldt had performed daring feats. During a dangerous boat journey—made all the more so because he could not swim—he first demonstrated that the Orinoco River drainage exchanged waters with the Amazon basin, a feature unique among the major rivers of the world. Along with Aimé Bonpland and Carlos Montúfar, he climbed to 5,875 meters on the tallest mountain in Ecuador, Mt. Chimborazo. Blocked by a crevice, they could not reach the 6,268-meter summit. At that time, Chimborazo was thought to be the tallest mountain on Earth. In fact, if one measures the heights of mountains as being the distance from the planet's center to the mountain's top (and not the conventional measurement of height above sea level), it is the world's tallest mountain. However one measures it, theirs was the highest ascent above sea level yet reached by any European mountaineer, a record that would stand for decades.

Alexander von Humboldt also had laid the scientific foundation that would make him the most celebrated intellect of his time. There were crates of scientific material brought back from South America. Humboldt and Bonpland collected sixty thousand plant specimens and discovered around 3,600 new species.[41] They settled in Paris and began to organize their vast collections and notes. The expedition would publish thirty volumes of findings.[42] Humboldt coauthored about one-third of these with Bonpland, who soon left the task of compiling field notes and specimens to take a position in charge of the Empress Josephine's gardens at Malmaison.[43]

When Josephine died in 1814, Bonpland married and took a professorship in natural history in Buenos Aires. He would never return to Europe. While on a field expedition near the Argentine border, he was wounded and placed under house arrest in a remote location in Paraguay until 1830. Deserted by his wife, he eventually ended up owning a ranch in Uruguay where he fathered several children with an Indian woman. He died in 1858.[44]

Meanwhile, Humboldt's old botanical tutor, Carl Ludwig Willdenow, had become the head of the Berlin Botanical Garden in 1801. He came to Paris in 1810 to work with his former student's material. Willdenow's Berlin herbarium today contains an inventory of specimens representing twenty thousand species, including a significant amount of the Humboldt/Bonpland South American material. Like Humboldt, he was interested in plant adaptation to the climate, and he noted that plants living in the same climate have common sets of characteristics.

Humboldt became one of the most productive scientists in history.[45] His work developed new vegetation/ecosystem-based plant geography and inspired others to follow this creative lead.[46] According to a recent biographer,[47] this paradigm shift to a vegetation-oriented plant geography was as significant as Lavoisier's "new chemistry" from 1789 or Lyell's "new geology" developed in his book *Principles of*

Geology, from 1830–1833.[48] The difference is that while Lyell and Lavoisier significantly realigned existing sciences in an enlightened age, Humboldt essentially invented a new science.[49]

Alexander von Humboldt saw his own accomplishments more modestly. When he was in his eighties, he wrote a letter, dated October 31, 1854, to his publisher, Georg von Cotta,[50] which listed his "only" three accomplishments as:

1. *Observations concerning geomagnetism that resulted in the establishment of magnetic stations throughout the planet.* These magnetic stations were the product of a direct appeal of Humboldt to the Royal Academy and to the British Society for the Advancement of Science. The station observations eventually tied changes in the Earth's magnetic field to sunspot activity.[51]

2. *The geography of plants, particularly of the tropical world.* This was a grand contribution. As but one example of his influence, his work on the biogeography of plants and his vivid description of tropical ecosystems led Darwin to sail to the tropics on HMS *Beagle* and was a principal influence on the development of Darwin's theory of natural selection. In a surviving letter responding to Darwin's stated appreciation of Humboldt's work, he wrote, "You told me in your kind letter that, when you were young, the manner in which I studied and depicted nature in the torrid zones contributed toward exciting in you the ardour and desire to travel in distant lands. Considering the importance of your work, Sir, this may be the greatest success that my humble work could bring."[52] Darwin carried copies of Humboldt's work on his own voyage of discovery on the *Beagle.*

3. *The theory of isothermal lines.* This accomplishment and previous one are particularly germane to this chapter. Isothermal lines connect locations with the same annual average temperature. As calculated by Humboldt, this temperature is obtained by averaging two daily observations, the temperature at sunrise and at 2 p.m. on each day over the course of a year.[53] Isothermal lines change with elevation and latitude. Humboldt noted that they predict the regular variations in the height

of snow on mountains (the higher the latitude, the lower in altitude is the line of permanent snow) and vegetation features (tree lines, transitions from evergreen to deciduous forests, etc.).

Humboldt collected copious measurements on the state of the environment both regularly and everywhere he went. From this myriad of observations, he decided that isothermal lines provided a general summary of the global pattern of environment and vegetation over the vast and biologically complex regions he explored. Humboldt probably would not have viewed isotherms as the sole "limiting factor" for vegetation interacting with climate. Humboldt saw the world as full of strong interconnections—"*Alles ist Wechselwirkung*" (everything is interconnected) was his slogan.[54] Isotherms were measurements that best captured the entire interconnected physical, chemical, and biological interactions in what might nowadays be called an ecosystem—even though that term was not formalized until well over a century later.[55] Humboldt inspired a succession of globally oriented plant geographers who worked to perfect world vegetation maps.

The first global-scale vegetation map was developed in 1872 by A. H. R. Grisebach.[56] This two-volume 1,300-page opus described the vegetation of different zones. The first section of each vegetation description was titled "*Klima*" and described the climate of the region. This was followed by detailed accounts of the species in different regions. He also designated some fifty-four plant forms to develop appearance-based vegetation categories. Many of these forms have a strongly ecological focus (bamboo, banyan, liana, steppe grass, annual grass, etc.); others are more taxonomic (cactus, palm, ferns, bromeliads, etc.). Despite all of this description, Grisebach's tome has but one illustration, the first world vegetation map, folded in the back of his book.

A. F. W. Schimper published a text in 1898 that also included a world vegetation map, this time with the biomes still used in modern vegetation: tropical rainforest, needle-leaved forest, savanna, steppe, heath, dry desert, tundra, cold desert, etc. His book, which was produced slightly more than twenty-five years after Grisebach's, displays hundreds of photographs illustrating the remarkable diversity of the planetary vegetation. Allowing the reader to see the vegetation observed on the voyages of discovery, Schimper's book combines an armchair explorer's photographic tour through the marvels of the planet's diversity with a scientific account of the vegetation.

The Danish ecologist Christen Christensen Raunkiaer developed a categorization of plant forms based on the height of what he called "perennating tissue."[57] Perennating tissue becomes inactive during cold (or dry) periods and, when conditions are favorable, then produces new plant structure. The buds on the ends of branches of trees and shrubs, which produce leaves and twigs in the springtime, are familiar examples of perennating tissue. Trees and shrubs with buds in the ends of their branches were called *phanerophytes* in Raunkiaer's classification. Plants surviving in harsh environments protect their perennating tissue in different ways, and these patterns are used to develop a classification of life forms. The proportions of different life forms in a local plant community are related to the environment and respond over time to, for example, experimental manipulation of shelter.[58] At the global scale, for example in tropical rainforests, phanerophytes with exposed perennating tissue at the end of branches (trees or vines) comprise virtually all of the plants. In moist tundra vegetation, *hemicryptophytes*, with their perennating tissues protected near the ground surface, predominate. In another similar classification, plants in different climates have unique physiological adaptations allowing them to survive minimum temperatures.[59]

Climate maps reinforced the importance of the interconnection between climate and vegetation.[60] For example, in the Köppen climate system,[61] climate is initially divided into main climate zones (equatorial, arid, warm temperate, etc.) that also relate strongly to broad categories of vegetation. There is not a lot of difference between these categorizations and the *klimata* of Aristotle and Ptolemy millennia ago. Subordinate categories in some cases have embedded vegetation formations (desert, tundra, etc.). Different climates are delineated by temperature, precipitation, and the seasonality of precipitation.

Seasonality in rainfall separates areas in which the rain does not occur during the most favorable season for growth from those in which it does. Savannas receive their precipitation in summer and are rain-green. They lose their leaves in the dry season. Mediterranean shrub-lands have winter rain and are mostly evergreen. Detailed climate maps often show climates for locations that have no weather stations (for example, the tops of mountains in remote locations). These climate maps are often based on observing the vegetation there and inferring an expected climate pattern.

CLIMATE-VEGETATION RELATIONSHIPS IN A CHANGING WORLD

Two hundred years after Alexander von Humboldt, we have learned a lot about the vegetation of the Earth and how it is related to climate, soils, and associated human society—all classic Humboldtian themes. Attempts to understand more clearly environmental interactions with the planet's vegetation have continued to occupy ecologists to the present day. Ecologists still have a lot to argue about and many theories to test. However, we have a substantial new problem, one with which the plant geographers from earlier eras did not concern themselves.

To paraphrase Alexander von Humboldt, *Alles ist Wechselwirkung aber einige Verbindungen neue sind*—everything is interconnected, *but some connections are new*. We are challenged to understand what the future vegetation might be on our planet with a different, human-altered climate and a level of carbon dioxide (along with other green-house gases) in the atmosphere not seen for millions of years.

Two interdependent, fundamental issues must be resolved to evaluate the effect of climate change on ecosystems. First, what aspects of the ecosystem does one wish to understand? How is the ecosystem resolved in time, space, or complexity? To understand climate change's effects on vegetation, must one understand the leaf response, the whole plant response, the population response, or somehow all of these and more? If so, how does one synthesize among these levels to understand the ecosystem's response to climate change?

Second, what environmental factors dominate the response of whatever aspect we wish to understand? What are the controlling factors? Most importantly, will the importance of these environmental factors change under novel conditions? In particular, will the importance of factors change with an increase in the CO_2 levels of the atmosphere— which has already happened to a degree from human actions and will continue to do so? Or, under novel climatic conditions, will new factors control the natural system response?

These questions stem from ecology's historical roots—from ancient philosophers such as Theophrastus; from the brave ecological explorers such as Joseph Banks, Alexander von Humboldt, and Darwin; from the plant geographers of the nineteenth century; and from the proto-ecosystem ecologists and vegetation scientists of the last century. The essential need for scientific understanding that drove centuries of scientists to explore the world is brought into fresh immediacy by present-day challenges of understanding the consequences of global environmental change.

THE PROBLEM OF SLICING AND DICING VEGETATION

The early challenge of mapping of vegetation over large areas and the current problem of evaluating the effects of climate change on vegetation both require identifying the best level of scale. How does one settle upon the ecological organization that will help us best understand vegetation patterns and changes in these patterns? How should one "slice and dice" the vegetation into its components? Alexander von Humboldt focused on the vegetation as a whole as his unit of interest. Physiologists focus on fundamental biochemical processes such as photosynthesis or respiration and how these basic processes respond to climate. Alternatively, an ecologist might group species into functional groups to reveal climate/vegetation relationships better. Perhaps amalgamating whole assemblages of species into recognizable plant communities is the key? These issues were part of the deeper history of vegetation/climate relations and are topics of research in global change today.

The understanding of the appropriate factors to consider, sometimes called the "scale problem," remains a challenge in formulating models to predict future vegetation patterns. Vegetation responses to future climate change are hard to predict. Because of the complexity of these issues, future predictions are often accomplished by applying computer models. The expectations of and suspicions toward computer models held by the general public—and by scientists as well—range from a belief in the models by some to a complete disbelief of the models by others. As was pointed out in chapter 1, "belief" is a loaded term when used in scientific discourse.

Models do have advantages over consulting experts, conducting shouting matches, or querying oracles to predict the future. The assumptions that are used to formulate ecological models may vary from one model to the next. What assumptions have been used to formulate the current suite of ecological models being used to assess the effects of a global change in climate? It is helpful to group these models into categories based on how and what their creators choose to emphasize in a given model's formulation.[62]

Different ecological models conclude that the climate changes associated with human-caused global warming could significantly change the vegetation of this planet. There is plenty we do not know and likely need to know. Models are abstractions, and they assume the important factors to understand at particular scales of time and space. Will there someday be the perfect model to predict how the vegetation will change? Perhaps. Currently, we have a multiplicity of ecological models. Some may be correct on a given scale, others correct at other scales. Hopefully, ecological modelers are taking different paths toward the same truth. When several different models converge upon what is called a "robust conclusion," an outcome that persists in spite of the differences in model formulation, one is inclined to take such an outcome seriously. What are these different modeling paths, and where do they seem to be taking us in our understanding of how climate change might produce vegetation change?

Vegetation and Climate Relationships

A straightforward method to assess the effects of climate change on global vegetation is to assume the current relationships among climate and vegetation types, such as those illustrated for deserts at the front of this chapter, will be the same in the future. The current recipe to project this assumption forward is to:

1. Calculate the expected vegetation at roughly ten thousand weather stations spread around the world.
2. Check to see if the predictions at each weather station are correct.
3. "Paint" (using graphics programs) the vegetation across the planet based on the weather station points with corrections for elevation and other factors to produce a world vegetation map.
4. Change the climate at each of the ten thousand weather stations and repeat the process.
5. Repaint a global vegetation map on a map of expected climates under a scenario of future global warming.
6. Compare to see how and where the maps differ.

The first global-scale application of this procedure used what is called Holdridge life-zones to assess greenhouse gas–related climate-change predictions using a database of about eight thousand weather stations.[63] In this case the expected vegetation is predicted based on a heat index, much like Humboldt's isotherms; the annual precipitation; and an estimate of how much water the vegetation could use, if water were in unlimited supply. Results showed a substantial change in global vegetation with greenhouse warming predictions. Subsequently, considering several different predictions of climate change,[64] between 35 and 45 percent of the Earth's vegetation was altered at the biome level (grasslands turning into deserts; arctic tundra turning into boreal forests, etc.). Change focused strongly on vegetation in the higher northern latitudes. There was also a substantial shift toward drier vegetation types across much of the Earth. Changes of 35 to 45 percent in the Earth's vegetation are worrisome numbers. They imply a considerable displacement of agricultural and grazing systems and a potential disruption of the conservation of species using the current suite of nature preserves and parks (discussed in more detail in the following chapter). In 1994, a Holdridge-model vegetation-covered terrestrial surface was included in a general circulation model, the first attempt to include changing vegetation directly in a global climate model.[65]

As one might expect, researchers used other theories of climate and vegetation relations to map climate change using the same procedures. The Russian scientist M. I. Budyko developed a bioclimatic theory on the relationship between climate and vegetation by using the radiation balance and a dryness index.[66] When this Russian climate/vegetation model was applied to predict climate changes, the effects were also significant.[67] They implied, among other things, the shifting of the northern boreal forest of Russia virtually out its current boundaries.

One disturbing aspect of these evaluations of the change is the possibility of a so-called runaway greenhouse effect—that a change in vegetation from warming produces increases in the rate of natural releases of greenhouse gases, which then promotes more warming. Over the long term, this appears not to be the case.[68] However, on the short term (hundreds of years) processes that release greenhouse gases (thawing of permafrost in the higher latitudes, destruction of forest by wildfires, increasing the rate that microbes break down carbon in the soil) outpace the processes that remove the greenhouse gases from the atmosphere (migrating trees onto the tundra, regrowing forests, storing carbon during ecological succession).[69] Thus, the initial response of a warming planet's terrestrial ecosystems is to release more greenhouse gases and produce a positive but certainly not desirable feedback.[70]

Painting the World with Plant Functional Types

A more detailed approach to assessing climate change is to predict where on Earth the components that make up the vegetation might be increased or decreased by climate change. That a forest is a type of vegetation mostly made of trees has a basic appeal to reason. Defining these components or life forms can be complex. Must a tree have branches? Must it have leaves? Does it have only one trunk, or can it have several? Many plant geographers have often related the shapes and sizes of leaves, plant sizes, and other aspects of plant form to systems of vegetation classification.

Elgene Box designed a set of plant forms designed to be related to climate variables.[71] He started with the premise that features of a plant that are involved in its water and energy balances are the key consideration, an idea strongly indicated by the sometimes striking convergence in the form of unrelated plants in similar environments. This is shown for a pair of desert plants at this chapter's start.

A full-blown application of Box's approach uses nineteen structural types primarily having to do with the size and pattern of branching of the plants (trees, shrubs, grasses, etc.) along with dimensions involving relative plant size, leaf type, leaf size, leaf structure, and leaf photosynthesis habit.[72] He kept it simple. For example, life forms vary in their height by being tall (large), normal, short, and dwarf.[73] Even so, in Box's scheme, the 19 structural types × 4 plant sizes × 4 leaf types × 4 leaf sizes × 6 leaf structures × 7 photosynthetic habits equals more than 500,000 theoretical plant life forms. Some of these combinations are illogical, such as leafless trees with large leaves, and can be eliminated, and other combinations either do not occur in nature or are very rare. Virtually all of the world's plants fall into around one hundred of Box's life forms. The responses to climate of each life form are determined with respect to several temperature variables, measures of aridity and rainfall, and the seasonal pattern of these variables. One can then apply the computer "painting" method described above to all of the locations for each of the hundred or so life forms. For a given place, the vegetation is the collection of the life forms that can survive there. Box's life-form classification produces descriptive, sometimes almost poetic, names for the different elements of the vegetation (for example, rain-green broad-leaved trees, xeric tuft-treelets, leaf-succulent evergreen shrubs). Change the base climate data to that expected in a greenhouse warming scenario, and the future vegetation changes significantly.[74] These changes confirm the patterns seen in the simpler vegetation mapping procedure already described.[75]

The life-form approach does have the advantage that it can predict novel vegetation not seen today with new climates. We know this happened in the past; it may be the case in the future. The overall implication of this alternative approach is for significant change in the Earth's vegetation given global warming.

Painting with Plant Species

One can map plant species according to their climate niches, in the sense of Grinnell. An initial study[76] mapped the changes in the distribution of four North American trees—eastern hemlock (*Tsuga canadensis*), American beech (*Fagus grandifolia*), yellow birch (*Betula lutea*), and sugar maple (*Acer saccharum*)—in response to two different climate models simulating climate change associated with a doubling of carbon dioxide in the atmosphere.[77] Very much in the spirit of Alexander von Humboldt's isotherms, the species' ranges were based on average January and July temperatures. The species' range changes are large, somewhat larger than those seen in the vegetation-type or life-forms methods. While the details are different, the larger patterns are consistent.

A similar European study[78] evaluated two important European tree species ranges—beech (*Fagus silvatica*) and Scots pine (*Pinus sylvestris*)—responding to climate change predicted for a doubling in CO_2 by two other climate models.[79] As seen in the North American case, the displacements in the ranges of the species are substantial. The potential future range of the species under such conditions of climate change would not overlap with the current species range over areas the size of large European countries. Several species of European birds also would be displaced—for example, the parrot crossbill (*Loxia pytyopsittacus*), which is highly dependent on Scots pine cones for food.[80] One would expect the bird to be completely shifted from its present range. A similar habitat evaluation in Australia of fifty-seven threatened vertebrate species found that a 1°C warming should reduce suitable habitat for 84 percent of these species.[81]

Such species-niche-based models have been compiled for large collections of plant and animal species. When species ranges are compiled into maps of expected communities, changes at the community level can be predicted as well. Significantly, these modeling approaches have been tested successfully in predicting vegetation from past climates—an implication that they might be similarly capable in their future predictions.[82]

Letting Individual Plants Assemble Themselves Into Dynamic Ecosystems

An even more detailed approach to evaluating climate change's effects on the Earth's vegetation is to predict how climate change might affect the birth, growth, and death of individual plants. Such a model is categorized as an individual-based model (IBM). With the development of increasingly powerful computers, several different scientific disciplines (physics, astronomy, ecology) developed IBMs that task computers to bookkeep changes and interactions of many individuals to produce a prediction of the overall system's change. For example, astronomers simulate hundreds of thousands of stars and their interactions to understand how galaxies form. Early versions of these models in ecology were developed by population ecologists interested in including animal behavior in population models,[83] and these led to a diverse array of applications for fish, insects, and birds.[84] In forests, models that computed the interaction of each tree with the others around it could predict the changes in the forests as the trees grew over time.

One class of individual-based models, which were developed for forests but have now been applied to grasslands and savannas, are called gap models.[85] These models feature relatively simple estimates of how trees function.[86] They use the considerable body of information on the more common temperate- and boreal-forest tree species (how fast they grow, what is required for seedlings to survive, how tall trees of a given trunk diameter are, how long they live, etc.) to develop the rules to grow trees in a simulated forest. Other simple rules include interactions among individuals (shading, competition for limiting resources, etc.) and equally simple rules for the birth, death, and growth of individuals. The simplicity of the relations in the models has positive and negative consequences. The positive aspects are largely involved in the ease of estimating model parameters for a large number of species, the negative aspects with a desire for more "correct" formulations.[87]

The entry of gap models into climate change prediction sprung from their reconstructions of the vegetation of past climates.[88] This particular class of models since has found wide application in the prediction of climate change effects.[89] Two special issues of the journal *Climate Change* have reviewed the gap-model applications for climate change assessment.[90] These models project significant changes in species composition, vegetation structure, productivity, and standing biomass in response to climate changes at regional and continental scales.[91] These results are in agreement with the other methods used to paint the Earth with a new vegetation induced by a climate change. They have the advantage both of predicting the rate of vegetation change and the formation of novel plant communities under different climates. Because of their detail, these models require powerful computers (or a lot of computer time) to be applied over large areas.

Different researchers have sought to modify these types of models and speed them up computationally. One simplification represents the species in the models by fewer "functional types" to help with the sometimes difficult problem of estimating the model parameters used for each species, particularly in tropical forests. This approach has been pioneered on tropical rain forests with applications in Borneo, French

Guiana, and several other tropical locations.[92] Other functional type approaches have been used on the deciduous forests of Tennessee and northern China.[93]

One of the products of a gap model is the statistical distribution of the sizes of all the trees. The Japanese ecologist Takashi Kohyama produced a sophisticated model predicting change in tree-size distributions over time for tropical rain forests in Borneo.[94] The Kohyama approach, fused with gap models and plant production models, has found application in predicting forest dynamics in the Amazonian rainforest.[95] However, without using these methods for reducing the computing demand, maps of climate change's effects on boreal forests have been produced by direct application of a gap model to simulate forest change at each of thirty thousand gridded locations across Eurasia.[96] In this study, a positive feedback effect was noted—changes in the mixtures of deciduous to evergreen vegetation under climate warming made the Russian landscapes reflect less of the Sun's radiation into space and increased the degree of surface warming.[97]

Whether one paints the Earth's vegetation in response to climate change using relations with climate and vegetation, climate and components of the vegetation, climate and plant species, or climate and individual plants, the overall patterns are similar. The 2°C to 5°C increase in average global temperature predicted by climate models that reflect our understanding of the response to the Earth's climate is large. Indeed, it is large enough to be highly disruptive to natural and human systems.

But these grim evaluations generally do not consider that, while CO_2 is a greenhouse gas being released in large quantity by human activities, mostly from increasing burning fossil gas, petroleum, and coal, CO_2 is also necessary for the process of plant photosynthesis. Will the so-called CO_2-fertilization effect stimulate plant performance as we introduce more CO_2 into the air? Will our grandchildren inherit a manmade Garden of Eden, with our crops growing at ever increasing rates? One answer might be, "Do you feel lucky?" Another might be to use an entirely different class of models to investigate the scientific state of understanding of this problem.

ECOSYSTEMS THAT CIRCULATE MATERIALS

Another representation of vegetation and its change is to emphasize the ecosystem's function as one of circulating materials through ecological processes. Models based on this viewpoint and much of their mathematical formulation initially came from procedures used in medicine and pharmacology to follow the movement of drugs and other tracers through the body.[98] These approaches were initially applied to ecosystems in the 1950s and 1960s, when concern over the transfer of radioactive isotopes through natural environments to people was sponsored by the U.S. Atomic Energy Commission (USAEC). Background research involved laboratory studies and field work as well as direct measurements of rates of transfer of isotopes injected experimentally into natural environments in whole-ecosystem experiments.[99]

The resultant models have been able to assess radioactive pollution problems both on long and short time scales. The half-life of a radioactive isotope is the time necessary for one-half of an initial amount of a radioisotope to be lost to radioactive decay. For isotopes with relatively short half-lives, these models can be used to identify the principal rapid pathways that would carry radioactive doses to humans in the event of a nuclear release.

For example, in the Chernobyl nuclear reactor incident, milk was immediately taken off of European grocery shelves. Iodine isotopes, mostly ^{131}I with an eight-day half-life, travels to humans through the rapid air→grass→cow→milk→grocery shelf→people pathway. This is a very fast transfer, and relatively little of the radioactive dose is lost to radioactive decay. The iodine radioactive isotopes concentrate in the human thyroid. Cheese, another dairy product, delivers isotopes to humans through a much slower pathway. It takes time to make cheese.

By the time cheese products reach people, much of the radioactive iodine isotopes have been lost to radioactive decay. So in Switzerland after Chernobyl, the milk came off of the shelves, but cheese remained for sale.

For longer-lived isotopes, these models of material transfer can determine where and to what levels particular radioactive isotopes would be most concentrated. One problematic isotope, ^{90}Strontium, tends to behave similarly to the element calcium, and like calcium, ^{90}Sr concentrates in bones. ^{90}Sr has a relatively long half-life (29.8 years). Once in the bone matrix, beta radiation from the ^{90}Sr bombards the bone marrow and can induce bone cancer and leukemia. Thus, the central issue in evaluating its potential hazard is the magnitude of the bone concentration—a problem computed today using material transfer models.

In global applications, recent versions of this class of models calibrate the flows of materials into, inside, and out of different types of vegetation cover (polar desert/alpine tundra, wet/moist tundra, desert, tropical savanna, etc.). These transfers alter with changes in monthly temperature and/or moisture conditions, allowing the models to respond to climatic change. This type of model can evaluate change in the uptake and storage of carbon and the associated carbon-based greenhouse gasses.

The current state of accounting for the CO_2 released into the atmosphere by fossil fuel combustion and concrete manufacture (9.1 Pg of carbon, or 9.1×10^{15} g in the year 2010)[100] shows that about a third of the carbon emitted is unaccounted for in the global carbon budget. This carbon is thought to be stored annually in the Earth's forests.[101] The worrisome problem is that, if it is being stored in forests, what mechanisms might cause this beneficent storage to cease or even change its sign from negative to positive and become a source. To understand this critical issue better, models that circulate materials and that have more detailed processes are deployed. These models paint the Earth with a covering of leaves to determine how carbon compounds might transfer in and out of the vegetation, particularly if the amount of CO_2 in the air is changed.

LEAFY PLANT CANOPIES

Leaves and plant canopies must perform a balancing act among the water leaving the plant, the temperature of the leaves, and the amount of carbon dioxide that enters the leaves to fuel photosynthesis. Water, carbon dioxide, and heat are not easily balanced in some cases. Canopy process models simulate these balancing processes in plants. Everything is interacting, but one can start by considering the inward diffusion or uptake of carbon dioxide into a plant leaf. Once in the plant, this carbon dioxide, along with water, is converted to sugars via the plant photosynthesis process. The carbon dioxide diffuses inward through pores, typically on the underside of the leaf surface, called stomata. The stomata open or close to different degrees (depending upon environment) and control the transfer of CO_2 to the leaf interior. When the leaf is open for CO_2 to come in, water from spaces inside the leaf can flow out. This is the challenge: to gain CO_2, leaves lose water, and the drier it is the more water they lose. Water and CO_2 fluxes are interwoven. Loss of water cools the leaf through evaporation and helps a leaf shed some of the heat that builds as a consequence of leaves capturing incoming radiation from the Sun.

Canopy process models represent the plant canopy as a single or multilayer collection of leaves. Photosynthesis and transpiration (water loss) are simulated by estimating the microscopic meteorological conditions near the leaf and the stomatal conductance (the plant-adjustable width of the pores in the stomata). Canopy process models simulate the carbon dioxide, water, and heat fluxes of leafy canopies over large areas.

Relatively simple models based on leaf physiology of plants can do a remarkable job of predicting the area of leaves expected over the Earth. One application[102] implemented a simple model of the energy and water balance of a plant canopy.[103] The model adjusted the leaf area so that over the course of the year the soil did not dry too much, and after the year the soil had as much water as when the model started. New leaves were added when all the year's precipitation was not used and subtracted when the canopy used too much water. This produced an estimate of the maximum sustainable area of leaves at a location. With the same paint-the-Earth approach used for other models, the model simulates the expected area of leaves globally and matches maps and satellite data on the vegetation leaf area.[104]

Why bother? If the other models that have already been discussed work to predict vegetation structure and species, why is it so important to add physiological detail merely to simulate the area of leaves? These approaches can assess the CO_2-fertilization issue.

We have been conducting an ad hoc global experiment by injecting large amounts of carbon dioxide into the atmosphere since the Industrial Revolution. The concentration of carbon dioxide in the atmosphere has crept up from the 280 parts per million before 1750 to the 400 parts per million as this chapter is being written in the summer of 2013.[105] Given the increasing human emission of carbon dioxide associated with energy consumption, which is increasing with population and consumption in developing economies, the concentration in the atmosphere should continue rising over the next century. Depending on economic development and energy conservation policies, minimal current projections would predict almost a doubling of historical carbon dioxide by 2100 (an increase to 535 parts per million). The last time the planet saw concentrations nearly this high was about three million years ago, when the Earth was 2° to 3°C warmer and the seas were fifteen to twenty-five meters higher than today.[106] Under less "green" policies, the future could see a three-and-one-half-fold increase to 983 parts per million of carbon dioxide in the atmosphere.[107]

Will more CO_2 in the atmosphere promote more plant growth, making crops and other vegetation grow better, delivering a Garden of Eden to humanity? One of the effects of having more CO_2 in the atmosphere that can be demonstrated in laboratories is to increase the water-use efficiency of plants. Water-use efficiency is the ratio of the amount of plant produced for a given amount of water lost from the plant. At higher concentrations of CO_2, more CO_2 diffuses into the plant per amount of water lost, so the plants have greater water-use efficiency. This implies that in a high-CO_2 future, plants might grow in places that are currently too dry. Indeed, the steady encroachment of "woody weeds" into grasslands and pastures worldwide is consistent with increased water-use efficiency, but there are a large number of other potentially contributing causes.

It is well documented that plants change their morphology when CO_2 concentration changes. The numbers of stomata per unit area of their leaves has increased by about 40 percent since the onset of the Industrial Revolution,[108] and plants are more efficient at using water today than they were 150 years ago.[109] This implies that in response to more CO_2 in the air over the past century and a half, plants have changed the morphology of their leaves mostly to lose less water, not to grow more.

The inclusion of the so-called direct effects of CO_2 (increased photosynthesis or increased water-use efficiency) calls for models that include leaf-related processes. Measurement of these direct effects of CO_2 is difficult, relatively local, and expensive. Free Air Carbon Dioxide Experiments (FACE) use complex instrumentation and release towers to hold a small patch of vegetation at double-CO_2 levels in the open environment. This avoids the complicating effects of putting tents or chambers over the plants in the field, which confounds interpretation of the measured effects. FACE studies have produced complex results. For young forest stands, the net primary productivity response to increased CO_2 levels (approximately 550 ppm) is an increase of around 23 percent,[110] a result consistent with a recent model-based study of the effect of increased CO_2 on global net productivity.[111] The steps to convert this net primary productivity increase into increased carbon storage (respiration, decomposition, tree mortality, etc.) could be expected to bring this number down considerably.

In one FACE experiment, the initial increased productivity from elevated CO_2, which was initially around 24 percent, declined to 0 percent after four years into the experiment.[112] The cause appeared to be related to shortages of nitrogen in the soil. Similar "down regulation" in photosynthesis rates has also been seen in a variety of longer-term laboratory experiments.[113] In summary, the evidence supporting a large positive effect of more CO_2 on vegetation is far from demonstrated.[114] Don't bet the planet on a CO_2-fertilization-based Eden just yet.

Dynamic Global Vegetation Models

The models of vegetation change described above are reminiscent of the story of the four blind men describing an elephant. One touches the trunk and declares an elephant is like a hose; another touches the tail and says an elephant is actually like a rope; the third touches the side to find an elephant is like a wall; the fourth touches a leg and thinks an elephant is like a tree. The models described thus far are a bit like this. Is the Earth covered with vegetation, with life forms, with different kinds of plants, with individual plants, with systems that process materials, or with leaves? Or is it covered with all of these things? One can ask on what do models based on these different views of Earth agree. In general, they agree that the climate changes that we are generating should change the living surface of our planet.

Another approach amalgamates the types of models so far described to try to know better the consequences of dynamic climate change—particularly the interaction between vegetation change and climate change. The resultant Dynamic Global Vegetation Models (DGVMs)[115] incorporate the mechanisms based on leaf processes in plant canopies with the biogeographical approaches used to paint vegetation on landscapes to simulate better the effects of climatic change.[116] These models have been used to determine the degree that increased plant productivity might slow the anthropogenic increase in CO_2 in the atmosphere.[117] Because they can include the effects of increased CO_2 on plant processes along with the effects of climate changes, DGVMs also can assess the likelihood that a warming might induce the positive feedback associated with a runaway greenhouse effect.

Large, complex models with many parameters have the appealing aspect of including the many mechanisms that might come into play in the runaway effect. As a downside, they are hard to test, and their solutions may accumulate errors in estimating their many parameters.[118] Joining DGVMs and complex climate models is also a demanding computer problem. The challenges are offset by the potential importance of better predicting the dynamics of climate change.

COULD WARMING THE EARTH EXCITE PROCESSES TO PRODUCE EVEN MORE WARMING?

The most computer-efficient way to couple a complex vegetation model such as a DGVM with a complex climate model is to use what is termed "asynchronous coupling" of the models—a climate model is run for a period of time, then its results are fed into a vegetation model, which is then run based on these new inputs; the vegetation model's results then are fed into the climate model, and so on. In 1997, an asynchronously coupled GCM and DGVM were found to produce small but positive feedbacks on climate.[119]

Just a few years later at the prestigious United Kingdom Meteorological Office Hadley Centre, a fully coupled climate model with an internal vegetation model called TRIFFID produced quite alarming results.[120] Under simulated future warming, the terrestrial biosphere initially stored carbon. After about fifty years of simulation, the terrestrial surface instead became a *source* for carbon dioxide. The effect was so strong that atmospheric carbon dioxide concentrations were 250 ppm greater by 2100. This increase in carbon dioxide added 1.5°C of extra warming in addition to the 4°C increase seen from the increased CO_2 but without the vegetation coupled to the atmosphere. This effect is a significant positive feedback, a self-reinforcing greenhouse effect, in which warming begets more warming. Recently, eleven different coupled GCMs/DGVMs showed vegetation dynamics from large carbon uptakes ranging from the vegetation reducing the greenhouse effect to large releases with the vegetation amplifying a greenhouse warming, all by the end of the century.[121] The TRIFFID model used in the initial coupled simulation described above produced the largest positive feedback over the twenty-first century.[122]

The results from other simpler approaches discussed earlier in this chapter produced positive feedback results as well. Regional warming in Siberia could produce vegetation composition change that may cause a positive feedback to generate even more warming.[123] Vegetation paint-the-Earth methods that track the carbon expected to be released when vegetation changes with climate change produces an emission of fifty to sixty-five gigatons of carbon into the atmosphere released for each 1°C degree of warming, also a positive feedback.[124] These numbers are close to those of the early TRIFFID experiment.[125]

If we look to the past to see if other natural climate changes produce similar results, the numbers are also close to model predictions for the yield of CO_2 per 1°C warming. One case is the so-called Little Ice Age between the years of 1450 and 1800. Over this period, the climate and atmospheric CO_2 data, the latter derived from ice bubbles trapped in glacial ice, show lower temperatures and also lower CO_2 levels than have occurred since.[126] As the Earth warmed from the Little Ice Age, changes in climate were followed by changes in atmospheric CO_2 concentration with a lag of about fifty years. The temperature sensitivity of CO_2 emissions from land and ocean sources was between 20 and 60 ppm of CO_2 released per 1°C of warming (with the assumption of no human-generated CO_2 emissions). These numbers all imply a similar sensitivity with positive feedback between the carbon cycle and climatic change.

However, not all of the DGVMs show this response, and many depict a planet where the increased rate of photosynthesis stores more carbon and reduces the warming. The result is a policy maker's nightmare, a situation where there are tremendous potential effects but relatively little certainty in the results. A recent review concludes by answering questions about the state of knowledge and certainty of climate change as a consequence of global alteration of the chemistry of the atmosphere by humans.[127] These questions (in quotes) and commentary are:

1. "Are the expected changes in climate large enough to have a meaningful effect on the world's vegetation?" The answer is yes. The

approaches, which were discussed earlier in this chapter (ranging from simple correlation to complex process-based models), virtually all show a sensitivity of vegetation to climate change.

2. "If the vegetation change is meaningful and detectable in the face of all the other things we are doing to the vegetation, how rapidly might it occur?" Models emphasizing plant physiology and leaf processes imply rapid responses. Other models, which include phenomena such as succession, soil development, and other slower processes, produce complex responses over longer time periods. DGVMs, which incorporate both sorts of processes, produce substantial responses in decadal time scales.

3. "How might the change in vegetation interact with the atmosphere with climate change?" Vegetation change makes the attempts to understand climate change much more complex. In longer-term dynamic vegetation models, processes that release greenhouse gases respond faster than those that store greenhouse gases. Complex dynamics of some models imply considerable difficulty in reversing the effects of climate change.

4. "Are there other factors that might significantly alter our conclusions in trying to answer the first three questions?" Both model predictions and results from field experiments on vegetation imply that elevated CO_2 increases vegetation net productivity similarly over a range of very different vegetation types.[128] This and the increased efficiency of plant use of water under elevated CO_2 could moderate vegetation response, but the uncertainty of these effects remains high. A more worrisome effect is noted in cases in which DGVMs are coupled directly with climate simulation models. In these cases, strong positive feedback is seen in the interactions between vegetation and climate. This warming-generating case needs further exploration both globally and locally.

Taken as a whole, what we have learned to date implies that vegetation operating at all scales interacts with climate. Of particular concern is the identification of positive feedback loops between the climate and vegetation. Such loops could make changes in the response to environmental change greater than one might initially expect from the climate models alone.

CONCLUDING COMMENTS

Many theologians see the whirlwind questions as a divine demonstration that humans are a part of the planet but not the sole divine focus of the creation.[129] Indeed, it has been interpreted as grounds for a belief that God's Earth was not constructed so much *for people* as *with people.* We may be but one of the many sorts of the Earth's creatures, with no intrinsic primacy over the others. Tucker notes,

> God brings rain not only to the just and unjust, but on the desert as well as the sown land. The wilderness is—quite literally—not Godforsaken. These lines [speaking of the same verses that head this chapter] bring a very important voice into the conversation about the environment and respond to what is all too commonly viewed as the main line of the Bible's understanding. The notion that all creation is to serve human interests is rejected.[130]

This represents significant grounds for, at the very least, discussions across science and religion.

Interestingly, we are in an era in which ecologists and notably conservation scientists are prone to point to "ecosystems services," in which natural systems when preserved and protected provide value to people, often calculated as monetary value. This economic motivation to protect the planet better by pointing out that we get things we need from it is certainly appropriate to modern times. It is true that watersheds with natural forests produce cleaner water, that coral reefs and mangrove trees protect coastlines, that cleaner rivers provide more fish, that purer air causes less respiratory health problems, and so on. A modern Job, unable to understand why it rains in the desert, which is empty of human life, could well be persuaded by this point of view, for it would match his own. This modern Job might well stunned when questioned from a modern whirlwind, "Who is this that darkens counsel by words without knowledge? Gird up your loins like a man, I will question you, and you shall declare to me. What on Earth do you think you are doing?"

This chapter starts with a complex Joban question and deals with but one facet of its implications: that the rain can "make the ground put forth grass." This part of the question only makes sense if the interrogated understands that deserts are associated with arid climates and the rare rain that falls there will not convert the desert to something else. However, we know that over long periods of time deserts have come and gone as the climate has changed. So have other ecosystems.

The nature of the relationship between vegetation and climate is a deep question in ecology. The daunting complexity of vegetation's responses at different scales of time, space, and biological resolution makes these issues a rich scientific challenge. Centrally, one must appreciate that climate and vegetation dance in a two-way interaction, each with the power to alter the other. We need to know is more about the nature of this land/atmospheric pas de deux. It is obvious we are changing the land surface. It is obvious we are changing the chemistry of the air. Our best understanding is that we are changing the climate. We simply need to know more, both regionally and globally.

"Dying Lioness". Detail of an alabaster mural relief from the North Palace of Ashurbanipal, Nineveh, Assyrian period, c. 650 BCE. *Source*: Photograph by Matt Neale, taken in room 10 of the British Museum.

9

Feeding the Lions

The Conservation of Biological Diversity on a Changing Planet

Can you hunt the prey for the lion, or satisfy the appetite of the young lions, when they crouch in their dens, or lie in wait in their covert?

—Job 38:39–40 (NRSV)

The lion in the Book of Job would have been the Asiatic lion (*Panthera leo persica*).[1] This lion subspecies was formerly found in the coastal forests of northern Africa and from northern Greece across southwestern Asia to eastern India.[2] The epitome of fierceness and power, it was used then and now as an emblem of these attributes, such as the Sphinx of ancient Egypt or the incorporation of lion motifs into the regalia of pharaohs. Lion hunts appear on ancient monuments from the origins of Western civilization. One striking example, "The Dying Lioness," from a 650 BCE alabaster panel from Nineveh, is shown at this chapter's start. Over two thousand years have not diminished the power of the carving or the beauty and fierceness of the animal represented, even at the point of her death.

Today, the awe-inspiring Asiatic lion's presence on Earth is represented by about 350 individuals that are found primarily in a biological reserve in India's Gir Forest and National Wildlife Park.[3] Some of the animals in this population, perhaps one hundred, are located outside the reserve.[4] Even though poaching is an issue, the population seems stable at this time.[5] The Asiatic lion is considered an endangered species because of its small population found in one location. The answer to the whirlwind question, "Can you feed the lion and take care of its young?" would likely be, "no," or, at best, "maybe" for the Asiatic lion.

"Can you take care of the lion?" is the first of a series of animal questions presented in the whirlwind questions. We have already treated three other creatures in this bestiary, the aurochs in chapter 3 and the onager and the horse in chapter 4, along with birds in chapter 6. Taken as a whole, the animal-based whirlwind questions treat different facets of animal survival: procreation, domestication, life cycles, behavior, and so on. The lion whirlwind question focuses on the feeding and care of wild creatures. Matitiahu Tsevat discussed the passage and noted, "Man is not wont to give much thought to the food supply of wild animals, but it is a problem all the same. Would Job assume the responsibility of providing food for one of their kind who normally has no difficulty in getting what he needs?"[6]

The lion question that challenged Job continues as a current question in conservation science. Modern society has so altered the planet that knowing how to "feed" the lion takes on additional importance. Let's rephrase the whirlwind question as, "Can you assure the survival of a dramatic, large predator, such as a lion, which occupies the top tier of the food chain?" This is a central issue in conservation ecology.

In a changing world, the preservation of creatures such as the lion is one of the hardest cases to solve. Such top-tier predators require vast landscapes. Strategies to preserve such animals are sensitive to all sorts of changes, including climatic change. The lion is such a striking creature that its survival commands the public's interest—these are symbolic and charismatic animals. For a modern conservation scientist, feeding the lion (or the tiger, or the jaguar, or any large, fierce predator) is the ultimate challenge. This chapter will discuss why this is the case and why this already difficult whirlwind question is becoming increasingly more so in a changing planet.

ON THE CONSERVATION OF THE BRIGHT
AND BEAUTIFUL, BIG AND DANGEROUS

Caring for large predators in the wild is challenging for several rather obvious reasons. First, although it may not seem that way to us when we step on a scale after the festivities of a holiday season, warm-blooded animals are relatively inefficient at converting food into animal tissue. This is particularly the case for herbivores converting plant tissues. Carnivores, animals that eat animals, are more efficient, but it requires a remarkable amount of plant production to feed the prey species that in turn are eaten by a predator and thus provide food energy for the production of predators.[7]

PATTERNS OF FOOD CHAINS AND FOOD WEBS:
WHY IT'S LONELY AT THE TOP

A classic study of the ecology of an abandoned farm field in southern Michigan calculated the energy transferred by feeding from plants to rodents to weasels.[8] The result is called a food-chain analysis. To produce a unit of weasel increase (least weasel, *Mustela nivalis*) requires the production of 45 units of rodents (meadow voles, *Microtus pennsylvanicus*), which in turn requires 377,000 units of production of plant material.[9] So the efficiency of producing weasels from rodents is 1 unit of weasel produced per 45 units of rodents eaten, a production efficiency of 1/45 or about 2 percent. Considering a range of mammalian and avian carnivores in a variety of places, this number varies from around 1 to 3 percent.[10] In so-called cold-blooded animals (*poikilotherms*), the efficiency is higher: around 10 percent on average for fish, 10 percent for social insects (bees, wasps, etc.), 25 percent for nonsocial insects, and 36 percent for other invertebrates.[11] These results imply that food chains with some or all of the links featuring cold-blooded animals potentially could have more links than food chains composed of vigorous warm-blooded animals, such as lions and their prey.

It requires a remarkable amount of plant productivity to support a population of warm-blooded, small carnivores such as weasels. A carnivore that feeds exclusively on weasels would be rare simply because it would take a very large area to support the plants that produce the food that feeds the mice that feeds the weasels that feeds the "weaselvore." Without launching into a chorus of "There was an old lady who swallowed a fly . . . ," a body of very interesting theoretical work in ecology illustrates just this point: the number of steps in a food chain toward some ultimate predator has limits.[12] As a rule of thumb, it takes the productivity of ten thousand kilograms of prey to sustain ninety kilograms of carnivore, regardless of the size of the carnivore species.[13]

The details of just how many steps a food chain might have in a particular location and the underlying causes for the different lengths and other features of food chains remain a topic of active discussion among ecologists.[14] One of the earliest concepts was that since the amount of food available for each link in the food chain was less at each step, as is the case in the Michigan old-field example, there necessarily would be fewer weasels than rodents. One would also expect that the size of the individual predators generally would be larger than that of their prey, with the exception of pack-hunting predators. Thus, the biological limitation on the body size of the top predators could also limit the length of food chains.[15] Food webs, a more complex rendering of the way food energy moves through natural systems, would appear to have the same features—both reduced energy inputs and larger animals higher in the food web.[16] Food webs provide greater detail on the transfers of food energy in cases in which predators feed on more than one prey or in which omnivores feeding on both plants and animals are important.

Who Runs the Ecosystem?

The patterns of food chains and the dominant factors controlling these patterns have significant implications in the responses of ecosystems to change.[17] One line of logic suggests that the predators at the top of a food chain control the numbers of herbivores so that they cannot become dense enough to overgraze the plants.[18] This has two significant implications: First, removing the top predator could permit overgrazing by the herbivores. When they have eaten all the vegetation, there is a collapse from starvation. Second, if herbivore grazing is controlled by the predation of the top carnivores in this way, then a "healthy" ecosystem features predators that keep the herbivore density relatively low. Consequently, the factors that control the vegetation would be mostly environmental factors such as climate or nutrient availability (and not grazing). This interpretation, referred to as "top-down" control, has been demonstrated in several ecosystem-level experiments.[19]

The alternative "bottom-up" control amplifies the observation that the plants control their being eaten by being relatively inedible, thorny, or poisonous to varying degrees. The world may appear to be green with plants, but it is not a gigantic tossed salad. Plants control the number of herbivores, and hence they control the food chains.[20] This would imply that the pattern of vegetation would be largely influenced by environmental conditions. There is a repertoire of theories that either elaborate or mix the concept of top-down versus bottom-up control.[21] It is not hard to imagine that one of these theories might be in action in one ecosystem but not in another or in the same ecosystem at one time and not another time.

In general, food-web theories point to the situation in which the abundance and productivity of plants respond mostly to environmental controlling factors. Does this imply that the techniques used to map the patterns of vegetation change in different climates (discussed in chapter 7) should work without consideration of the changes in the animal populations? The bottom-up theorists would say, "Yes," and the top-down theorists, "No." The understanding how food webs work becomes a question of importance in global change—how much of the vegetation is controlled by climate and how much by other local factors? This issue implies potential uncertainty in our capacity to predict the consequences of global environmental change on the vegetation.

Ecosystems with Lions

For protected ecosystems (parks) that have the African lion as a significant top carnivore, such as those located in Tanzania, Zimbabwe, and the Republic of South Africa,[22] there is evidence for top-down control of the carnivores on the herbivores in some locations but not others.[23] For these examples, in savanna ecosystems that support lions, the small- to medium-sized herbivores appear to be limited by predation, which would be the top-down control case. Large herbivores, the elephants and rhinoceroses, are *not* controlled by predation to any great degree or only in exceptional circumstances.[24] Presumably their maximum densities are controlled by the amount of suitable vegetation available, the bottom-up control case. These megaherbivores are so large that they have minimal predation from even the top carnivores.[25]

For lions, sitting at the top of the food chain in the ecosystems in which they occur, rarity is a consequence of the amount of land required to convert energy through food chains to lions. To the degree that they control of the ecosystems in which they occur with top-down control, they are truly the "king of beasts"—a keystone species with great leverage on the overall integrity of the ecosystems in which they are found.

Potentially, a large decline of very large herbivores or large predators could have drastic effects on the African savanna ecosystem. Basically, the vegetation-controlled-by-climate case flips to being a predators-in-control case. Such a flip-flop effect has been postulated in past ecosystems in North America, causing what appears to have been a mass extinction of the "megafauna," a diverse assemblage of very large herbivores and carnivores extant over ten thousand years ago.[26]

LIVING ON THE EDGE: ON THE CHALLENGES OF BEING A REALLY LARGE CARNIVORE

For lions, there is another aspect of food webs that must be considered in their management. In general, the animals at the higher levels of a food web need to be larger than their prey, but how much larger? If one considers mammal carnivores, two different groups that differ by size also differ strongly in the manner in which they feed. These differences in foraging behavior also influence the way they function in the ecosystems in which they occur:

1. Small carnivores, which weigh less than twenty kilograms per animal and feed on small prey (invertebrates and small vertebrate animals). Because they feed upon animals that are quite a bit smaller than they are, the relative energy costs for finding and capturing prey for small carnivores is relatively low. The animals cruise at a relatively slow pace through their feeding area. When they find small insects or mice, they capture and eat them with dispatch. Small carnivores that feed primarily on prey similar to their size are rare.[27]

2. Large carnivores, which weigh over twenty kilograms per animal and eat large vertebrate herbivores.[28] They prey on animals that are much closer to their own size. Some of their prey are quite dangerous—armed with horns or antlers and disposed to kicking and fighting when captured. Many of them can outrun large carnivorous predators once they get up to speed. Large carnivores expend considerable energy running down prey animals. Subduing and killing the prey and, in some cases, dragging it back to their lairs costs even more energy.

Large and small carnivores use two essentially different strategies to obtain food. Obtaining a dependable, even if small, payoff using a small energy output is the strategy of the small carnivores. High-rolling large carnivores wager large energy expenditures in capturing their prey in hopes of getting substantial payoffs when they bring down large prey animals.

There is little choice in these matters. Basically, it would be difficult or impossible for lions to make a living by solely hunting mice and insects given the typical densities of these small prey creatures. On theoretical grounds, the payoff going after larger prey versus capturing small prey switches when the carnivore species weighs around fifteen kilograms.[29] This prediction is close to what appears to be the body-size boundary (at around twenty kilograms), which separates large- and small-carnivore feeding strategies.

One possible limitation to the number of lengths in a food web is the maximum possible size of the top-of-the-food-chain supercarnivore, such as the lion. Lions are animals weighing 142 kilograms on average.[30] Their daily energy expenditures are such that they have to kill an animal(s) of relatively similar mass as theirs on average once a day.[31] Lions have a relatively narrow margin between the food they must consume and the energy they must burn to obtain their sustenance.[32]

The attractiveness of domesticated animal herds with relatively more docile prey to large predators is obvious. However, lion predation on pastoral herds carries substantial risks. It has been estimated that in Kenya ranchers alongside the Tsavo National Park each lose US$290 each year to livestock eaten by lions.[33] The comparable number for Waza National Park in Cameroon is US$370.[34] These numbers are about 35 percent of the average per capita income in each country.[35] From these simple economics, it is small wonder that one of the principal sources of lion mortality is persecution by humans.

Lions partly maintain their energy budgets with behaviors that conserve their energy. They spend over 90 percent of their time inactively: sleeping and looking regal.[36]

CAN WE FEED THE LIONS?

The Joban whirlwind question could have hardly focused on a more challenging creature for humans to feed, protect, and preserve than the lion. One must pump a remarkable amount of energy through a food web to maintain a lion at the top. If the area of a wildlife preserve is too small, the plant production at the base of the food web is only large enough to support a few lions. Small populations of any animals have several problems. If the populations are too small, there is a chance that "luck of the draw" will produce fewer females than males. This in turn reduces the number of reproductive females, reduces the potential rate that lion cubs are produced over the entire population, and thus reduce the numbers in the population.[37] Small, inbreeding populations also tend to lose their genetic diversity over time. This produces a number of effects, such as the increased chance of recessive deleterious genes decreasing the overall hardiness of individuals in the populations. Lion prides on average have four to six adult animals. Lions need populations of fifty to one hundred lion prides to maintain the population's genetic diversity. Inbreeding becomes a significant issue in populations of fewer than ten prides.[38] These conditions are now met in only seventeen large reserves found across Africa, the "lion strongholds."[39]

As to our current state of tending of the lions in natural settings, the population seems to be decreasing over Africa. The numbers of lions are thought to have decreased by about 30 percent over the past two decades, and the species is listed as "vulnerable" on the International Union of Conservation Networks "Red List."[40] This is despite the efforts of some remarkably capable and dedicated wildlife ecologists and range managers working to maintain lions, which are major tourist attractions for visitors to African national parks. The source of the decrease appears to be the killing of lions to avoid loss of human life and livestock as well as decreases in the prey species that feed the lions.

THE MEGAFAUNAL EXTINCTIONS AT THE END OF THE PLEISTOCENE

Extinction of species is a part of the Earth's biological history. What appear to be more-or-less periodic catastrophes have been associated with mass extinctions of higher forms of life. These cycles are associated with exceptional occurrences (volcanic eruptions and other tectonic events, changes in sea level, reversals in the magnetic poles, and large prehistoric meteor impacts). These cataclysmic events' periodicity may derive from the movements of our solar system with respect to the Milky Way's galactic plane.[41]

The largest extinction "event" occurred 245 million years ago (in the Permian period) when known genera of marine animals dropped by around 80, and perhaps as much as 96, percent.[42] At the end of the Cretaceous period, sixty-five million years ago, the "End of the Age of Dinosaurs" involved the extinction of approximately half of the living genera at the time, including microscopic aquatic plants and animals of a variety of kinds, marine and flying reptiles, and dinosaurs.[43] However, other land plants, crocodiles, snakes, mammals, and many kinds of invertebrates managed to survive.[44] There will likely continue to be differences of opinion as to the cause. An asteroid impact or violent volcanic eruptions have been suggested.[45]

The diversity of the species of plants and animals on our planet is the heritage of thousands of millions of years of evolution. We would be wise to preserve this heritage. We have precedents for the potential of humans to eliminate remarkable number of species both in prehistoric and historical times—so much so, that the actions of our species have been termed the "Sixth Asteroid," likening human action on planetary biodiversity to the great extinction events in geological history. Often in conjunction with a change in climate, humans colonizing and expanding across landscapes have generated substantial drops in the biotic diversity of the Earth. The impacts of humans on lions and the difficulty guaranteeing their survival as a species exemplify a microcosm of the difficulties in preserving biotic diversity in a world with shrinking habitats and other human actions on natural systems. Climate changes working in concert with human pressures on Earth's systems only exacerbates this problem. The role of humans and the complexity of understanding the causes of decreases in biotic diversity can be seen in examples from the recent past and from the recent geological epoch, the Pleistocene.

The Pleistocene epoch started about 2.5 million years ago and lasted until about 12,000 years ago. Continental movement was likely less than one hundred kilometers over the entire epoch, and the continents were more or less in the same position then as now for the entire Pleistocene. Nonetheless, this was a very dynamic time in the Earth's history in terms of climate. Glaciations ("ice ages") that covered as much as 30 percent of the planet's surface with ice alternated with interglacial periods when the extensive ice retreated. There were also minor fluctuations in the glaciers, called stadials for advances and interstadials for retreats. During the most recent of these glaciations, extensive continental glaciers up to three kilometers thick covered much of the Northern Hemisphere. Mountain glaciers in the Southern Hemisphere were much more extensive than today. So much water was tied up in this ice that the seas stood one hundred meters lower than they do at present.

The end of the last glaciation produced a substantial extinction of large animals in Eurasia, the Americas, Australia, and elsewhere.[46] Because of their size, these creatures are called the "megafauna." For example, in North America about ten thousand years ago, almost forty genera of large North American mammals became extinct during what appears to be a relatively short interval of time.[47] Accounts of the megafaunal species stir the imagination.[48] Carnivores including huge bulldog-faced bears, saber-toothed tigers, cave lions, dire wolves, dog- and hyena-like animals of various sorts were distributed through Europe, Asia, and North America. An equally diverse array of herbivores, such as long-necked camels, beavers the size of modern bears, horned giraffes, giant armored armadillos, and ground sloths as large as elephants, all grazed on vegetation in the Americas and Eurasia that today supports nothing even resembling such creatures.

A zoo full of the species of large Pleistocene mammals would need to be over twice the size of today's zoos. Near the end of the Pleistocene, over half of the species of large mammal herbivores in North America, South America, and Australia became extinct. Thirty-seven percent of the large herbivores were lost from Eurasia. However, Africa, where humans evolved in the presence of these large animals, saw only about a 10 percent loss of species.[49] Climate change, actions of prehistoric people, and the coincidence of the evolutionary turnovers of species are all candidate causes for this extinction,[50] but these could very well work in concert with one another.[51]

Modern megafauna, notably the elephant, strongly alter the habitats in which they occur. From direct observations in Africa and Asia, the effect of elephant herds and other large herbivores on vegetation structure and composition is significant.[52] Elephants in Africa are a major factor in altering dominant tree species and the very nature of the vegetation. Grasslands are converted to closed canopy woodlands where elephants are excluded. It is likely that the effect of the extinct megafauna on the past vegetation in other regions was equivalently substantial. The recent disappearance of so many large mammals would have altered the Earth's vegetation and the kinds of plants that one would expect to be successful.

The large mammal extinction is called the Late Quaternary Extinction.[53] It occurred worldwide. However, the extinction occurred at different times depending on the location. Overall, half of the 167 genera of large animals (animals heavier than forty-four kilograms) found on the Earth's continents became extinct.[54] The loss of the megafauna was quite abrupt in North and South America as compared to Europe. The temporal variation in the location of megafaunal extinction supports a premise that the migration of humans out of Africa had an important role in the extinction of these creatures. The extinction occurred at different times on different continents and islands and correlate with the successful migration of humans to a particular place. In what has been termed a "dreadful syncopation," [55]Homo sapiens arrives in a location, and large mammals and other animals become extinct there.

Examples of this syncopation give an impression of the magnitudes of these changes and endorse the power of Stone Age people to modify species diversity over very large areas. There has been what a long prehistory of human expansion across the world and a disappearance of large animals correlated with this expansion. The details of how this might work are topics of new discovery and debate and a marvelous example of the scientific process.

EXTINCTION OF THE NORTH AMERICAN MEGAFAUNA

In a scientific debate over the interpretation of complex data and with new data constantly being acquired, it is sometimes useful to step back and take inventory of what is known. Globally, and clearly in North America, for the past two million years climate variations have driven ice ages with continental-scale glaciers a couple of kilometers thick moving south to as far as 40°N latitude. These glaciations were periodically terminated with significant glacial retreats called interglacials. We are living in such an interglacial now. We have no expectation from geological considerations that this long-term periodic cycle of change will end in the next million years or so.

The last retreat of the glaciers about 12,500 years ago was attended by a colossal extinction of very large land animals, an event that did not happen during any of the previous glacial retreats. There were large-animal-hunting people in North America before the apparent time of this extinction and at the end of this glaciation. This was a unique aspect of the last glaciation/deglaciation cycle. Thus, humans are a logical suspect as being the cause of the unique megafaunal extinction.

The Classic Explanation of the North American Megafaunal Extinction

Before discussing the complications and excitement caused by recent archaeological findings, it is useful to consider the "classical" explanation of the role of humans in the decimation of the Pleistocene fauna.[56] The Clovis culture, named after Clovis, New Mexico, where their artifacts were first found, hunted the megafauna, including the elephantine megaherbivores. Bones, often lots of them, of mammoths, extinct species of bison, mastodons, giant sloths, tapirs, paleo-llamas, and other large herbivores have been found in Clovis sites. Several factors are thought to have aided the spread of these hunters, notably, that the animals they hunted were naïve to the hunting practices that the Clovis people or their ancestors in Siberia had perfected in becoming mammoth hunters. Crossing through a newly formed corridor through the ice sheets that covered Canada about 13,500 years ago, they brought with them a willingness to hunt very large animals and the Clovis stone technology, which included large, doubly fluted spearheads. Clovis culture spread over North America, where most of their artifacts are found. There have been a few Clovis artifacts discovered in Central and South America, as well.[57] Recent recalibration of carbon-14 dates for North American Clovis sites indicate that the material culture was in place starting between 13,200 and 13,100 years ago and lasted until 12,900 or 12,800 years ago.[58] This has two significant implications. First, if these data are correct,[59] the remarkable spread of the Clovis people across North America could have occurred in as little as two centuries and at the most four centuries. Second, the dates are close enough in time that it is difficult to interpret the pattern or direction of the spread of the Clovis.[60]

The Clovis hunters were seen as originating as a killing force in the far north of North America and then spreading southward in a wave toward increasingly better hunting grounds.[61] They would prey on the local megafauna while increasing their own population density.[62] The creativity of paleoecologists in unraveling past change is quite remarkable. *Sporormiella* fungi of different species grows on the dung of large terrestrial herbivores. The spores of these fungi spread through the air and are deposited in lake sediments, where they can be interpreted as indicators of large herbivore densities. In upstate New York, the spores are common in lake sediments until about 12,000 years ago, at which time they virtually disappear. They reappear with the advent of cattle with European settlement.[63] This indicates a sudden demise of large herbivores—or at least of their droppings, although it is hard to imagine one without the other.

Computer models were constructed using data from Africa for large-animal densities and reproductive rates to approximate the recovery of the North American megafauna to hunting. From these, a small population of one hundred Clovis people originating in what is now in Edmonton, Canada, could theoretically roll south for a distance of 3,200 kilometers in under 350 years while eliminating megafaunal populations as they went.[64] This is termed the "Overkill" or "*Blitzkrieg*" theory.[65] According to the model, they could be at the tip of South America in as little as one thousand years, leaving a megafaunal extinction in their wake.[66] This is the now "classic" conception of the role of human hunters in dispatching the North American megafauna.

As we have noted, the human role in the extinction of the North American megafauna is being debated scientifically in ongoing complex arguments. These are exciting times for understanding this event. There are also complications that have developed as we have learned more about the nature of the megafaunal extinction and particularly about the arrival of people to the Americas.

Coming to America

Human arrival to the Americas appears to be more complex than originally was thought.[67] The seas were well over one hundred meters lower, and a large land area called Beringia connected Asia and North America. Rising seas from the meltwater of the continental glaciers flooded Beringia and formed the Bering Strait that now separates Asia from North America.

In the Pleistocene, two ice sheets, the Cordilleran and Laurentide ice sheets, joined and covered much of Canada.[68] These ice sheets would have blocked entry to much of North America for any people who had crossed into Beringia from Asia. As it became warmer, the ice sheets retreated and separated. This created ice-free corridors: one corridor in the plains east of the Canadian Rockies, the other along the Pacific coast. The corridor between the Rockies and the plains opened about 13,500 years ago and allowed people to move south. Classically, this was thought to be the opening that allowed a land passage and the point of origination of the Clovis megafaunal-hunting culture.

Halfway down the coast of Chile about 13,000 years ago, a Paleoindian people produced fluted projectile points but not Clovis points. They hunted large animals at a lakeside site on the Andean Pacific.[69] Even earlier, 14,000 years ago, at the Monte Verde site far into southern Chile in northern Patagonia, people built a complex of twelve rectangular rooms made of planed logs with a roof of mastodon hides.[70] An adjoining egg-shaped structure may have had a religious function. From their refuse, the people ate wild potatoes, perhaps with salty seaweed, brought from the coast thirty kilometers away. They used (and discarded) medicinal plants chewed as quids between the cheek and gum. Wooden mortars and grinding stones were used to process plants. Bones of six mastodon-like gomphotheres found at the site imply that the people obtained meat by either hunting or scavenging large animals. There are several other sites in South America that date before the time of the Clovis culture and before the mid-Canadian corridor was open.[71] The same case holds for early North American sites.

In North America, there are also sites that date prior to the Clovis culture. Recently, excavations at an archaeological site in central Texas, the Debra L. Friedkin Site on the Edwards Plateau, was shown to be an extensive pre-Clovis site with 15,528 human stone artifacts overlain by younger strata with Clovis tools.[72] The importance of the find is that the Clovis and pre-Clovis material were found together. This pre-Clovis "Buttermilk Creek Complex" dates between 12.8 and 15.5 thousand years ago. Artifacts with similar Clovis/pre-Clovis stratification and dates have also been found at the Gault site, which is also on the Edwards Plateau.[73]

These findings lend credence to the antiquity of archaeological material from other sites with pre-Clovis strata found scattered over North America: stone tools and debris dating to 14.2 and 14.8 thousand years ago occurring in Kenosha County, Wisconsin, found along with the remains of two mammoths;[74] tools dated between 13.4 and 15.2 thousand years ago from the Meadowcroft Rockshelter in Pennsylvania;[75] human coprolites found at Paisley Caves, Oregon, dated from 14.1 thousand years ago;[76] and a small number of flakes appearing at Page-Ladson, Florida, in strata with associated organic matter dating to 14.4 thousand years ago.[77] Other sites in Cactus Hill, Virginia, and Miles Point, Maryland, provide additional dates and put human presence on the continent a few more millennia earlier.[78]

From such data, people seem to have occupied the area that is now the continental United States for at least 15,500 years. The earliest cultures known to date produced bifaces (two-sided stone tools), stone blades, and bladelets. Their early occupation of North America provided them with ample time to settle into the environments of North

America and to have colonized South America by at least 14.1 to 14.6 thousand years ago.[79] This pre-Clovis, earlier human presence also opens the possibility of other arrival pathways. These early humans in the Americas could not have come to America through the corridor between the Cordilleran and Laurentide ice sheets. It opened too late, 13,500 years ago.

As was mentioned earlier in this section, there is another route, which would have been open as early as 15,000 years ago. It lay along the northern Pacific coast, but traveling this route would have required the use of boats.[80] We know that people in North America used boats as early as 13,000 years ago because of the human remains from that time found at the Arlington Springs site on Santa Rosa Island off the coast of California.[81]

If the pre-Clovis people used boats to travel the northern Pacific route, this would solve the problem of finding evidence of people in the Americas too early. However, it also opens up other possibilities of ocean-based colonization routes. The technology of the production of Clovis fluted points is similar to that of the European Solutrean culture, the culture associated with the cave paintings of aurochs and other large mammals in the French Lascaux caves (discussed in chapter 2). The Solutrean inhabitants of northern Spain made stone tools that have a strong resemblance to Clovis tools. Significantly, they used the same technology in their production.[82] The Solutrean culture is more than six thousand years older than the dates of Clovis material, and they were living on a different continent.

In era of the Solutreans, who lived during the harsh Late Glacial Maxima that occurred between 22,000 and 16,000 years ago, there was an ice bridge that connected Europe and North America. If some Solutreans were skin-boat, ice-edge hunters with a lifestyle similar to that of modern Inuits, Aleuts, and Eskimos, then they conceivably could have crossed the distance from Europe to America.[83] The distance to North America from Europe across this ice front is comparable to the distances (c. 2,500 kilometers) that the Thule Inuit people now travel in skin boats between Greenland and Alaska.[84]

Such travel across the North Atlantic would put people in North America very early. It would conform to the still controversial reports for the oldest dates of several eastern North American sites (such as Cactus Hill and Meadowcroft Rockshelter).[85] These people would then have had ample time to perfect the Clovis technology and rapidly emerge, perhaps by a cultural transfer of the Clovis technology, as a widespread culture. As will be discussed below, the genetic evidence for such colonization is, at least, not yet found in the genes of Amerindians.

One aspect of very early humans in the Americas is that their appearance seems not to have been particularly like that of modern Amerindians. An example is found in the analysis of human skulls found in caves and rock shelters at a site in central Brazil.[86] These skulls are dated older than 11,000 years ago. When their measurements are compared to the measurements on a reference collection of skulls from all over the world,[87] the most similar modern skulls are from Easter Island in the Pacific. The skulls of "paleo-Americans" in general are often different from those of more recent Native Americans[88] However, the genetics of modern Native American people indicate that they share a common genetic origin.[89]

From the current genetic analysis of a large sample of Native American people from North and South America, it is believed that they spring from a single population originating in Siberia. This group then moved toward the Americas across Beringia, likely less than 22,000 years ago.[90] The subsequent dispersal of these people south from Beringia into the Americas seems have involved a founding population of fewer than five thousand individuals.[91] There is no genetic evidence found in Native American people for a European ancestry. However, the genetic record does detect dispersal of Eskimo-Aleut people from northeastern Asia at a later time.

People appear to have been living in western Beringia 32,000 years ago. There were no glacial sheets to block their movement through western Canada during this relatively warmer time. However, there is no current evidence that they moved south at this time.[92] Current patterns indicate that people came to America via the Pacific sea route around 15,000 years ago. The Clovis culture could have been a second entry of people of the same genetic stock through the interior passage about 13,500 years ago.

The Clovis also may have represented the perfection of a hunting technology among the people already living in North America. The Clovis technology could have spread as an innovation transferred to established human populations. Or it could be the result of the population growth of the successful Clovis culture resulting from their success as megafaunal hunters. The Clovis culture very likely was involved in the extinction of the North American megafauna. The details of the way that all of this developed remain a topic of very active research and discussion. Before concluding this section, it is useful to outline some of the other theories that have been posited as to the cause.

Other Theories on the Pleistocene Extinction

There are alternative theories as to what might have happened to cause the megafaunal extinctions as well as new refinements to the classic blitzkrieg concept. The use of fire as a hunting and management tool, which was described in chapter 2 as "fire-stick farming" in its use by Australian Aboriginal people, could have had a role in the disappearance of the megafauna.[93] Fire appears to have had that role in the earliest colonization of Australia about 45,000 years ago, which will be discussed below. Another theoretical consideration would involve the changes in the nitrogen cycle at the end of the Pleistocene, putting the megafauna under nutrient stress.[94]

One of the early alternatives to the overkill theory is that there was a "hyperdisease" that arrived with the humans or dogs in North America. This conjectural disease wipes out the megafauna or enough of them that their extinction becomes highly likely. One can reason that the megafauna would be struck more strongly than other animals is because their relatively lower birth rates would slow their recovery from the disease.[95] The disease would have a high infection rate and a high lethality. It would also need to be able to infect animals of several different and not closely related species. Within a species it would infect both sexes and all age classes. It would also need a reservoir species (humans?) that could harbor the disease after it had infected and thinned the local host populations.[96] The disease organism that might have caused this extinction is not known. If it existed, it is unlike any extant disease. Rabies might be a possible example of such a disease, but it has a low infection rate.[97]

In addition, the effects of human hunting could have been amplified by food chain effects. Worldwide, human pressures have caused the extinction of elephants and their kin.[98] The extinction of the largest of the megaherbivores could have changed the grazing patterns and provided an initiator for further extinctions.[99] Human hunting also could have caused other predators to switch to other prey as the larger animals became reduced in number, with an associated collapse in the herbivore community followed by a collapse of the predator population. The loss of the elephantine keynote species, in conjunction with climate change and modest levels of human hunting, could have then produced a diversity collapse.[100] Currently, there appears to be a worldwide "shortening" of food webs (loss of the predators at the top of the food web) and an increase in herbivores and associated increased grazing pressure.[101]

A highly interdisciplinary and controversial theory postulates that an extraterrestrial event, a collision with an asteroid cratering into the melting continental glacier somewhere north of the current U.S./ Canadian Great Lakes caused the megafaunal extinction.[102] This asteroid impact is postulated to have occurred 12,900 years ago, wiping out the megafauna and disrupting the Clovis people that hunted them. This latter theory provides a large number of testable consequences that involve rather precise timing, feeding the scientific controversy over dating of various finds from the past.[103]

Complications from Climate Change

One obvious consideration that occurred at the time of the megafaunal extinction was the significant change in the climate as the world transitioned from a glacial to an interglacial condition. Several scientists have proposed this change as the underlying cause of the megafaunal extinction event.[104] The end-of-Pleistocene climate change would produce changes in vegetation and in the ecological systems and would stress the megafaunal populations. The wild horses and onagers in Alaska (*Equus ferus* and *E. hemionus*) showed a decrease in body size, indicating an environmental stress effect on the population while headed for extinction in North America.[105] Beringean steppe bison (*Bison bison occidentalis*) in Alaska had a similar decline in size.[106] Would these species and others have recovered and survived without human pressures? This is really hard to know. There are events that look like natural experiments that perhaps provide insight but still do not answer the question. For example, dwarfed versions of the North American giant ground sloths living on islands survived well past the extinctions of the continental species,[107] only to perish with the arrival of humans to the islands. This certainly demonstrates the power of people as agents of extinction, but it does not prove that the extinctions on the continent would have happened without climatic change.[108]

Of course, it is possible that the influx of human hunters along with a major climate change was the "double whammy" that wiped out the megafauna, which had been exposed to global climate change at the ends of previous ice ages without the massive extinctions.[109]

The classic interpretation of a wave of Clovis hunters arriving across the Beringian land bridge from Asia and "overharvesting" the megafauna in a wave of colonization is subject to some conditions. There were people in North America before the appearance of the Clovis hunters, and they were hunting (or perhaps scavenging) large mammals. This complicates the idea that the Clovis people were hunting a naïve fauna that allowed them to close the distance on their prey. From observations of the reintroductions of wolves in the western United States, there is evidence suggesting that prey populations alter their feeding habits and habitats fairly quickly when a predator population is introduced.[110] Significant ecosystem perturbations from people and/ or from the climate changes that were occurring could have produced feedbacks that amplified the negative effects of human hunting. Other aspects of human presence on the landscape, such as increased wild-fires, could have made a bad situation for the megafauna worse. At some point, domesticated dogs that would have also come over from Asia with people (see chapter 2) could have had a role in the mega-faunal extinction. Generally, hunters with dogs are remarkably more effective than hunters without them.[111]

Taken in its entirety, the evidence indicting humans as agents of the megafaunal extinction are strong, all the more so if the human effects are accompanied by other factors.[112] There may not be enough evidence to convict humans of causing the megafaunal extinctions, but there is certainly enough evidence to hold them without bail while rounding up the other suspects.

THE EXTINCTION OF THE AUSTRALIAN MEGAFAUNA

Forty-six thousand years ago in Australia, another continental-scale extinction event involving large mammals along with large birds, giant carnivorous lizards, and big snakes seems to have occurred.[113] A large and genuinely amazing collection of creatures are not found after this date. Some of these extinct mammals were monotremes, egg-laying mammals that are today represented solely by the platypus and the echidna (spiny anteater). The rest were marsupials—mammals that carry their babies through the latter stages of their development in pouches. The largest of these was *Diprotodon opatum*, a large rhinoc-eros-like marsupial. Males were three meters long, two meters tall at the shoulder, and weighed up to 2,800 kilograms. Females were quite a bit smaller, about half the size of the males. It is a bit difficult to imag-ine these giants carrying their babies in pouches, but female skeletons have been found with babies located where the mother's pouch would have been. Related to huge extinct *Diprotodon*, there were a smaller (but still large) now-extinct species. *Zygomaturus trilobus* measured two meters long and one meter tall at the shoulder and may have had a short trunk. Its habitat was thought to be wetland ecosystems, where they foraged by shoveling up reeds and sedges using two forklike incisor teeth. Another diprotodontoid was *Palorchestes azael* (sometimes called the marsupial tapir). It was similar in size to the *Zygomaturus* and had long claws and a long trunk. The nearest living relatives to all these creatures are the koala (*Phascolarctos cinereus*) and the three living wombat species (*Vombatus ursinus*, *Lasiorhinus krefftii*, *L. latifrons*).

Along with these large (over a thousand kilograms) marsupials were many species of large and strange kangaroos. *Procoptodon goliah* (the giant short-faced kangaroo) was the largest of these, standing at two to three meters tall and weighing up to 230 kilograms. *P. goliath* had a shortened flat face with forward-looking eyes, which provided it with stereoscopic vision. Its jaw and teeth were adapted for chewing tough semiarid vegetation. *P. goliath* was one of the seventeen species of kangaroos in the Sthenurine kangaroo subfamily (all now are extinct). Sthenurines were open-woodland creatures. All the toes on the leg (or hindlimb) were vestigial except for an extremely well-developed, hooflike fourth toe. Their forelimbs were long and had two extra-long fingers and claws (unlike the small, stiff arms of modern kangaroos). These may have been used for pulling branches nearer for eating and for moving on all four limbs for short distances. Along with a brace of other extinct kangaroos, one of the stranger, now extinct marsupials was *Propleopus oscillans* (the "carnivorous kangaroo"), a large (seventy kilogram) rat-kangaroo with large shearing and stout grinding teeth. It may have been an opportunistic carnivore that could eat insects, vertebrates (possibly carrion), fruits, and soft leaves. There was an assortment of large wallabies, leaf-eating kangaroos, large versions of modern kangaroos, and oversized koalas.

There were also marsupial carnivores. Notable among these was a leopard-sized marsupial "lion" (*Thylacoleo carnifex*) with bladelike slicing teeth.[114] It had the ability to use its tail and hind legs to stand as a "tripod," leaving free its powerful forelimbs, which were equipped with opposable thumbs and retractable, sharp thumb-claws. The marsupial lion was almost certainly carnivorous and likely a tree-dwelling predator. From the geometry of its jaws, it had an extremely powerful bite, one comparable to that of a modern lion.[115] Other marsupial carnivores included *Thylacine cynocephalus*, or the Tasmanian tiger, and a large version of the Tasmanian devil (*Sarcophilus harrisii laniarius*).

There were other very nasty carnivores as well, notably a large predatory lizard, *Varanus priscus*. It may have been over six meters long. Predictions of this lizard's size, based on its fossil vertebra, have produced different estimates of just how big this animal was, depending on which living varanid lizard is used as a model to interpret its morphology.[116] If *V. priscus* was built like the largest of its living relatives (the Komodo dragon lizard, *V. komodoensis*), then a large one would have weighed 1,940 kilograms.[117] It is likely that this monster also was poisonous; the Komodo dragon is, and so are other related large varanid lizards.[118] Other strange and dangerous reptiles included a terrestrial crocodile, *Quinkana fortirostrum*, with long legs, a length of five meters, perhaps more, and serrated teeth for slashing flesh.[119] There were three species of omnivorous, flightless birds related to ducks. The largest, *Dromornis stirtoni*, was three meters tall and weighed five hundred kilograms. The other two, *Bullockornis planei* and *Genyornis newtoni*, were ostrich sized. If one throws in giant pythons six meters long (*Wonambi naracoortensis*) and a genus of huge terrestrial turtles (*Meiolania*) up to 2.5 meters long with a horned heads and spiked tails, then one has an Australian menagerie that would excite the most jaded zoo visitor.

These creatures are not found in the fossil evidence after 46,000 years ago, the time of the migration of people to Australia.[120] There is a relative shortage of good sites for fossil preservation in Australia, and many of these species last were found at times that are quite a bit earlier than 46000 BP. This raises the possibility that they were already extinct when humans arrived. However, several species are found in fossil deposits dated from around 46000 BP but not later. Thus, the estimated age and the rapidity of the extinction of Australia's megafauna should be considered a reasoned theory but not an established fact.[121]

Enter Humans, Stage Right

If it seems that getting people to the Americas *might* have involved the use of boats, the getting of people to the Australian continent *certainly* involved sea crossings by humans over long enough distances to require boats. Perhaps because of stereotypes in movies and television, there is a tendency to imagine that prehistoric "cave people" would have spent a lot of time banging rocks together, speaking in monosyllables, and generally slapping one another around. These people may not have had the Internet, but they were our species (*Homo sapiens*). They were fully equipped with the same cognitive capacities that we possess.

For people to get to Australia from southeastern Asia on the coastal route out of Africa, they would have had to make multiple ocean crossings. The longest of these is over ninety kilometers of ocean, even taking into account the lower sea level from glaciation. For *Homo sapiens*, these crossings are generally thought to have happened about fifty thousand years ago.[122] At that time the oceans were one hundred meters or more shallower than today. Borneo, Sumatra, Java, and Bali were joined to the Malay Peninsula to form a continental land mass called Sunda. Similarly, Australia and New Guinea were joined by the lowered seas to form the Sahul landmass. Between these two mainlands was an archipelago of islands (Lombok, Timor, Buru, Sulawesi, Halmahera, Ceram, plus many more small islands). To get to Australia, *H. sapiens* must "island hop" across this archipelago. Indeed, *H. sapiens* was not the first human species to make some of these maritime crossings. Hundreds of thousands of years ago, *Homo erectus* manage to cross the ocean over the same archipelago at least as far out as the Flores Islands.[123] *H. erectus* does not appear to have gotten to Australia.

The human arrival to Australia (or, more properly, Sahul) likely occurred between 42,000 and 45,000 years ago.[124] This is a difficult time in the past to gauge, because carbon-14 dating methods produce variable results this deep in the past. These prehistoric sites are often dated by a suite of alternative methods. Reconstruction of past events from bits of fossil evidence is rarely a clear-cut enterprise. There is latitude

for debate in specific cases of humans' role in the demise of any megafaunal species for several reasons. Implication of humans in the megafaunal extinctions involves relatively precise timing of events in the
deep past. Is the date for human presence in an area the earliest, or are
there other remains of earlier humans not yet found? Similarly, the last
fossil is unlikely to be the very last individual of the species.

So how long did a particular species "hang on" after its last fossil
was produced? The processes that cause a fossil to form are not random. How does this affect the interpretation of the numbers of fossils and the timing of their deposition? Also, what are the details of
how these extinctions occur? Were they a product of human hunting
of the animals? Not just in the Australian case but in general, were the
extinctions attributable to the animals brought by humans (dogs, rats,
etc.)? Were the extinctions caused by the human technologies (ignition of wildfires, bows and arrows, etc.) that made humans a novel and
effective predator? Perhaps virulent disease vectored to a native fauna
by humans (or their animals) serving as a disease reservoir wiped out
the megafauna. Potentially, could complex combinations of any these
work in conjunction with climate change and other events?

The time of the disappearance of the Pleistocene megafauna was
thought to have occurred with the arrival of humans. One hypothesis
that fuses a variety of opinions about the causes of change in Australia,
postulates an initial blitzkrieg-type extinction followed by a reduction
of grazing and a buildup of plants that then provided fuel for wildfires, causing larger, hotter fires and significant long-term change in the
vegetation.[125] Supporting this hypothesis was the understanding that
the major Australian megafaunal extinction occurred around 46,000
years ago,[126] and there are indications of a change in the pattern of
wildfires and a reorganization of the vegetation at about this time.[127]
Several scientists have argued against this model while at the same time
endorsing a long-term deterioration of the Australian climate as the
cause of the decline in Australian megafaunal diversity.[128]

In some senses, these issues turn upon timing of events quite a long
time ago. There is evidence that in some regions in Australia humans
and various megafaunal species seem to have lived on the same land-

scapes for quite long periods of time, perhaps thousands of years.[129] There is and likely will continue to be argument on the causes of the loss of the Australian megafauna.

NO MOA, NO MOA IN OLD AOTEAROA:
THE NEW ZEALAND EXTINCTIONS

People in hunting-and-gathering societies can produce manifest changes to their environment, probably to a greater degree than many people might imagine.[130] Species extinctions that have occurred on remote islands in fairly recent times provide insight into the ways extinctions may have occurred in the past. Consider this account from Taylor, reporting on his observations in New Zealand:

> Resting on the shore near the Waingongoro Stream I notice the fragment of a bone which reminded me of the one I found at Waiapu. I took it up and asked my natives what it was. They replied "A Moa's bone, what else? Look around and you will see plenty of them." I jumped up, and to my amazement, I found the sandy plain covered with a number of little mounds, entirely composed of Moa bones; it appeared to me to be a regular necropolis of the race.[131]

Taylor was reporting on finding the bones of moas, ancient flightless birds of the genus *Dinornis* and native to New Zealand. The tallest moas were three meters in height and weighed up to 250 kilograms. By comparison, the ostrich (*Struthio camelus*) is the tallest living bird at 2.5 meters. The smallest moa species were about the size of a large domestic turkey, about a meter tall and weighing twenty kilograms. Moas are known from a diverse array of remains including eggshells, eggs, a few mummified carcasses, great numbers of bones, and some older fossilized bone.[132] The eleven moa species currently recognized occupied ecological niches usually filled by large mammalian browsing herbivores elsewhere.[133] There is some evidence that they may have had relatively low reproductive rates. They usually laid only one egg at a time.[134]

Uniquely, they had neither wings nor even any residual wing bones. Moas were herbivores and had muscular gizzards for grinding plant material. They swallowed small stones into their gizzards to aid in this food grinding before digestion. These polished stones, which were up to five centimeters (two inches) in diameter, are called "gastroliths" and often occur in groups along with moa bones. Bones of the larger moas were confused with ox bones by nonscientist observers, at least on initial inspection.

The moas evolved after New Zealand broke away from a giant Southern Hemisphere continent called Gondwanaland between eighty and eighty-five million years ago. As the Tasman Sea formed as a growing separation between New Zealand and Australia, New Zealand's fauna and flora were launched on an evolutionary trajectory independent from the rest of the world's. Creatures that could not cross considerable distances over ocean waters could not spread to New Zealand. Snakes and mammals, which evolved after the separation, are not indigenous to New Zealand. Bats, which could fly there, and seals, which could swim there, were the only mammals on New Zealand before the arrival of Polynesian settlers.

The Maori, New Zealand's Polynesians, arrived around one thousand years ago. They intensively exploited moas for food, feathers, bone, and skin from about nine hundred to six hundred years ago. Moa hunting subsequently declined, but opportunistic hunting of the animals continued until about 500 BP on the east coast of the South Island and on the western interior of the South Island until as recently as 300 to 200 BP.[135] It seems possible that when Captain James Cook first visited New Zealand in 1769, moas (or at least one of the moa species—the upland moa, *Megalapteryx didinus*) may have still survived in the remote areas in the western part of New Zealand's South Island.[136] If so, these individuals would have been the last relics of their kind.

Because they were so unique and because they became extinct so quickly, moas and their sad recent history have piqued the interest of ecologists, paleoecologists, and paleontologists. Climatic conditions in New Zealand appear to have been relatively stable over the period that moas became extinct. Different causes could have worked to some degree in concert with one another to account for the abrupt disappearance of moas:

1. *Forest Burning*: Vegetation changed with the Polynesian occupation of New Zealand. This change is not easily explained by climate variations but appears to be a product of burning. Forest and shrubland burning appears to have reduced the prime habitat of many of the moa species. The main forest burning started around seven hundred years ago, so that it followed what current archeological evidence indicates as the most intensive stage of moa hunting.[137] The east side of the South Island of New Zealand was burned most extensively. This habitat destruction seems to have occurred after the moa populations already were depleted. Large forest tracts remained in the most southern part of the island. Presumably, the presence of some suitable habitat could have sheltered small populations of moas.

2. *Human Hunting*: On the South Island of New Zealand, hunting appears to have been a significant factor in depleting the moa populations. In one location alone, six railway carloads of moa bones were sent to the bone mills at Dunedin.[138] Polynesian settlements and artifacts increased at the time of the most intensive moa hunting (950 to 650 years ago). This was followed by an apparent decline in the Maori population and a societal transition to smaller, less numerous settlements.[139] This pattern fits that expected for overexploitation of moas as food resource.

3. *Introduced Animals*: The Maori introduced the dog (*Canis familiaris*) and the kiore or Polynesian rat (*Rattus exulans*) to New Zealand. Either of these could have reduced moa populations by eating eggs from moa nests. The Maori may have brought pests and disease organisms from their chickens, which could have crossed over to eradicate moa populations. The possibility of using ancient DNA to identify past diseases of extinct animals is being explored.[140] However, evidence of such diseases is difficult to determine directly from paleoecolgical or archeological remains.

The moas and the rate of their demise raise ecological issues on the vulnerability of species to human-caused changes—including altered vegetative cover of the landscape, change in the physical environment, and modification of the biota from eliminating some species (such as the moa) and introducing others (such as the dog and rat).

The number of moas killed by Polynesians and the numerous bones stored in known archeological sites are sufficient to indicate that hunting alone could have accounted for the disappearance of the moa.[141] Modest rates of hunting can significantly reduce the population of an animal with a low rate of reproduction, as technological human societies have demonstrated by virtually eliminating the great whales of the oceans in about half the time interval postulated for the moa extinctions. Computer models of hunting human populations and moas indicate that the moa extinction could have occurred in as little as one hundred to 160 years.[142]

The demise of the moas on New Zealand is mirrored on other islands and is associated with initial human colonization. The prehistorical human colonization of the islands in Oceania is accompanied by compelling evidence for the extinction of vertebrates.[143] The extinction of 20 percent of the world's bird species were thought to be lost in the human colonization of the Pacific Islands.[144] Around 350 BCE, Iron Age people migrated to Madagascar. Their arrival was followed by a decline in the megafauna, which included seventeen giant lemurs, some the size of gorillas; three species of pigmy hippopotami; large grazing tortoises; and "elephant birds" (*Aepyornis* spp. and *Mullerornis* spp.), the largest birds ever to have lived.[145] The fungal spores indicating herbivore dung on the landscape dropped in lake sediments between 230 and 410 CE. This drop was followed by a substantial increase in charcoal in lake sediments, indicating a change in fire regime, a pattern found worldwide following human colonization.[146] Some of the megafauna, including pygmy hippos, elephant birds, giant tortoises, and large lemurs, coexisted for one thousand years or more with human colonizers. As late as 1661, Etienne de Flacourt, the French colonial governor, reported eyewitness accounts of animals that could have been giant lemurs, elephant birds, and hippopotami. Indeed, hippopotami were reported in other accounts as recently as the early twentieth century.[147] There is little argument that the advent of humans on islands can have a profound effect on the species found on islands.

$$\approx$$

What is to be learned from this? First, there is a refreshingly brisk discussion of the role of humans in the extinctions of a diverse array of animals in relatively recent geological time and at varied locations. From the point of view of the workings of science, this represents an excellent parable of the process of scientific discovery and debate.

Second, in the later Pleistocene epoch and in the epoch in which we now live, the Holocene, the record has featured a substantial increase in the extinction of animal species. Species are disappearing much, much faster than new ones are evolving to take their places. We have lost some species, such as the moas, that have been evolving for tens of millions of years. Human alteration of landscapes, hunting, and the variations in the world's climate, along with other possible factors, are at the root of these extinctions.

The Joban whirlwind question, "Can you hunt the prey for the lion?" would have to be answered thus: we have eliminated the prey for the lion and with that elimination, we have put lions and lionlike creatures in grave jeopardy.[148] This past in which human action and climate change act in concert and produce species extinctions does not convey pleasant omens for a future—with human activities altering the climate and human population growth increasing the pressures on the land's resources.

CONSERVING DIVERSITY ON A CHANGING PLANET

Francis Galton produced an essay in 1865 (see chapter 2) that listed the attributes necessary to allow different animals to be domesticated: "I will briefly restate what appear to be the conditions under which wild animals may become domesticated; 1. They should be hardy . . ." In a chilling paragraph that immediately followed, he opined,[149]

It would appear that every wild animal has had its chance of being domesticated, that those few which fulfilled the above conditions were domesticated long ago, but that the large remainder, who fail sometimes in only one particular, are destined to perpetual wildness so long as their race continues. As civilisation extends they are doomed to be gradually destroyed off the face of the earth as useless consumers of cultivated produce.

If Galton were alive today, he might well be surprised that his prediction has moved so rapidly toward realization. The pieces of this sad future world are the wheat fields, corn farms, rice paddies, cattle feedlots, and so on that can be seen across much of the modern landscape. It does not require a powerful imagination to assemble these pieces to see into the monotonous but productive "Galton World." The biotic diversity in Galton World would be the weeds, pests, and assorted vermin that are able to adapt and prosper in a land totally dedicated to cultivated production. It would require quite an innovative piece of salesmanship to create a travel brochure able to entice tourists to visit Galton World. Humanity needs to ask itself if we really want to go there; we are certainly on the way.

Somewhere in the trip through time toward Galton World, species begin to be lost.[150] The past patterns in diversity loss are understandable in the context of human exploitation of the environment, and we have considerable evidence that climate change, particularly climate change on an already altered human landscape, can have catastrophic consequences for the biota.

HOW MANY SPECIES MIGHT WE LOSE?

We are far from a complete inventory of all the species on Earth, and the species that we have not named are often rare, inconspicuous, or found in very isolated locations. One estimate based on the projected changes in climate that are expected to occur by 2050,[151] given our current and expected releases of carbon dioxide into the atmosphere, is that the number of species going extinct will be greater than 10 percent.[152] It is important to note that these changes are not expected to occur by 2050 but are an estimate of the long-term consequences that might develop if the expected 2050 climate were to persist over a long period of time.[153] The risk of extinction was calculated in several different ways, each representing theories on the expected distribution of abundances of species. For terrestrial species, the risk of extinction was estimated between 18 to 34 percent of the estimated five million terrestrial species based on the statistical models of the abundance of terrestrial species. This estimate amounts to a projected loss of 900,000 to 1.7 million species. This estimate has inspired a book compiling the scientific aspects of this projection under the title *Saving a Million Species*,[154] in which a number of different scientific experts analyze the potential of such a monstrous loss of diversity. Needless to say, these extinction estimates are the subject of hot scientific debate,[155] but other scientists have come up with relatively similar results using different methods.[156]

What Is Being Done?

At present, our efforts to conserve biotic diversity have taken a couple of essential directions. One strategy is to conserve species by identifying and protecting critical areas or hot spots for biotic diversity. There are locations that thanks to a combination of biogeography, history, and lack of human utilization contain a remarkable proportion of the planet's biotic diversity. These locations are "living arks" that could carry the planet's biotic diversity forward into the future. Such locations are potentially vulnerable to climate change, particularly the drying and warming that is being predicted to be the response of the atmosphere to the increased loading of greenhouse gases as a consequence of human activities. According to the International Panel on Climate Change 2007 report, the last time the greenhouse gases in the atmosphere were as high as we expect to have by 2050 was three million years ago, during the Mid-Pliocene epoch.[157] At that time global temperatures were 2° to 3°C warmer than today. Such temperatures occurring today could be expected to shift vegetation up mountains and necessarily shrink the area available for higher-elevation plants and animals. High-alpine ecosystems could be lost if the mountain tops became too warm to support them.[158] The disruption to our systems of biological reserves under a 2° to 3°C climate warming could be considerable.[159] These numbers are on the low range for the climate changes expected to come from human emissions of greenhouse gases in the coming century.

A second strategy for preserving biotic diversity is to focus on charismatic animals that appear to be keystones to different ecosystems and hope that the conservation of these will bring along the other associated species. The Joban whirlwind question regarding the lion implies this strategy. Perhaps our efforts with the conservation of lions are a microcosm of the challenges of this approach. It is difficult to guarantee the continued survival of lions as it is for many other large, beautiful, and dangerous species. These more species-oriented plans share the vulnerability to climate change.

There is also a mixed strategy in which one preserves species-rich locations hopefully to save the species found there and manages other areas to assure the survival of the species that we do know how to manage. One of the problems with rare species is that we often do not know enough about them to make ecological decisions on how to increase the likelihood of their continued existence. From the megafaunal extinctions of our immediate past we can easily infer what a laissez-faire approach to maintaining the biodiversity that surrounds us will bring—it will bring extinctions and lots of them.

CONCLUDING COMMENTS

By any reckoning, the Joban question about lions is as pertinent today as it ever was. Considering the story in its entirety, the Book of Job begins with Job as the owner of tremendous herds of domesticated livestock and as an inhabitant of an agricultural world, all of which he controls. Job is a denizen, a very important one, of an ancient version of Galton World, a landscape of human creation under human management. By the end of the account in the whirlwind speech, Job is chastised by God for his lack of understanding of the workings of nature and shown without a doubt that he is but a small piece of nature and not its centerpiece. As was discussed in the introduction, this "man-in-nature" motif is in sharp contrast with the "man-in-charge-of-nature" theme in other creation accounts in the Bible.

In this context, "Can you hunt the prey for the lion, or satisfy the appetite of the young lions, when they crouch in their dens, or lie in wait in their covert?" becomes a seminal question. If the answer is "No," then we must recognize ourselves as being in nature, one more group of passengers on a planetary bubble whirling through an uninhabitable void, a bubble that must be maintained to survive. We necessarily must lower our planetary footprint, consume more efficiently, and err toward caution when we put our hand to change the planet and alter its parts. If the answer is "Yes," then we must make certain that our care of the lions and of the nature as a whole that they symbolize is done in a way that can sustain future generations. We cannot tolerate errors in judgment over this hopefully long future. We must manage well and wisely. The past human history is *not* prologue for this enlightened management. The planet has lost and is losing species at a remarkable rate. The changes that we have made to our planet speak to how difficult it has been for us to shepherd the planet in past times.

Realistically, we probably do not have a good answer in either direction, a point that God hammered home to Job in the whirlwind speech. At our population density, saying "No" and hoping the planet will take care of us seems risky at best. Saying "Yes" implies a wisdom that we do not have and a will not manifested in our actions. We must find better answers than we have. Our future depends upon it.

Masks of Chaac. Masks of the Mayan rain god, stacked along a staircase leading up the Pyramid of the Magician in Uxmal, near Mérida, Yucatán, Mexico. *Source*: Photograph by the author.

10

Making Weather and Influencing Climate

Human Engineering of the Earth

Have you entered the storehouses of the snow, or have you seen the store-houses of the hail, which I have reserved for the time of trouble, for the day of battle and war?

—Job 38:22–23 (NSRV)

Can you lift up your voice to the clouds so that a flood of waters may cover you? Can you send forth lightnings, so that they may go and say to you, "Here we are"?

—Job 38:34–35 (NRSV)

Who has the wisdom to number the clouds? Or who can tilt the water-skins of the heavens, when the dust runs into a mass and the clods cling together?

—Job 38:37 (NRSV)

As one can see from the verses from the whirlwind speech above, God's power over the weather and wisdom on its workings differentiates Himself from mere mortals such as Job. This association of God with weather is neither unique to the whirlwind speech nor to the Book of Job. In the Torah (or the Pentateuch), the first five books of the Bible (Genesis, Exodus, Leviticus, Numbers, Deuteronomy), one finds:

If you follow my statutes and keep my commandments and observe them faithfully, I will give you your rains in their season, and the land shall yield its produce, and the trees of the field shall yield their fruit.

Leviticus 25:3–4 (NRSV)

and

If you will only heed his every commandment that I am commanding you today—loving the Lord your God, and serving him with all your heart and with all your soul—then he will give the rain for your land in its season, the early rain and the later rain, and you will gather in your grain, your wine, and your oil; and he will give grass in your fields for your livestock, and you will eat your fill. Take care, or you will be seduced into turning away, serving other gods and worshipping them, for then the anger of the Lord will be kindled against you and he will shut up the heavens, so that there will be no rain and the land will yield no fruit; then you will perish quickly from the good land that the Lord is giving you.

Deuteronomy 11:13–17 (NSRV)

as well as other similar verses.[1] In addition to instructions from the Torah, one finds praise for God's power over rain. For example:

When he utters his voice, there is a tumult of waters in the heavens, and he makes the mist rise from the ends of the earth. He makes lightnings for the rain, and he brings out the wind from his storehouses.

Jeremiah 10:13 (NRSV)

and

He it is who makes the clouds rise at the end of the earth; he makes lightnings for the rain and brings out the wind from his storehouses.

Psalms 135:7 (NRSV)

Clearly, provenance over the weather, particularly thunder, lightning and rain, is a divine attribute. Indeed, the Indo-European root of the word "divine" derives from a rain and storm sky god.[2]

Early agricultural people generally had a keen interest in favorable growing seasons and a fear of drought and famine. The façade of the masks of Chaac, the Mayan god of rain,[3] shown in the illustration at the head of this chapter is but some of the many that adorn the temples and other structures of the Puuc Mayan site located south of Mérida on the Yucatán peninsula of Mexico. It is small wonder that the people of Uxmal would regard the favor of the rain god with great importance. The Puuc Maya region is an area that has no natural surface water. Rainfall quickly drains into the porous limestone bedrock underlying the region as soon as it falls. From 600 to 1250 CE (the Late Classic to Early Postclassic Mayan periods), the inhabitants of Uxmal and other centers in the region survived by constructing two different kinds of water-capture systems: *aquadas* were open, still-water reservoirs; *chultunob* were small underground cisterns with fresh water for household use.[4]

Recent data based on the oxygen isotopes in the layers of stalagmites from limestone caves in the Puuc region provide a climate record of droughts and wet periods for the past 1,500 years.[5] During the Terminal Classic period, about 800 to 950 CE, a series of horrific droughts struck the region. These droughts had estimated rainfall reductions between 38 and 52 percent of the current average monthly rainfall and lasted for from three years up to as many as eighteen years. The regional population, which was thought to be around four million people, was decimated and fell to a few hundred thousand.[6] The system of divine kingship virtually disappeared, and the classic Mayan civilization went into a 150-year trajectory of depopulation, social unrest, warfare, and eventual disintegration.[7]

The Aztec rain god analogous to Chaac was called Tlaloc, the god of hail, thunder, and lighting. Tlaloc required sacrifices, primarily of children, whose tears were thought to encourage the god to bring rain.[8] These sacrifices intensified during times of drought. There is a long tree-ring record of drought for the region starting in about 850 CE, and this record has been reconciled with the Aztec calendar.[9] The Aztecs had a 260-day religious calendar (the *tonalpohuallii*) with a count of one to thirteen of the day numbers as well as twenty day signs (Rabbit, Death, Snake, Deer, etc.). They also had a 365-day solar calendar (the *xihuitl*), with eighteen months of twenty days each and an extra five days that are counted but are outside the calendar and considered unlucky. A solar year takes its name from the festival *tonalpohuallii*, held on the calendar day of the last (the 360th) day of the solar, *xihuitl* year. This complicated calendar is loaded with cycles (a thirteen-day *tonalpohuallii* cycle, a twenty-day sign-cycle interlacing though both calendars, an eighteen-month *xihuitl* cycle along with the five missing days). Given the complex combinatorics of these different cycles, years can only have four day names—Rabbit (*tochtli*), Reed (*acatl*), Flint Knife (*tecpatl*), and House (*calli*). It takes fifty-two years to cycle through all the combinations of years. Each combination (One Rabbit, Two Reed, etc.) only comes up once in this fifty-two-year cycle.

The year One Rabbit is the first year in the counting of the fifty-two-year cyclical Aztec calendar. There was a "Curse of One Rabbit Year" because this year was strongly associated with famine and catastrophe. The period between 882 and 1558 CE had thirteen One Rabbit years for which the climatic conditions have been determined.[10] Based on climate reconstructions derived from analyzing tree-ring width variations in the region, ten of these thirteen years were immediately proceeded by droughts and were a time of the resulting famines—a remarkable coincidence that is sufficiently unlikely as to be statistically significant.[11] One can speculate whether "predictable" catastrophe, even when it is a statistical fluke, is less destabilizing than more random disasters. Apparently the Aztecs, while suffering a fifty-two-year drought-cycle curse of the One Rabbit—along with other additional random droughts—saw the periodic calamity as reinforcing their belief in their calendar and its cycles.[12]

Along with Chaac and Tlaloc were other analogous rain gods in the Mesoamerican cultures: Cocijo for the Zapotec in southern Mexico, "Jaguar" in Olmec culture, and Dzahui in the Mixtec religion. There is also a strong convergence in other religions in other regions, for example, between Mesoamerican Olmec religious icons and those of the ancient Chinese Shang, the San people of southern Africa, and a long history of rainmaking in Jewish history.[13]

Indeed, the word "deity" ultimately derives from the Indo-European word *dieuos god of the daytime sky wielding thunder and lightning. *dieuos is the source of the word for "god" common to the most Indo-European languages.[14] In Greek, *dieuos is seen as Zeus, the god of sky and thunder. Jupiter, Zeus's Roman equivalent, was implored to bring rain and break droughts in a sung ceremony, the *aquaelicium* (Latin: calling the waters), which involved pouring water on a stone moved from its location near the Roman Temple of Mars.[15] The words from the root *dieuos represents "god" in Sanskrit, Latin, Oscan, Volscian, Umbrian, Old Irish, Old Welsh, Modern Welsh, Old Cornish, Breton, Old Icelandic, Old High German, Anglo-Saxon, Lithuanian, Lettic, and Old Prussian.[16]

MAKING WEATHER

Ceremonies, rain dances, and offerings continue to be used to break droughts and control the rain. The modern Mayans still give offerings to ancient gods that have merged with Catholic saints in the hopes of controlling rain, but the gifts are now cacao rather than human sacrifices.[17] There is a longstanding tradition of rain dances among Amerindians, notably Hopi, Navaho, and Zuni tribes in the U.S. desert southwest. The *paparuda* is a Romanian and Bulgarian rain ritual of considerable antiquity and possibly Thracian origin.[18] Within the last five years, two governors of U.S. states have proclaimed days of prayer to break droughts. "Sonny" Perdue, governor of Georgia, on November 14, 2007, stated, "I'm here today to appeal to you and to all

Georgians and all people who believe in the power of prayer to ask God to shower our state, our region, our nation with the blessings of water." Texas's governor Rick Perry proclaimed the three-day period from Friday, April 22, 2011, to Sunday, April 24, 2011, as "Days of Prayer for Rain in the State of Texas": "I urge Texans of all faiths and traditions to offer prayers on that day for the healing of our land, the rebuilding of our communities, and the restoration of our normal and robust way of life." Be they the Puuc Mayans, Romans, Georgians, Texans, or any other peoples, drought and lack of rainfall strain the very fabric of a society.

RAINMAKERS

In the United States, particularly in agricultural regions of the American West, rainmakers plied their trade in times of drought. This practice probably reached a crescendo in the "Dust Bowl" era of the Great Depression, in the 1930s. Among this rich collection of mountebanks, dreamers, and confidence tricksters, let us consider but two individuals as examples.

Charles Mallory Hatfield, "The Moisture Accelerator" (c. 1875–January 12, 1958)

Mr. Hatfield transitioned from working as a sewing machine salesman in southern California to rainmaking. By 1902, he had developed a secret mixture of over twenty chemicals that he stored in large evaporating tanks, which he claimed caused rain. By 1904 he was collecting fees of fifty to a hundred dollars for using his tanks and chemicals to produce rain and eventually obtained an assignment to produce eighteen inches of rain over Los Angeles for a thousand dollars.[19] By 1906, he had shifted to the Canadian Yukon Territory, where he contracted to create rain for the Klondike goldfield for ten thousand dollars. He failed but still collected $1,100. Then he went back to California, where, as documented in the January 5, 1916, *Los Angeles Times*:

The Southland was hit by the first in a wave of storms that dumped 11.4 inches of rain. The resulting floods killed 20 people and sparked a flurry of lawsuits, many of them aimed at rain-maker Charles Mallory Hatfield, who billed himself as a "Moisture Accelerator." The city of San Diego had hired Hatfield to coax precipitation from the clouds.

The city of San Diego reneged on a ten-thousand-dollar contract, which they had made with Mr. Hatfield. Working with his brother Paul, Charles Hatfield claimed 503 successful rainmaking attempts over his career. He eventually retired to Eagle Rock, California, where he continued to sell sewing machines.[20] His career inspired a Broadway play that became the 1956 hit film *The Rainmaker*, starring Katherine Hepburn and Burt Lancaster as the rainmaker. The eighty-one-year-old Hatfield attended the Hollywood premiere.[21]

Wilhelm Reich (March 24, 1897–November 3, 1957) and His "Cloudbuster"

Wilhelm Reich was born in Dobzau, Austria. He graduated from the University of Vienna in 1922 and became a psychiatrist and psychoanalyst. He studied under the Nobel laureate Julius Wagner-Jayregg and worked as the deputy director of Sigmund Freud's psychoanalytic polyclinic. With the rise of Hitler in 1933, Reich moved from Berlin to Scandinavia (to Denmark, where he was accused of corrupting Danish youth, then briefly to Sweden, and in 1934 to Oslo, Norway) and then to New York in 1939. Out of his controversial career in psychoanalysis he developed a general concept of a sexual primordial energy called "orgone." In 1940, he began to construct devices that could collect and concentrate atmospheric orgone and to experiment with these devices on cures for cancer and promoting plant growth. On January 13, 1941, Reich met with Albert Einstein, who agreed to experiment with an orgone concentrator but lost interest after initial experiments failed.[22] Reich eventually proposed a second, antiorgone energy called Deadly Orgone Radiation. He felt that DOR caused desertification, which led to his development of a "cloudbuster" device. In theory, the cloudbusters emitted orgone and channeled it into clouds to destroy DOR and cause rain.[23] He conducted several experiments with the device and was eventually hired with apparent success to produce rain to save Maine's blueberry crop from drought.[24] Things ended badly for Reich. Five days after the Japanese attack on Pearl Harbor, he was arrested by the FBI on December 12, 1941, largely because he was an immigrant with a communist background. He was held for about a month. Following his release, he enjoyed five very successful years practicing his version of psychotherapy and bought a 160-acre farm in Maine, which he named Orgonon.[25] There he constructed a facility that now houses the Wilhelm Reich Museum. In 1947, he was inspected by the Federal Drug Administration following exposés of his work in *Harper's* and *The New Republic*.[26] After an unscheduled inspection of Orgonon in 1952, the FDA decided that he was a fraud and obtained an injunction in 1954 to prevent him from interstate shipment of his publications and orgone accumulators.[27] In 1956, he was sentenced to two years in federal prison for violating the federal injunction. He died in prison a year later, at the age of sixty.[28]

WEATHER MODIFICATION AS A SCIENTIFIC ENDEAVOR

Clark C. Spence summarized the history of rainmaking in the United States thus:

> Especially in the 19th and 20th centuries, it has excited the imagination of Americans and brought forth hundreds of rainmaking schemes for private and public consideration. Some of these suggestions have undoubtedly been the offspring of wags or crackpots; some have been advanced by ignorant, but sincere, persons

who clung tenaciously to utterly hopeless beliefs; some have been seriously proposed by men of reputation as scientific solutions to the problem of drought; some—a great, many, unfortunately—have been offered by charlatans and "rain fakirs" interested only in lining their own pocket.

The patina of charlatanism from some of the commercial rainmakers of the early- to mid-twentieth century and the veneer of mysticism from rain ritual in a wide variety of cultures has tended to produce negative or, at the very least, suspicious interpretations of scientific rainmaking research efforts as well.

Early Rainmaking Attempts in the United States

In the United States at the turn of the nineteenth century, James Pollard Espy (1785–1860) advanced a theory of storms. The description of his theory sounded very much like that of a steam engine—inrushing winds, heat-induced uplift of air, and condensation of moisture to release atmospheric "steam power" all combined to drive storms.[29] Espy was a serious and well-known scientist in his day, and his ideas of storms as engines were consistent with Carnot's concept of winds and ocean currents as heat engines (discussed in chapter 7). Starting in 1817, Espy was a part-time teacher of mathematics and classics at the Franklin Institute in Philadelphia and served as the chairman of the Joint Committee on Meteorology of the American Philosophical Society and the Franklin Institute. He established a network of weather observers in each county of Pennsylvania as well as a national network of volunteers. He moved to Washington, D.C., in 1942 to become the first federally funded U.S. government meteorologist. He was at the center of meteorological research in his time, and his steam-energy theory of storms was accepted by other influential scientists of the time, including William Ferrell, who was also discussed in chapter 7 with regard to his seminal contribution on the circulation of the atmosphere.[30]

Espy's enthusiasm for his theory led him to the concept of generating artificial rains by igniting huge fires. In 1849, he contracted for twelve acres of timber in Fairfax County, Virginia, to be set alight to promote artificial rain. The experiment ended in failure.[31] His friends were concerned that his notion of lighting up forests in patches up to seven miles long to generate rain was impractical and potentially embarrassing to him.[32] On good advice, after this his rainmaking efforts faded.

The interest in artificial rainmaking certainly did not end with James Espy. Edward Powers produced a book, *War and the Weather; or, The Artificial Production of Rain.*[33] It begins with what might be the theme of scientific rainmaking: "The idea that rain can be produced by human agency, though sufficiently startling, is not one which, in this age of progress, ought to be considered as impossible of practical realization." Powers's theory was based on the analysis of observational data. He found upon analysis that rainfall often followed the major battles of the U.S. Civil War. Powers felt this was a consequence of artillery fire inducing rain. It is possible the observed relationship may instead be attributable to a correlation with the tendency of the generals to choose to fight during breaks in the weather when possible.[34] The U.S. Congress authorized $2,500 to test the theory.[35]

In 1880, the Confederate brigadier general Daniel Ruggles (1810–1897), who famously commanded "Ruggles' Battery" of sixty-two cannons against the Union line known as the "Hornet's Nest" in the Battle of Shiloh, patented a rainmaking "concussive theory."[36] In 1890, Robert St. George Dryenforth collected the ten thousand dollars appropriated by Congress to break a drought in Texas by using this concussive theory. His attempt included explosives on kites and balloons at different altitudes to compliment ground-level explosions. At the time, the news magazine *The Nation* characterized this as the "silliest performance that human ingenuity could devise."[37]

EARLY RUSSIAN AND AMERICAN EXPLORATIONS
IN SCIENTIFIC RAINMAKING AFTER WORLD WAR II

The Leningrad Institute of Rainmaking began field experimentation with rainmaking in 1932 with attempts to produce rain by seeding clouds with calcium chloride. This work lasted until 1939, when the experiments were curtailed because of World War II.[38] After the war, the work resumed using dry ice and silver iodide to seed clouds.

Despite a long history of theorization and application of rainmaking methods such as Espy's concepts, the initiation of scientific research on the modification of local weather conditions is often attributed to Vincent Schaefer's initial work with freezer chambers and using dry ice to cause ice crystal formation.[39] Pure water can be supercooled to temperatures slightly below $-40°$ (which is the same temperature in both Celsius and Fahrenheit), and it will freeze instantly if nuclei of condensation are introduced. In the summer of 1946, while working at the General Electric Research Observatory Laboratories, Schaefer dropped some dry ice into a cold chamber of supercooled water vapor. His breath into the chamber caused it immediately to fill with a cloud of millions of ice crystals. It was soon discovered that silver iodide could also promote explosive growth of ice crystals in chambers filled with supercooled water.[40] By November of the same year, Schaefer rented an airplane and spread six pounds of dry-ice pellets in a cloud over the Berkshire Mountains. This created of a streak of ice and snow three miles long. His boss and mentor, the Nobel laureate Irving Langmuir, was watching from the control tower and was on the telephone to the *New York Times* before the plane landed.[41] In the subsequent *Times* article, Langmuir opined, "a single pellet of dry ice, about the size of a pea . . . might produce enough ice nuclei to develop several tons of snow," thus "opening [the] vista of moisture control by man."[42] General Electric became wary of lawsuits that might attach to side effects of cloud-seeding experiments and transferred its work to the U.S. military. This was to emerge from secrecy during in the Vietnam War twenty-five years later and will be discussed in the section that follows.

The General Electric research was the first case of a continued interest in rainmaking by seeding clouds. By the early 1950s, about 10 percent of the U.S. land area was under arrangements with commercial cloud-seeding firms.[43] Basically, cloud seeding works on the basis of three related assumptions.[44]

1. To make rain, either relatively large water droplets must be present in a cloud to collide with one another and form raindrops or there are ice crystals in a supercooled cloud that can release snow and rain.[45] Schaefer observed this latter process firsthand in his laboratory's cold chambers.

2. Some clouds do not produce rain because of a lack of water droplets or ice crystals.

3. The lack of water droplets or ice crystals can be remedied by artificially introducing solid carbon dioxide (dry ice) or silver iodide to induce the formation of ice crystals, by introducing water droplets, or by introducing different materials that absorb water.

A reprise in *Science* of the success of cloud seeding to produce precipitation based on these assumptions lists only one successful experiment in cloud seeding in the thirty-five years after Schaefer's initial work.[46] One of the reasons for the low success rate derives from the rigor required in the *Science* review to be considered successful. To evaluate a cloud-seeding experiment statistically, one must develop an experimental design in which there are experimental and controlled cases, usually assigned at random. This means one must not simply apply cloud-seeding procedures to clouds and see that rain occurs; one must also observe cases in which equivalent clouds are not seeded and are scored on whether or not there also is rain. The cases must be randomly selected.

The one successful experiment tabulated in the *Science* article was an Israeli experiment, which compared rainfall on 388 days when clouds over the watershed of the Sea of Galilee were seeded (or not) in a predetermined random order with silver iodide.[47] Rainfall increased 18 percent in a smaller target area and 13 percent in the larger area (the Sea of Galilee watershed). This later finding is important because it goes to a political issue: if more rain is induced by rainmaking in Israel, will less rain fall in Jordan? From this experiment, the answer to this politically loaded question appears to be "no."

A scientific review of forty-seven years of rainfall modification in Australia concluded that cloud seeding is ineffective in the plains area of Australia but that it may work under certain special conditions in Tasmania.[48] A randomized rain enhancement experiment carried out in the Australian plains for 260 days in the years 1988 through 1994 found no significant effect of the rainmaking procedures.[49] The experiment was terminated in 1995. One somewhat recent review covers research and application projects up until about 1997. It points to the difficulty in establishing a statistically rigorous verification that the methods work. But it also identifies the great increase in information about the functioning of rain clouds and holds out some promise for the future using new methods.[50]

Two of the regular themes in the history of scientific rainmaking would have to be: it is harder than it seems, and more information about the workings of clouds and storms is necessary. One sees this in the initial work on cloud seeding springing from Schaefer's cold chamber observations. It is also the case for the ambitious Project STORM-FURY, which attempted to reduce the strongest winds of a hurricane by seeding the area next to the hurricane's eye. Three hurricanes were experimentally seeded. One promising result, Hurricane Debbie, had wind speeds that dropped but then strengthened after seeding the storm in two instances. The problem is that such behavior also can occur when hurricanes have not been seeded.

Unfortunately, as more was learned about the internal structure of hurricanes, it became clear that the amount of supercooled water in them was not sufficient for the seeding to be successful. Also, the variability of the behavior of hurricanes would have required an infeasible number of tests to be statistically rigorous. The good news is that the STORMFURY studies motivated the acquisition of WP-3D Orion meteorological research airplanes, which have collected invaluable information about the atmospheric sciences and about hurricanes in particular. The Orion data and the other information from the experimental study of rainmaking still provide basic data to test theories on the workings of clouds and storms today.

Indeed, the testing of theories in rainmaking experiments and identifying the need for better scientific knowledge may be the important long-term contribution of large-scale experiments on rainmaking. A large international cloud-seeding experiment in drought-stricken Queensland, Australia, seeded 127 clouds. The statistical variability of the results and the relatively small sample size did not allow for a statistically significant conclusion.[51] The results, at the same time, pointed to the importance of the cloud experiments to gain a physical understanding of cloud and rain dynamics. This could be said of many of the other scientific rainmaking experiments of the past.

Notably, early studies in cloud seeding in Missouri found that seeding clouds can possibly *reduce* the amount of rainfall.[52] Possibly this effect arises from having too many particles coalescing water or ice, so that a large number of small droplets form instead of a lesser number of larger droplets that are still large enough to fall to the surface as rain. It highlights how much more we need to know and would like to understand. The U.S. National Research Council in 2003 outlined important issues to be considered in weather modification,[53] as did a symposium held by the Tyndall Centre for Climate Change Research on "Macroengineering Options for Climate Change Management and Mitigation" at Cambridge.[54] A synthesis among the National Research Council writers and members of the Weather Modification Association aired differences in points of view but largely agreed on the need for an increased program in weather modification.[55] One significant point aired in this report was that if the inadvertent changes in global environment can be associated with global-scale climate change, then one might reasonably expect intended modifications to be equally effective.

MAKE MUD, NOT WAR: WEATHER MODIFICATION AS A MILITARY ASSET

Almost as soon as the General Electric Research Laboratory Observatory began its exploration of cloud seeding, its research was transferred to the U.S. military. The military implications of the weather modification technology compelled an enthusiastic response. General George C. Kenney, commander of the U.S. Strategic Air Command, stated, "The nation that first learns to plot the paths of air masses accurately and learns to control the time and place of precipitation will dominate the globe."[56] Rear Admiral Luis De Florez: "Start now to make control of weather equal in scope to the Manhattan District Project which produced the first A-bomb."[57]

During October 1969, Daniel Ellsburg and Anthony Russo photocopied a secret history of the U.S. Vietnam War requested by Secretary of Defense Robert McNamara with the title *United States–Vietnam Relations, 1945–1967: A Study Prepared by the Department of Defense*, better known as the Pentagon Papers. In 1971, Ellsberg gave forty-three volumes of the material to the *New York Times*. Release of the Pentagon Papers ignited street protests, controversy, and lawsuits. Senator Mike Gravel (Democrat, Alaska) read 4,100 pages of the material into the *Congressional Record* using the immunity from the threat of trial for treason for senators on the floor of Congress.[58]

One of the many things revealed from the Pentagon Papers was the documentation of a military rainmaking program, Project POPEYE, in Southeast Asia, which was designed to hamper the opponents of the United States in the Vietnam War by turning road surfaces to mud, causing landslides along roads, washing out bridges and river crossings, and extending the duration of soggy conditions.

The investigative journalist Jack Anderson had reported on Project POPEYE, which he called Project Intermediate-Compatriot in his column in March 1971. Operation POPEYE was briefly mentioned in the Pentagon Papers, which began to be printed by the *New York Times* on June 23, 1971. On June 29, Senator Gravel read them into the *Congressional Record*. But it was not until almost a year later, on July 3, 1972, that the existence of Project POPEYE was disclosed on the front page of the *New York Times*.[59]

Project POPEYE was not a small operation. During the rainy season from March to November, whenever possible, three C-130 Hercules aircraft (four-engine turboprop military transport airplanes) and two F-4C Phantom long-range fighter/bomber/reconnaissance airplanes flew two sorties daily out of Thailand to seed cumulus clouds with silver iodide. Based on a subsequent Senate subcommittee investigation of the project,[60] it was revealed that the Project POPEYE cloud-seeding flights amounted to 2,602 sorties flown over North and South Vietnam, Laos, and Cambodia between March 20, 1967, and July 5, 1972. The last flight was two days after Operation POPEYE details were reported in the *New York Times*. Operation POPEYE was considered a success in meeting its objectives.[61]

The Senate subcommittee hearings concluded that there was a need for an international agreement to prohibit the use of weather modification as weapons of war.[62] The Convention on the Prohibition of Military or Any Other Hostile Use of Environmental Modification Techniques was adopted by resolution by the UN General Assembly on December 10, 1976, opened for signature on May 18, 1977, in Geneva, and entered into force on October 5, 1977. As of January 2012, it was ratified by seventy-six countries. The convention bans weather modification for purposes of war. It is now called the Environmental Modification Convention (ENMOD).[63] In the United States, federal funding for applied weather modification has evaporated since about 1979.[64]

GEOENGINEERING: CLIMATE MODIFICATION

Even if modification of the environment for military purposes violates ENMOD, there is still the consideration of using environmental modification to ameliorate the effects of other changes in the environment. Stemming from concern about the possibility of a potentially irreversible planetary warming as a consequence of human activities,[65] learned societies (the Royal Society, U.S. National Academy of Sciences, American Association for the Advancement of Sciences, American Geophysical Union, etc.) have convened special conferences and published special issues of scientific journals on the topic.[66] There are two interrelated topics to consider. The first involves the historical precedents of nations contemplating changing the climate to better their situation in some way. The second involves how actually to modify the climate to "solve" the problem of global warming. The two topics converge on several critical questions that amount to the question: "If we 'fiddle' with the climate, who gets to set the thermostat?"

Early Geoengineering Ideas

U.S. postwar research on atmospheric modification was primarily focused on rainmaking. The USSR's program had a broader focus and included intentional climate modification as a major component.[67] In 1948, Stalin developed a plan to expand the Soviet economy by transforming nature and controlling the weather and climate.[68] By 1961, the Twenty-Second Congress of the Soviet Communist Party listed control of the climate as one of the USSR's most urgent problems.[69] The USSR occupied a major international position thanks to its development of thermonuclear weapons. The rocket delivery systems for these weapons were a significant byproduct of the Cold War and of Soviet competition with the United States in a space race. The USSR had the research and scientific infrastructure to deliver large, technologically difficult projects. Further, it had a harsh climate that perennially produced problems for Soviet agricultural productivity and development of its wild northern lands.

The proposed climate-modification projects included a plan to orbit metallic particles in a "rings of Saturn" arrangement to reflect heat and light to northern Russia and simultaneously to shade the tropics to produce a more temperate climate there.[70] The cover of a book published in 1960 and titled *Man Versus Climate* featured the Earth surrounded by a planetary ring.[71] This book finishes with what well might be the theme of the Soviet era of climate modification:

> We have described those mysteries of nature already penetrated by science, the daring projects put forward for transforming our planet, and the fantastic dreams to be realized in the future. Today we are merely on the threshold of the conquest of nature. But if, on turning the last page, the reader is convinced that man can really be the master of this planet and that the future is in his hands, then the authors will consider that they have fulfilled their purpose.

Proposals in the book included such plans as damming and diverting the Congo River to irrigate the Sahara, dispersing cloud cover over the Arctic Basin, blocking the Gulf Stream with a dam between Florida and Cuba, and damming the Bering Strait and then pumping the cold Arctic water into the Pacific and drawing warmer Atlantic water into the Arctic Ocean to melt the polar ice.

Melting the Arctic ice was seen by several Soviet scientists as a large-scale application to warm the Russian climate.[72] The presidium of the USSR Academy of Sciences met in November 1959 to discuss controlling the Arctic climate. There were two follow-on meetings in Leningrad in 1961 and 1962.[73] M. I. Budyko, who proposed a human role for the extinction of the megafauna just discussed in chapter 9, was a major figure in Russian climatology in this era. Budyko felt that the Arctic Sea ice, if melted, would not refreeze and would warm and hydrate northern Russia.[74] Water in a melted Arctic absorbs incoming solar radiation to heat the region. Conversely, ice reflects incoming solar radiation, and its presence has a cooling effect.

At about the same time in America, the Nobel laureates John von Neumann and Edward Teller advocated manipulation of the weather as a Cold War weapon.[75] Two decades later, President Lyndon Johnson's scientific advisory committee was concerned with options to mitigate the possible planetary warming that could result from fossil fuel burning.[76] Opposition on whether to engineer a warmer world or a cooler world between the two Cold War opponents brims with irony.

GEOENGINEERING TO COMPENSATE FOR ANTHROPOGENIC GLOBAL WARMING

The eruption of Mount Pinatubo in the Philippines in June 1991 represents something of a natural experiment on chemistry-based geoengineering. In volcanic eruptions, sulfur dioxide blown into the atmosphere forms sulfate particles that can remain in the stratosphere for months. The Pinatubo eruption lowered air temperatures and reduced the water vapor in the atmosphere. The Nobelist Paul Crutzen in 2006 suggested injecting sulfur into the stratosphere to produce sulfate clouds, which would whiten the Earth from space. The additional clouds would reflect more incoming sunlight back into space and represent a possible way to offset the planetary warming caused by human greenhouse gases.[77] Budyko made a similar proposal in 1982 and calculated that ten million tons a year of sulfur dioxide released in the stratosphere would cancel out the expected warming from a doubling of atmospheric carbon dioxide from human fossil fuel use.[78] Analysis of the Pinatubo eruption event indicated that the shielding of the sun by increased clouds reduced the rate of evaporation of water.[79] Injecting sulfur dioxide into the stratosphere is only one of several means to eliminate the possible warming brought upon from increases in greenhouse gases in the atmosphere. Imagine a board game for "science nerds" that involves guessing whether a particular scheme for engineering the Earth's climate was dreamed up in the USSR climate-modification era in the 1960s or elsewhere more recently. The key to winning could stem from realizing that the early Soviet proposals emphasized (mostly) the transportation of heat to new places using megascale public works projects; more recent strategies (also mostly) have involved engineering the Earth's energy balance. For many of the geoengineering proposals there is a parallel, inadvertent potential climate change that is ongoing.[80]

Changing the Earth's Energy Balance on the Incoming Side

One might be able to cool the Earth by reducing the amount of incoming solar radiation. The injection of sulfur dioxide into the stratosphere to reflect incoming sunlight is but one of the methodologies proposed to compensate for global warming. Others, starting from outside the Earth and moving in are:

- *Installation of reflecting materials in orbit.* The ideas of orbiting reflective materials around the planet to reflect sunlight were proposed in the early Soviet space era.[81] Putting reflectors in space to beam sunlight to Earth to light up solar power facilities in Earth orbit was discussed in 1981 in a U.S. National Academy Committee but was thought to be prohibitively expensive.[82] Reflecting shields in low-Earth orbits act as solar sails. Incoming sunlight pushes them out of orbit.[83] One can get around this by putting the reflectors in orbit at the L_1 Lagrange point, the location in space between the Earth and Sun where the two gravitational attractions are exactly in balance. The shields would need small positional adjustments to remain perched on the Sun-Earth gravitational fence.[84] The capability to correct position also would mean one could adjust the shields.[85] In 1997, the best guess at the cost of such a system of Lagrangian reflectors was $50 to $500 billion.[86] A somewhat similar proposal is to put a cloud of reflective particles around the L_1 Lagrange point.[87]

- *Using atmospheric aerosols to cause greater reflection of incoming light.* These plans almost invariably involve injection of material into the stratosphere.[88] Putting material into the atmosphere nearer the Earth, the troposphere, is vexed by the problem that rainfall washes it out in precipitation. The whitening of the Earth using sulfur dioxide to cause sulfate clouds in the stratosphere is but one of the proposed methods of reflecting more of the incoming solar radiation. Another possibility is the injection of ten million tons of metallic aerosols (probably alumina) into the stratosphere. These could be in the form of a micromesh or microballoons.[89] Still another is the injection of soot or black carbon into the stratosphere, but this seems riskier than a sulfur injection.[90] Sea salt particles emitted from wind-powered vessels using artificial seawater sprayers could provide the condensation nuclei to produce increased marine stratocumulus clouds, but again there is much uncertainty about how efficient this might be.[91] A field test of a government project to inject stratospheric particles of reflective aerosols in the United Kingdom (called SPICE: Stratospheric Particle Injection for Climate Engineering) was cancelled in 2012 over unresolved issues of patent ownership and intellectual property rights over the potentially lucrative capability to mitigate climate change.[92]

- *Changing the reflectivity of the Earth's surface.* The engineering of an ice-free Arctic for the benefit of Russia's north is a longstanding geoengineering concept intended to warm the Earth by altering its reflectivity. Other concepts involve changing the proposals to change the surface to cool the planet and neutralize global warming include, in one way or another, changing the surface albedo of the ocean by using various floating objects to reflect incoming light. There are obvious downsides to such applications, including littering shorelines and harming commercial fisheries. Clearing large tracts of the boreal forests of the Northern Hemisphere have planetary cooling effects by ways of several different feedbacks.[93] Reroofing cities with white roofs would also make urbanized areas reflect more light and also reduce air conditioning demand, conserving energy. The contrast in clearing boreal forests versus white roofs is one of scale. Plans to alter the reflectivity of the Earth's surface (or to reduce the emissions of greenhouse gases) demonstrate a continuous gradient of local "green" energy conservation strategies to more grandiose geo-engineering projects.

There are at least two broad and problematic considerations in reducing the amount of incoming sunlight by whatever means, at regional or global scales. First, the complexity of planetary interactions is considerable. Do we have the wisdom to avoid the unintended consequences of our geoengineering plans? Certainly, there is a need to know much more than we know do. Recently, two atmospheric scientists at the National Center of Atmospheric Research, Kevin Trenberth and Aiguo Dai, concluded their analysis and review of the issue of injecting sulfate into the stratosphere to increase the Earth's reflectivity by writing, "creating a risk of widespread drought and reduced freshwater resources for the world to cut down on global warming does not seem like an appropriate fix."[94]

Second is the persistent issue of who determines the beneficiaries and those who must make sacrifices on a geoengineered planet. Recall that as far back as the early 1960s, with the Soviet plan to put a reflective ring around the Earth there was a rationalization that the reflecting of radiation onto the Russian north would shade and make more temperate the sweltering tropics.[95] Whether this is a correct assertion is not obvious. It is rather difficult to imagine that the consequences of geoengineering will benefit everyone on Earth.

Changing the Earth's Energy Balance on the Outgoing Side

The greenhouse warming of the Earth is a product of the interception of some portion of the outgoing short-wave radiation from the Earth's surface. The control of outgoing radiation involves reducing greenhouse gases from the atmosphere. The principal greenhouse gas is water vapor, but modification of the amount of water vapor in the atmosphere has not been proposed.[96] Most of the plans involve reducing the amount of carbon dioxide in the atmosphere. Fossil fuel use can be reduced. The carbon dioxide generated by fossil fuel use can be captured. Geoengineering can remove carbon dioxide from the atmosphere. For terrestrial ecosystems, carbon can be removed using forestry practices that store carbon in the soil and in trees.[97] If trees are harvested, the wood can be stored,[98] burned as fuel, or even burned as fuel with recovery of the carbon dioxide. One can also develop agricultural practices that store more carbon in the soil.

In the oceans, iron may be the nutrient limiting plant growth over large regions,[99] with each atom of iron causing the uptake of about ten thousand atoms of carbon. Field experiments have shown short-term large increases in ocean productivity,[100] but it is not clear that these increases can be sustained over a longer period of time.[101] Further, there is reason to be concerned over any large-scale change in the chemistry of the oceans. The same can be said for using the oceans as a repository for carbon dioxide that has been captured from fossil fuel combustion.[102]

CONCLUDING COMMENTS

It is no surprise that the power to control the weather is a principal dimension of divine omnipotence. Does the sensitivity of simulations of the Earth's climate to inadvertent human changes in the atmosphere and the planet's surface imply that geoengineering could be effective to manifest planetary-scale changes? In other words, if we can change the climate by accident, just think what we could do if we really put our minds to it. The stakes to control the weather have always been high.

Certainly, control of weather has both tactical and strategic warfighting implications. Choosing to fight battles under favorable conditions has been an aspect of warfare since time immemorial. Predicting these conditions is intrinsic to modern warfare. Modifying the environment to favor one military opponent over another has been deployed in the past but is currently under international injunction through treaties.

If to intensify storms, blizzards, hurricanes, and hail is the ultimate weapon, then to moderate these same calamities is the ultimate magnanimity. Breaking or causing droughts could control the fates of regions and cultures. Simply being able to produce rain at critical times during the growth and maturation process of crop plants could determine economic success or failure of agriculture at a myriad of scales. Issues associated with the geoengineering of the Earth have parallels with these issues. One problem is to know when and how geoengineering might favor one people or one nation over another. This was a persistent concern with respect to the USSR's climate modification plans. The melting of the Arctic Sea was one of the preferred Soviet schemes.[103] The possibility of this event worsening climate elsewhere in the Northern Hemisphere was a worrisome consequence of this action. Ironically, at the time of writing there is a decline in Arctic Sea ice attributed to a general warming of the Arctic.[104]

Can we really geoengineer the Earth? Some scientists would see humans as having had a profound change on the climate with the dawn of agriculture.[105] A majority of atmospheric scientists feel that human activities are altering climate today.[106] Given the potential consequences of a climate change, geoengineering is becoming a popular technological solution to a difficult policy problem.[107] Several holistic scientists see geoengineering as the challenge from which we cannot walk away, particularly at our population density and per capita use of materials.[108] A reenvisioning of our energy production capacity, an intrinsically complex and costly option with nothing, including nuclear energy, off the table, is likely to be as contentious a debate in the future as it was in the past.[109] Others express concern with the feasibility, ethical appropriateness, and likelihood of success of geoengineering.[110] Is it cheaper to reduce greenhouse gas emissions than to modify the planet somehow to compensate for our actions? The methodologies involved can be divided into those for managing solar radiation or for managing the atmospheric inputs that effect outgoing radiation (notably carbon dioxide). It appears that scrubbing carbon dioxide from industrial processes may cost as much as a thousand dollars per ton of carbon dioxide removed from the atmosphere, implying a multitrillion-dollar cost each year. Still others are keenly aware of the importance to consider the options wisely: encouraging research but stopping large-scale field experiments to develop scientific oversight.[111]

We have the challenge of navigating the moral and ethical issues that attend climate modification. There is a significant concern that rogue geoengineers might take it upon themselves to change the planetary climate. We must at the same time assess the feasibility, risks, and costs. The history of weather and climate modification portends considerable difficulty in negotiating these difficulties, but it is an assessment that must be undertaken wisely and soon.

The Moon in orbit around the Earth. Imaged by the NASA *Galileo* spacecraft looking back eight days after its last encounter with Earth. *Source*: NASA, http://www.nasa.gov/offices /oce/appel/ask/issues/42/42s_galileo_rocky_road_jupiter.html.

11

Conclusion

Comprehending the Earth

Have you comprehended the expanse of the earth? Declare, if you know all this.

—Job 38:18 (NRSV)

At the very start of the whirlwind speech, Job, who has been asking for an explanation for his terrible misfortunes occurring in spite of his piety, receives an abrupt comeuppance from God: "Who is this that darkens counsel by words without knowledge? Gird up your loins like a man, I will question you, and you shall declare to me" (Job 38:2–3). This is the introduction to a towering set of questions about planetary creation, function, and biodiversity. It is a revelation of the creation and the most elaborate of the creation accounts in the Bible.[1] At some level, the whirlwind speech is about Job's legal standing to ask God, "Why?" "Why," if Job was a good man and more than complied with what he was supposed to do, had he fallen into such a pitiable condition? Why does a benevolent, all-powerful deity allow bad things to happen to good people such as Job? "Why me, God?"

Such "why" questions arise in any culture whose religion centers on a protective, omnipotent deity. Accordingly, direct interactions between humans and gods to ask "why" are themes in many religious traditions. In these interactions, an immediate issue is whether a mere mortal should be allowed even to speak to the divine. This determination often involves a test. Job was not asked slay a horrific dragon along with its kith and kin, nor to tussle with various demigods, nor to bring back sundry magical objects from some intensely malevolent lady. Instead Job was challenged to comprehend the Earth and its workings. For an environmental scientist, the idea that understanding the way the world works somehow constitutes a challenge as daunting as slaying Grendel or capturing the Erymanthian Boar has an obvious appeal.

The quest to comprehend the Earth sweeps in a remarkable breadth of human knowledge, including the natural sciences of physics, astronomy, earth science, chemistry, and biology. Given the growing density of human beings on Earth and their propensity to change things around themselves as the planetary keystone species, the charge to comprehend Earth also has a sense of urgency. It is a scientific "good question"—difficult, well posed, synthetic, and important.

Answering this particular good question has been a scientific quest for much of the history of science. Certainly a high point was reached in the remarkable insights gained by the early scientists and naturalists of the seventeenth and eighteenth centuries. The voyages of exploration and discovery, the insights of the polymath scientists of the age of Enlightenment, and the simultaneously detailed and expansive conjectures of ecologists, anthropologists, and archeologists have provided the foundations upon which we build. In 1676, Isaac Newton wrote to Robert Hooke, "If I have seen further it is by standing on the shoulders of giants,"[2] humbly referring to Newton's exceptional contribution to science and his debt to the scientists of the past. Scientific discoveries of the present build on and reanalyze previous findings. This is certainly the case for scientists striving for increased comprehension of Earth. Surely, we stand upon shoulders of giants: Alexander von Humboldt attempting to understand the interconnected nature of living and physical systems; Svante Arrhenius calculating the effect of changing the amount of greenhouse gases in the atmosphere by expanding ideas from Joseph Fourier eighty years earlier; George Hadley and William Ferrel trying to understand the cause behind the patterns of sea winds that had allowed navigators from centuries earlier to traverse unknown waters of the oceans; along with many others great and small, famous and unappreciated.

A second Newtonian reflection made near the end of his life captures another essential aspect of the forward progress in global environmental sciences: "I do not know what I may appear to the world, but to myself I seem to have been only like a boy playing on the seashore, and diverting myself in now and then finding a smoother pebble or a prettier shell than ordinary, whilst the great ocean of truth lay all undiscovered before me."[3] Like Newton, the past giants of Earth-system science were driven by their own curiosity to understand better the connectivity of nature. They were smart, driven, and often willing to take great risks in hazardous situations.

Think of Andrés de Urdaneta deciding to use a *volta do mar* to navigate the Pacific route from the Philippines to Mexico, or Joseph Banks stepping on board the HMS *Endeavour* bound for the South Pacific and Australia with Captain Cook. Many of them were wealthy or backed by supportive patrons. They were essentially doing what they enjoyed, what they had to do, what was their life's quest. They resembled Newton's boy playing on a seashore collecting pebbles and shells, but they and Newton were simultaneously much more. They likely would have not have had an easy time writing an incremental grant proposal to some nation's science funding agency. A promotion and tenure committee at a university would have counseled them to try to get out some early results in a good journal—the more the better.

We live in a time of great need to understand our planet. We have the challenge of comprehending Earth as we simultaneously change the Earth. Are we creating the intellectual environment for creative, synthetic, and revolutionary researchers that can push us across the old boundaries into new paradigms? It is a not a question of letting scientists do what scientists do. It is a question for us all. Sadly, the politicization and the businessification of science may be taking the intellectual and creative environment in the opposite direction.

The tools at our disposal for the challenges in Earth-system science would be the envy of the environmental researchers who have come before us. We have satellite systems capable of remarkable measurements, along with a repository of innovative new systems on the shelf. Products of several of these are shown as illustrations in this text. However, the satellite constellation of the U.S. space agency, NASA, is falling into a state of disrepair. Some of this lost capability is being replaced by the orbiting instruments developed by other nations or by international and even commercial consortia. But overall, there is a loss of capability at this critical time. The conversion from satellite data provided free to researchers of any nation by NASA to a more nationally oriented, pay-as-you-go system may have a negative effect on creative, small-budget exploratory research.

Other capabilities include sophisticated chemical-analytical methods that can trace differences in the stable isotopes of elements and even isolate individual chemical compounds in microscopic samples for isotope analysis. These and the sophisticated analyses of DNA and other biochemical compounds using modern molecular biology provide new data to challenge old theories and produce new ones. These methodologies seem to have penetrated the public consciousness through their sometimes magical use in the fully instrumented forensic laboratories of TV police shows. Results of these chemical and physical instruments have also shown up in several of the chapters here, ranging from applications involving the details of the creation of the solar system to documenting fermented mare's milk in pottery of the Botai culture in southern Ukraine and Kazakhstan six thousand years ago.

The Earth is a place that life has made much different from what it otherwise might be. In the past, different organisms have changed the planet. Archeons altered the Earth's ocean and atmospheric chemistry. Photosynthetic species manufactured an oxygen-rich atmosphere far from its natural chemical equilibrium. These same photosynthetic species may have pulled enough carbon dioxide out of the atmosphere to reduce the greenhouse effect and produce a frozen "snowball" Earth. The effects of modern technological human culture, even with its population expected to reach nine billion by the year 2050, pales in comparison to these agents for change on Earth.

Given available resources, the human per capita use of energy and materials also has edged up over time. Because of the increased population and increased use of the resources of Earth, we have begun to modify the internal feedbacks of the planet. Not as much as the changes from the archeons or from the photosynthesizers, as the collision of asteroids, or as the eruptions of an unusually active volcanic era, but we are changing the Earth nonetheless. To think otherwise is to ignore a considerable body of fairly easily understood information.

That the Earth previously has changed radically in the past is cold comfort. Life on Earth in some form has held on through billions of years. The question really is whether our species or our complex society will continue to find a comfortable home on Earth, our home for which we clearly have no other habitable alternative. Comprehending Earth was the subject of the whirlwind speech and its questions. In considering these questions, this book has attempted to provide some insight on the functioning of this interactive planetary system; this tiny blue and white marble in the vastness of space; this, our galactic home. We truly need to comprehend the Earth—its physical systems, its biota, its internal feedbacks, and our place in this marvel. The lives of the generations to come depend upon it.

Notes

1. W. P. Brown, *Seven Pillars of Creation: The Bible, Science, and the Ecology of Wonder* (New York: Oxford University Press, 2010).

2. *Transformation of Larch-Dominated Forests and Woodlands Into Mixed Taiga* (NASA Proposal 06-IDS06-0078); *Optimal Dynamic Predictions of Semiarid Land Cover Change* (NASA Proposal 07-LCLUC07-0029); *Climate- and Fire-Induced Vegetation, Agricultural, and Albedo Change in Northern Eurasia—Consequences to Gases, Aerosols, and Radiative Fluxes* (NASA Proposal 09-IDS09-0116); *Carbon Monitoring Science Definition Team* (NASA Proposal 10-CMSSDT10-0003); *Synthesis of Forest Growth, Response to Wildfires and Carbon Storage for Russian Forests Using a Distributed, Individual-Based Forest Model* (NASA Proposal 10-CARBON10-0068); *Elephant-Habitat Dynamics in the Serengeti Ecosystem—A Predictive Modeling Approach Using Multiscale Optical and Radar Satellite Imagery* (NASA Proposal 11-Earth11R-0092); *LCLUC Synthesis—Forested Land Cover and Land Use Change in the Far East of Northern Eurasia Under the Combined Drivers of Climate and Socioeconomic Transformation* (NASA Proposal 10-LCLUC10-2-0031); *A Program for Computational Education and Internship Training for Environmental Science Students* (NASA Proposal 11-CMAC11-0006).

1. Introduction

1. There is some ambivalence in this interpretation of the Hebrew text. The literal translation of the Hebrew word, ברך or brk (to bless), would be "bless God and die," but, because writing "curse God" would be inappropriate in a holy text, "bless" was used as a gloss on this potentially offensive phrase. Recall the idiomatic phrase in English, "I *blessed* him out for his foolishness," in which "blessed" actually means the

opposite. An alternate view would be that the Hebrew word was used literally and that his wife's advice was essentially to pray and then die honorably. See T. Linafelt, "The Undecidability of ברך in the Prologue of Job and Beyond," *Biblical Interpretation* 4 (1996): 154–172.

2. Zophor does not speak in the third cycle.

3. G. B. Dalrymple, "The Age of the Earth in the Twentieth Century: A Problem Mostly Solved," *Geological Society of London*, Special Publications, 190 (2001): 205–221; G. Manhesa et al., "Lead Isotope Study of Basic-Ultrabasic Layer Complexes: Speculations About the Age of the Earth and Primitive Mantle Characteristics," *Earth and Planetary Science Letters* 47 (1980): 370–382.

4. W. P. Brown, *Seven Pillars of Creation: The Bible, Science, and the Ecology of Wonder* (New York: Oxford University Press, 2010).

5. Ibid.

6. K. Schifferdecker, *Out of the Whirlwind: Creation Theology in the Book of Job*, Harvard Theological Studies 61 (Cambridge, Mass.: Harvard University Press, 2008).

7. B. McKibben, *The Comforting Whirlwind: God, Job, and the Scale of Creation* (Cambridge, Mass.: Cowley, 2010); Schifferdecker, *Out of the Whirlwind*; Brown, *Seven Pillars of Creation*.

8. BCE stands for "Before Current Era," a term equivalent to the BC in the BC/AD (Before Christ/*Anno Domini*) dating system. It is appropriate for multireligious discussions. BP is also used in this book to denote "Before Present." This is a term often used in conjunction with radiocarbon dating.

9. Schifferdecker, *Out of the Whirlwind*.

10. Ibid.

11. G. Wigoder, *The Illustrated Dictionary and Concordance of the Bible* (Jerusalem: Jerusalem Publishing House; New York: MacMillan, 1986).

12. O. Lipschits and T. L. Thompson, *Early History of the Israelite People: From the Written and Archaeological Sources* (Leiden: Brill, 1992).

13. J. Blenkinsopp, "Bethel in the Neo-Babylonian Period," in *Judah and the Judeans in the Neo-Babylonian Period*, ed. O. Lipschits and J. Blenkinsopp (Winona Lake, Ind.: Eisenbrauns, 2003), 93–108. A. Lemaire, "Nabonidus in Arabia and Judea During the Neo-Babylonian Period," in *Judah and the Judeans in the Neo-Babylonian Period*, ed. O. Lipschits and J. Blenkinsopp (Winona Lake, Ind.: Eisenbrauns, 2003), 285–300.

14. J. A. Middlemas, *The Troubles of Templeless Judah* (New York: Oxford University Press, 2005), 10.

15. Schifferdecker, *Out of the Whirlwind*.

16. A. Korotayev, *Ancient Yemen* (Oxford: Oxford University Press, 1995).

17. Schifferdecker, *Out of the Whirlwind*.

18. N. Sarna, "Epic Substratum in the Prose of Job," *Journal of Bible Literature* 75 (1957): 13–25.

19. Ibid.

20. Brown, *Seven Pillars of Creation*.

21. B. R. Foster, "Epic of Creation (1.111) (*Enūma Elish*)," in *The Context of Scripture*, ed. W. W. Hallo and K. L. Younger (Leiden: Brill, 1996), 390–402.

22. Martien Halvorson-Taylor, Religious Studies Department, University of Virginia, Charlottesville, personal communication.

23. E. L. Greenstein, "Texts from Ugarit Solve Biblical Puzzles," *Biblical Archaeology Review* 36 (2010): 48–53, 70.

24. Sarna, "Epic Substratum in the Prose of Job."

25. A. E. Killebrew, *Biblical Peoples and Ethnicity: An Archaeological Study of Egyptians, Canaanites, and Early Israel, 1300–1100 BCE* (Atlanta, Ga.: Society of Biblical Literature, 2005), 10–16.

26. Ugarit was a city from c. 8000 BCE. It was walled c. 6000 BCE.

2. Laying the Foundation of the Earth

1. K. Schifferdecker, *Out of the Whirlwind: Creation Theology in the Book of Job*, Harvard Theological Studies 61 (Cambridge, Mass.: Harvard University Press, 2008).

2. S. N. Kramer, *Sumerian Mythology: A Study of Spiritual and Literary Achievements in the Third Millennium BC*, rev. ed. (Philadelphia: University of Pennsylvania Press, 1997).

3. Schifferdecker, *Out of the Whirlwind*.

4. Ibid.

5. Cherokee Creation Story. J. W. Powell, *Nineteenth Annual Report of the Bureau of American Ethnology, Part 1, 1897–98* (Washington, D.C.: Government Printing Office, 1900), 242.

6. Bushongo (Central Africa) Creation Story. E. A. W. Budge, *Osiris, or the Egyptian Religion of Resurrection, Part 2* (1911; repr. Whitefish, Mt.: Kessinger, 2003), 364.

7. S. Sturluson, *Edda*, trans. A. Faulkes (London: J. M. Dent & Sons, 1987).

8. E. Chaisson, *Epic of Evolution: Seven Ages of the Cosmos* (New York: Columbia University Press, 2006).

9. C. M. Linton, *From Eudoxus to Einstein—a History of Mathematical Astronomy* (Cambridge: Cambridge University Press, 2004); M. A. Finocchiaro, *The Galileo Affair: A Documentary History* (Berkeley: University of California Press, 1989).

10. C. R. Chapman, "Surface Properties of Asteroids: A Synthesis of Polarimetry, Radiometry, and Spectrophotometry," *Icarus* 25 (1975): 104–130.

11. B. E. Clark, "New News and the Competing Views of Asteroid Belt Geology," *Lunar and Planetary Science* 27 (1996): 225–226.

12. M. Mueller et al., "21 Lutetia and Other M-Types: Their Sizes, Albedos, and Thermal Properties from New IRTF Measurements," *Bulletin of the American Astronomical Society* 37 (2005): 627.

13. C. T. Cunningham, *The First Asteroid: Ceres, 1801–2001* (Surfside, Fla.: Star Lab, 2002).

14. C. T. Russell et al., "*Dawn Discovery* Mission to Vesta and Ceres: Present Status," *Advances in Space Research* 38 (2006): 2043–2048.

15. M. Brown, *How I Killed Pluto and Why It Had It Coming* (New York: Random House, 2010).

16. http://web.gps.caltech.edu/~mbrown/dwarfplanets/.

17. A. P. Boss, "Evolution of the Solar Nebula: VI. Mixing and Transport of Isotopic Heterogeneity," *Astrophysical Journal* 616 (2004): 1265–1277.

18. ^{60}Fe is radioactive with a half-life of about 2.6 million years. The abundance of ^{60}Ni, the stable decay product of ^{60}Co, which in turn is the radioactive decay product of ^{60}Fe, found in iron meteorites is the remnant of the initiating supernova.

19. A. P. Boss and S. A. Keiser, "Who Pulled the Trigger: A Supernova of an Asymptotic Giant Branch," *Astrophysical Journal Letters* 717 (2010): L1–L5.

20. J. E. Chambers, "Planetary Accretion in the Inner Solar System," *Earth and Planetary Science Letters* 223 (2004): 241.

21. S. Barabash et al., "The Loss of Ions from Venus Through the Plasma Wake," *Nature* 450 (2007): 650–653.

22. Au, Co, Fe, Ir, Mn, Mo, Ni, Os, Pd, Pt, Re, Rh, and Ru.

23. Al, At, B, Ba, Be, Br, Ca, Cl, Cr, Cs, F, I, Hf, K, Li, Mg, Na, Nb, O, P, Rb, Sc, Si, Sr, Ta, Th, Ti, U, V, Y, Zr, W, and the lanthanides.

24. B. Wood, "The Formation and Differentiation of Earth," *Physics Today* 64 (2011): 40–45.

25. J. Papike, G. Ryder, and C. Shearer, "Lunar Samples," *Reviews in Mineralogy and Geochemistry* 36 (1998): 5.1–5.234.

26. A. N. Halliday, "Terrestrial Accretion Rates and the Origin of the Moon," *Earth and Planetary Science Letters* 176 (2000): 17–30; U. Wiechert et al., "Oxygen Isotopes and the Moon-Forming Giant Impact," *Science* 294 (2001): 345–348; R. Canup and E. Asphaug, "Origin of the Moon in a Giant Impact Near the End of the Earth's Formation," *Nature* 412 (2001): 708–712.

27. Wood, "The Formation and Differentiation of Earth."

28. Ibid.

29. GRAIL stands for Gravity Recovery and Interior Laboratory.

30. D. L. Pinti, "The Origin and Evolution of the Oceans," in *Lectures in Astrobiology*, ed. M. Gargaud et al. (Berlin: Springer Verlag, 2005), 1:83–211.

31. Uranium has two relatively common isotopes (^{235}U and ^{238}U) that are radioactive. Their radioactive decay eventually produces lead (symbol Pb): ^{207}Pb comes

from the decay sequence of ^{235}U; ^{206}Pb from the ^{238}U sequence. The half-life, the time for one-half of the material to undergo radioactive decay is 704 million years for the ^{235}U pathway and 4.47 billion years for the ^{238}U pathway. These two clocks start running when a rock forms, and the ratio of each uranium isotope and its lead decay products can be used as a clock. Zircons are particularly useful for such timing because the crystals easily include uranium but exclude lead. Thus the lead found in a zircon very likely comes from radioactive decay of uranium after the rock has formed.

32. S. J. Mojzsis, M. T. Harrison, and R. T. Pidgeon, "Oxygen-Isotope Evidence from Ancient Zircons for Liquid Water at the Earth's Surface 4,300 Myr Ago," *Nature* 409 (2001): 178–181; S. A. Wilde et al., "Evidence from Detrital Zircons for the Existence of Continental Crust and Oceans on the Earth 4.4 Gyr Ago," *Nature* 409 (2001): 175–178.

33. Wilde et al., "Evidence from Detrital Zircons."

34. Ibid.; T. Ushikubo et al., "Lithium in Jack Hills Zircons: Evidence for Extensive Weathering of Earth's Earliest Crust," *Earth and Planetary Science Letters* 272 (2001): 666–676.

35. N. H. Sleep, "The Hadean-Archaean Environment," *Cold Spring Harbor Symposium Perspectives on Biology* 2 (2010): a002527.

36. Pinti, "The Origin and Evolution of the Oceans."

37. L. A. Frank, J. B. Sigwarth, and J. D. Craven, "On the Influx of Small Comets Into the Earth's Atmosphere. I: Observations," *Geophysical Research Letters* 13 (1986): 303–306; D. Deming, "On the Possible Influence of Extraterrestrial Volatiles on Earth's Climate and the Origin of the Oceans," *Palaeogeography, Palaeoclimatology, and Palaeoecology* 146 (1999): 33–51; L. A. Frank and J. B. Sigwarth, "Trails of OH Emissions from Small Comets Near Earth," *Geophysical Research Letters* 24 (1997): 2435–2438.

38. C. E. J. de Ronde et al., "Fluid Chemistry of Archaean Seafloor Hydrothermal Vents: Implications for the Composition of Circa 3.2 Ga Seawater," *Geochim. Cosmochim. Acta* 61 (1997): 4025–4042; C. G. A. Harrison, "Constraints on Ocean Volume Change Since the Archaean," *Geophysical Research Letters* 26 (1999): 1913–1916.

39. Pinti, "The Origin and Evolution of the Oceans."

40. N. H. Sleep, K. Zahnle, and P. S. Neuhoff, "Initiation of Clement Surface Conditions on the Earliest Earth," *Proceedings of the National Academy of Sciences* 98 (2001): 3666–3672.

41. W. W. Rubey, "Geologic History of Seawater: An Attempt to State the Problem," *Geological Society of America Bulletin* 62 (1951): 1111–1147.

42. J. Geiss and G. Gloecker, "Abundance of Deuterium and Helium in the Protosolar Cloud," *Space Science Review* 84 (1998): 239–250.

43. D. Gautier and T. Owen, "Cosmological Implication of Helium and Deuterium Abundances of Jupiter and Saturn," *Nature* 302 (1983): 215–218.

44. A. Drouart et al., "Structure and Transport in the Solar Nebula from Constraints on Deuterium Enrichment and Giant Planets Formation," *Icarus* 140 (1999): 59–76.

45. The much more common ^1H, called protium, and ^2H, called deuterium or "heavy hydrogen."

46. D. Bockelée-Morvan et al., "Deuterated Water in Comet C/1996 B2 (Hyakutake) and Its Implications for the Origin of Comets," *Icarus* 193 (1988): 147–162; R. Meier et al., "A Determination of the DH_2O/H_2O Ratio in Comet C/1995O1 (Hale–Bopp)," *Science* 279 (1998): 842–844; P. Eberhardt, D. Krankowski, and R. R. Hodges, "The D/H and $^{18}O/^{16}O$ Ratios in Water from Comet P/Halley," *Astron. Astrophys.* 302 (1995): 301–316.

47. N. Dauphas, F. Robert, and B. Marty, "The Late Asteroidal and Cometary Bombardment of Earth as Recorded in Water Deuterium-to-Protium Ratio," *Icarus* 148 (2000): 508–512.

48. A. H. Delsemme, "The Deuterium Enrichment Observed in Recent Comets Is Consistent with the Cometary Origin of Seawater," *Planet. Space Sci.* 47 (1999): 125–131.

49. A. Morbidelli et al., "Source Regions and Timescales for the Delivery of Water to the Earth," *Meteorit. Planet. Sci.* 35 (2000): 1309–1320; N. Dauphas, "The Dual Origin of the Terrestrial Atmosphere," *Icarus* 165 (2003): 326–339; Pinti, "The Origin and Evolution of the Oceans."

50. N. H. Sleep et al., "Annihilation of Ecosystems by Large Asteroid Impacts on the Early Earth," *Nature* 342 (1989): 139–142.

51. K. K. Takai et al., "Cell Proliferation at 122°C and Isotopically Heavy CH_4 Production by a Hyperthermophilic Methanogen Under High-Pressure Cultivation," *Proceedings of the National Academy of Sciences* 105 (2008): 10949–10954.

52. N. R. Pace, "Time for a Change," *Nature* 441 (2006): 289.

53. Y. Koga and H. Morii, "Biosynthesis of Ether-Type Polar Lipids in Archaea and Evolutionary Considerations," *Microbiology and Molecular Biology Reviews* 71 (2007): 97–120.

54. E. F. DeLong, "Everything in Moderation: Archaea as 'Nonextremophiles.'" *Current Opinion in Genetics and Development* 8 (1998): 649–654.

55. E. F. DeLong and N. R. Pace, "Environmental Diversity of Bacteria and Archaea," *Systematic Biology* 50 (2001): 470–478.

56. G. Schleper et al., "*Picrophilus* gen. nov., fam. nov.: A Novel Aerobic, Heterotrophic, Thermoacidophilic Genus and Family Comprising Archaea Capable of Growth Around pH 0," *Journal of Bacteriology* 177 (1995): 7050–7059.

57. E. V. Pikuta et al., "*Carnobacterium pleistocaenium* sp. nov., a Novel Psychrotolerant, Facultative Anaerobe Isolated from Fox Tunnel Permafrost, Alaska," *International Journal of Evolution and Systematics in Microbiology* 55 (2005): 473–478.

58. E. V. Pikuta, R. B. Hoover, and J. Tang, "Microbial Extremophiles at the Limits of Life," *Critical Reviews in Microbiology* 33 (2007): 183–209.

59. Sleep, "The Hadean-Archaean Environment."

60. Sleep et al., "Annihilation of Ecosystems."

61. J. W. Schopf, "Fossil Evidence of Archaean Life," *Philosophical Transactions of the Royal Society Series B* 361 (2006): 869–885, lists forty-eight stromatite fossils, fourteen Archaean geological units with forty different types of microfossils and organic geochemicals, and thirteen ancient (3.2 to 3.5 billion years old) Archaean geological units.

62. E. J. Nisbet and N. H. Sleep, "The Habitat and Nature of Early Life," *Nature* 409 (2001): 1083–1091.

63. M. Gogarten-Boeckels, E. Hilario, and J. P. Gogarten, "The Effects of Heavy Meteorite Bombardment on the Early Evolution—The Emergence of the Three Domains of Life," *Origins of Life and Evolution of Biospheres* 25 (1992): 251–264.

64. Nisbet and Sleep, "The Habitat and Nature of Early Life."

65. N. H. Sleep and K. Zahnle, "Refugia from Asteroid Impacts on Early Mars and the Early Earth," *Journal of Geophysical Research* 103 (1998): 28529–28544; Sleep et al., "Annihilation of Ecosystems."

66. The National Advisory Committee for Aeronautics (NACA) was founded in 1915 to encourage aeronautical research. On July 18, 1958, in the post-*Sputnik* era, NACA's National Integrated Missile and Space Vehicle Development Program reported that their "Type IV" launch vehicle (equivalent to the later Saturn C-3 rocket) would be used to send a 2,300 kg "probe" to Mars in January 1967. In October 1958, NACA was dissolved to become part of NASA. The NACA report evolved to be the Voyager and later the Viking missions. See M. Erickson, *Into the Unknown Together—The DOD, NASA, and Early Spaceflight* (Maxwell Air Force Base, Ala.: Air University Press, 2005).

67. J. E. Lovelock, "A Physical Basis for Life Detection Experiments," *Nature* 207 (1965): 568–570.

68. J. E. Lovelock, *Gaia: A New Look at Life on Earth* (Oxford: Oxford University Press, 1979); J. E. Lovelock, *The Ages of Gaia* (New York: Norton, 1988).

69. E. G. Nisbet et al., "The Age of Rubisco: The Evolution of Oxygenic Photosynthesis," *Geobiology* 5 (2007): 311–325.

70. The Plantomycetes phylum of bacteria are nitrifying and denitrifying bacteria that can convert nitrite and ammonium to nitrogen gas and water in anaerobic (oxygen-free) environments.

71. J. Zalasiewicz and M. Williams, *The Goldilocks Planet: The Four-Billion-Year Story of the Earth's Climate* (Oxford: Oxford University Press, 2012).

72. A. A. Pavlov et al., "Greenhouse Warming by CH_4 in the Atmosphere of Early Earth," *Journal of Geophysical Research* 105 (2000): 11981–11990; D. E. Canfield,

K. S. Habicht, and B. Thamdrup, "The Archaean Sulfur Cycle and the Early History of Atmospheric Oxygen," *Science* 288 (2000): 658–661; D. C. Catling, K. J. Zahnle, and C. P. McKay, "Biogenic Methane, Hydrogen Escape, and the Irreversible Oxidation of Early Earth," *Science* 293 (2001): 839–843; C. Sagan and C. Chyba, "The Early Faint Sun Paradox: Organic Shielding of Ultraviolet-Labile Greenhouse Gases," *Science* 276 (1997): 1217–1221; M. G. Trainer et al., "Haze Aerosols in the Atmosphere of Early Earth: Manna from Heaven," *Astrobiology* 4 (2004): 409–419.

73. J .D. Haqq-Misra et al., "A Revised, Hazy Methane Greenhouse for the Archean Earth," *Astrobiology* 8 (2008): 1127–1137.

74. Nisbet et al., "The Age of Rubisco."

75. L. J. Rothschild, "The Evolution of Photosynthesis . . . Again?" *Philosophical Transactions of the Royal Society Series B* 363 (2008): 2787–2801.

76. W. P. Brown, *Seven Pillars of Creation: The Bible, Science, and the Ecology of Wonder* (New York: Oxford University Press, 2010).

77. M. S. Smith, *The Early History of God: Yahweh and the Other Deities in Ancient Israel*, 2nd ed. (Grand Rapids, Mich.: Eerdmans, 2002).

78. M. S. Smith, *God in Translation: Deities in Cross-Cultural Discourse in the Biblical World* (Tübingen: Mohr Siebek, 2008).

79. Brown, *Seven Pillars of Creation*.

3. Taming the Unicorn, Yoking the Aurochs: Animal and Plant Domestication and the Consequent Alteration of the Surface of the Earth

1. The Septuagint is a translation of the Hebrew Bible into Greek developed over the third to second century BCE in Alexandria. The Vulgate is a Latin translation from the fourth century CE.

2. *Persica* and *Indica* are abstracted in Photius's *Bibliotheca* or *Myriobiblon*, a ninth-century CE work. Photius I was the patriarch of Constantinople from 858 to 867 CE. In *Bibliotheca*, he abstracted 279 ancient books that he had read.

3. T. S. Brown, "The Reliability of Megathenes," *American Journal of Philology* 76 (1955): 18–33.

4. J. H. Freese, *The Library of Photius*, vol. 1 (Suffolk: Richard Clay and Sons, 1920), 117.

5. Ibid.

6. Called the *Physiologus* because its chapters began with the phrase, "The physiologus says . . ."; "physiologus" means naturalist or natural philosopher.

7. F. McCulloch, *Medieval Latin and French Bestiaries* (Chapel Hill: University of North Carolina Press, 1962).

8. The *Aberdeen Bestiary*, Historical Collections, King's College, University of Aberdeen, University Library MS 24. http://www.abdn.ac.uk/bestiary.

9. http://en.wikisource.org/wiki/The_Notebooks_of_Leonardo_Da_Vinci.

10. Males with two tusks occur but they are rare; females also rarely can develop a tusk.

11. The coat of arms referred to here is that of the monarchs of Scotland used until the Acts of Union in 1707. The Royal Coat of Arms of the United Kingdom, among other changes, replaces one of the unicorns with an English lion. The Scottish unicorn is *sinister* (or on the left side when looking out from the coat of arms) in the United Kingdom but is *dexter* (right side) for its use in Scotland.

12. Horses are in the order Perissodactyla, the odd-toed ungulates, versus the order Artiodactyla, the even-toed ungulates.

13. Males ranged from 180 to 160 cm; the smaller females were around 150 cm. C. T. Van Vuure, *Retracing the Aurochs: History, Morphology, and Ecology of an Extinct Wild Ox* (Sofia/Moscow: Pensoft, 2005).

14. E. Thenius, *Grundzüge der Faunen–und Verbreitungsgeschichte der Säugetiere. Eine historische Tiergeographie* (Jena: V.E.B. Gustav Fischer, 1980).

15. W. von Koenigswald, "Palökologie und Vorkommen des pleistozänen Auerochsen (*Bos primigenius* Bojanus, 1827) im Vergleich zu den grossen Rindern des Pleistozäns," in *Archäologie und Biologie des Auerochsen*, ed. G.-C. Weniger (Mettman: Neanderthal-Museum, 1999), 23–33.

16. G. Curtis, *The Cave Painters: Probing the Mysteries of the World's First Artists* (New York: Knopf, 2006), 102.

17. See discussion in chapter 1.

18. Translation from W. A. McDevitte and W. S. Bohn, *The Gallic Wars* (New York: Harper and Brothers, 1869). See http://classics.mit.edu/index.html.

19. M. Rokosz, *History of the Aurochs (Bos Taurus primigenius) in Poland* (2006). http://users.aristotle.net/~swarmack/aurohist.html.

20. F. Galton, "The First Steps Toward the Domestication of Animals," *Transactions of the Ethnological Society, London* n.s. 3 (1865): 122–138; J. Clutton-Brock, *A Natural History of Domesticated Mammals* (Austin: University of Texas Press, 1989).

21. S. J. Crockford and Y. V. Kuzman, "Comments on Germonpré et al., *Journal of Archaeological Science* 36, 2009 'Fossil Dogs and Wolves from Palaeolithic Sites in Belgium, the Ukraine, and Russia: Osteometry, Ancient DNA, and Stable Isotopes,' and Germonpré, Lázkičková-Galetová, and Sablin, *Journal of Archaeological Science* 39, 2012 'Palaeolithic Dog Skulls at the Gravettian Předmostí site, the Czech Republic,'" *Journal of Archeological Science* 39 (2012): 2797–2801; A. S. Druzhkova et al., "Ancient DNA Analysis Affirms the Canid from Altai as a Primitive Dog" (2013), PLOS One *(3): e57754. doi:10.1371/journal.pone.0057754.

22. R. Coppinger and L. Coppinger, *Dogs: A New Understanding of Canine Origin, Behavior, and Evolution* (Chicago: University of Chicago Press, 2001); C. A. Driscoll, D. W. Macdonald, and S. J. O'Brien, "From Wild Animals to Domestic Pets, an Evolutionary View of Domestication," *Proceedings of the National Academy of Sciences* 106 (2009): 9971–9976.

23. W. E. Roth, "An Introductory Study of the Arts, Crafts, and Customs of the Guiana Indians," in *Accompanying Paper to the Thirty-Eighth Annual Report of the Bureau of American Ethnology to the Secretary of the Smithsonian Institution 1916– 1917*, transmitted by F. W. Hodge (Washington, D.C.: Government Printing Office, 1924), 25–745.

24. Ibid., 551–556.

25. Bear cubs in North America: F. Galton, "The First Steps Toward the Domestication of Animals," *Transactions of the Ethnological Society of London* n.s. 3 (1865): 122–138. Bear cubs in China: J. G. Frazer, *The Golden Bough: A Study in Magic and Religion* (London: Macmillan, 1922). J. Serpell, "Pet-Keeping and Animal Domestication: A Reappraisal," in *The Walking Larder*, ed. J. Clutton-Brock (London: Unwin Hyman, 1989), 10–21. This article reviews the role of women in taming a wide variety of animals for pets in a wide range of cultures. J. Macrae, *With Lord Bryon in the Sandwich Islands in 1825: Being extracts from the MS Diary of James Macrae, Scottish Botanist* (Honolulu: W. F. Wilson, 1922); M. Titcomb, *Dog and Man in the Ancient Pacific*, Bernice P. Bishop Museum Special Publication 59 (Honolulu, 1969).

26. H. H. Shugart, *How the Earthquake Bird Got Its Name and Other Tales of an Unbalanced Nature* (New Haven, Conn.: Yale University Press, 2004).

27. A. Newsome (personal communication), who conducted an important series of CSIRO studies on the biology and genetics of dingoes, relates that dingo pups taken into captivity after their eyes opened were always difficult to handle. None ever became tame enough to answer to commands. When given to the public after the lab tests were finished, they were always returned with complaints. Pups with their eyes closed at initial captivity and subsequently given to the public were never returned.

28. J. K. Gollan, *Prehistoric Dingo* (Ph.D. thesis, Australian National University, 1982).

29. There are earlier reported records (c. 8500 BP) for dingoes in Australia for isolated teeth, which may have dropped from a shallower level during the excavation of the archeological site. Olsen feels the evidence is too meager for a definitive determination, and subsequent scientists have largely upheld this opinion. S. J. Olsen, *Origins of the Domesticated Dog: The Fossil Record* (Tucson: University of Arizona Press, 1985).

30. Corbett favors the Thai dog as a source for the dingo; Gollan, the Indian pariah dog. L. K. Corbett, *The Dingo in Australia and Asia* (Sydney: University of New South Wales Press, 1995); Gollan, *Prehistoric Dingo*.

31. Corbett, *The Dingo in Australia and Asia*.

32. T. F. Flannery, *The Future Eaters: An Ecological History of the Australian Lands and People* (New York: Grove, 2003).

33. Titcomb, *Dog and Man in the Ancient Pacific*.

34. C. Vilà et al., "Man and His Dog," *Science* 278 (1997): 206–207; C. Vilà et al., "Multiple and Ancient Origins of the Domestic Dog," *Science* 276 (1997): 1687–1689.

35. P. Savolainen et al., "A Detailed Picture of the Origin of the Australian Dingo, Obtained from the Study of Mitochondrial DNA," *Proceedings of the National Academy of Sciences* 33 (2004): 12387–12390.

36. A. E. Newsome and L. K. Corbett, "The Identity of the Dingo. 3. The Incidence of Dingoes, Dogs, and Hybrids and Their Coat Colors in Remote and Settled Areas of Australia," *Australian Journal of Zoology* 33 (1985): 363–373; M. J. Daniels and L. Corbett, "Redefining Introgressed Protected Mammals: When Is a Wildcat a Wild Cat and a Dingo a Wild Dog?" *Wildlife Research* 30 (2003): 213–218.

37. Clutton-Brock, *A Natural History of Domesticated Mammals*.

38. C. S. Darwin, *On the Origin of Species by Means of Natural Selection, or the Preservation of Favoured Races in the Struggle for Life* (London: Murray, 1859).

39. J. Z. Wilczynski, "On the Presumed Darwinism of Alberruni Eight Hundred Years Before Darwin," *Isis* 50 (1959): 459–466.

40. D. K. Belyaev, "Destabilizing Selection as a Factor in Domestication," *Journal of Heredity* 70 (1979): 301–308.

41. L. Trut, "Early Canid Domestication: The Farm-Fox Experiment," *American Scientist* 87 (1999): 160.

42. Ibid.

43. L. Trut, I. Oskina, and A. Kharlamova, "Animal Evolution During Domestication: The Domesticated Fox as an Example," *BioEssays* 31 (2009): 349–360.

44. G. Nobis, "Der älteste Haushund lebte vor 14,000 Jahren," *Umshau in Wissenschaft und Technik* 19 (1979): 610; M. Street, "The Archaeology of the Pleistocene/Holocene Transition in the Northern Rhineland, Germany," *Quaternary International* 50 (1998): 46–67. S. J. M. Davis and F. R. Valla, "Evidence for Domestication of the Dog 12,000 Years Ago in the Natufian of Israel," *Nature* 276 (1978): 608–610; T. Dayan, "Early Domesticated Dogs of the Near East," *Journal of Archeological Science* 21 (1994): 633–640.

45. References and other documentation for a variety of sites can be found in Olsen, *Origins of the Domesticated Dog*. For the New World c. 8,400 years ago, see D. F. Morey and M. Wiant, "Early Holocene Domestic Dog Burials from the North American Midwest," *Current Anthropology* 33 (1992): 224–229. For genetic evidence that the dogs of the New World originated from Asian domesticated wolves that were subsequently brought by humans to the New World, see J. A. Leonard et al., "Ancient DNA Evidence for Old World Origin of New World Dogs," *Science* 298 (2002): 1613–1616.

46. J.-F. Pang et al., "mtDNA Data Indicate a Single Origin for Dogs South of Yangtze River, Less Than 16,300 Years Ago, from Numerous Wolves," *Molecular Biology and Evolution* 26 (2009): 2849–2864.

47. Ibid.; C. F. W. Higham, A. Kijngam, and B. F. J. Manly, "An Analysis of Prehistoric Canid Remains from Thailand," *Journal Archeological Society* 7 (1980): 149–165; S. Ren, "Important Results Regarding Neolithic Cultures in China Earlier Than 5000 B.C.," *Kaogu* 1 (1995): 37–49 (in Chinese); F. J. Simoons, "Dog Flesh," in *Food in China: A Cultural and Historical Inquiry* (Boston: CRC, 1991), 200–252.

48. See J. P. Scott, O. S. Elliot, and B. E. Ginsburg, "Man and His Dog," *Science* 278 (1997): 205; and N. E. Federoff and R. M. Nowak, "Man and His Dog," *Science* 278 (1997): 205; which are in response to Vilà et al., "Multiple and Ancient Origins of the Domestic Dog."

49. C. S. Troy et al., "Genetic Evidence for Near-Eastern Origins of European Cattle," *Nature* 410 (2001): 1088–1091.

50. F. J. Simoons and E. S. Simoons, *A Ceremonial Ox of India: The Mithan in Nature, Culture, and History* (Madison: University of Wisconsin Press, 1968).

51. Clutton-Brock, *A Natural History of Domesticated Mammals.*

52. Ibid.

53. M. A. Levine, "Botai and the Origins of Horse Domestication," *Journal of Anthropological Archaeology* 18 (1999): 29–78.

54. F. Wendorf et al., "The Earliest Pastoralism in Egypt," *Rechere* 21 (1990): 436–445.

55. D. Perkins, "Fauna of Çatal Hüyük: Evidence for Early Cattle Domestication in Anatolia," *Science* 164 (1969): 177–179.

56. J. L. Angel, "Early Neolithic Skeletons from Çatal Hüyük: Demography and Pathology," *Anatolian Studies* 21 (1971): 77–98.

57. Perkins, "Fauna of Çatal Hüyük."

58. The Anatolian mufflon (*Ovis orientalis anatolica*) was the wild sheep. The other wild animal, the onager (*Equus hemionus*), resembles a donkey. It is slightly larger than the donkey but has other characteristics more like those of horses.

59. J. Mellaart, "Çatal Hüyük: A Neolithic Town in Anatolia," in *New Aspects of Archaeology*, ed. M. Wheeler (New York: McGraw-Hill, 1967).

60. J. L. Angel, "Porotic Hyperostosis, Anaemias, Malarias, and Marshes in the Prehistoric Eastern Mediterranean," *Science* 153 (1966): 760–763.

61. Ibid.

62. Mellaart, "Çatal Hüyük."

63. R. T. Loftus et al., "Evidence for Two Independent Domestications of Cattle," *Proceedings of the National Academy of Sciences* 91 (1994): 2757–2761.

64. D. G. Bradley et al., "Mitochondrial Diversity and Origins of African and European Cattle," *Proceedings of the National Academy of Sciences* 93 (1996): 5131–5135;

C. S. Troy et al., "Genetic Evidence for Near-Eastern Origins of European Cattle," *Nature* 401 (2001): 1088–1091. H. Mannen et al., "Independent Mitochondrial Origin and Historical Genetic Differentiation in North Eastern Asian Cattle," *Molecular Phylogenetic Evolution* 32 (2004): 539–544.

65. J. Zilhao, "The Spread of Agropastoral Economies Across Mediterranean Europe: A View from the Far West," *Journal of Mediterranean Archaeology* 6 (1993): 5–53; A. Beja-Pereiraa et al., "The Origin of European Cattle: Evidence from Modern and Ancient DNA," *Proceedings of the National Academy of Sciences* 103 (2006): 8113–8118.

66. R. Kyselý, "Aurochs and Potential Crossbreeding with Domestic Cattle in Central Europe in the Eneolithic Period. A Metric Analysis of Bones from the Archaeological Site of Kutná Hora-Denemark (Czech Republic)," *Anthropozoologica* 43 (2008): 7–37.

67. Loftus et al., "Evidence for Two Independent Domestications of Cattle."

68. J. Kantanen et al., "Maternal and Paternal Genealogy of Eurasian Taurine Cattle (*Bos taurus*)," *Heredity* 103 (2009): 404–415.

69. R. J. Timmins et al., "*Bos javanicus*," in IUCN Red List of Threatened Species, version 2010.4. http://www.iucnredlist.org.

70. Clutton-Brock, *A Natural History of Domesticated Mammals*.

71. Ibid.

72. R. B. Harris and D. Leslie, "*Bos mutus*," in IUCN Red List of Threatened Species, version 2010.4. http://www.iucnredlist.org.

73. H. Epstein, *Domestic Animals of China* (England: Commonwealth Agricultural Bureaux, 1969).

74. Clutton-Brock, *A Natural History of Domesticated Mammals*.

75. F. E. Zeuner, *A History of Domesticated Animals* (London: Hutchinson, 1963).

76. S. Hedges et al., "*Bubalus arnee*," in IUCN Red List of Threatened Species, version 2010.4. http://www.iucnredlist.org.

77. Clutton-Brock, *A Natural History of Domesticated Mammals*.

78. Ibid.

79. S. Bökönyi, *History of Domestic Animals in Central and Eastern Europe* (Budapest: Akadémiai Kiadó, 1974).

80. Z. Naveh and J. Dan, "The Human Degradation of the Mediterranean Landscape in Israel: Ecological and Evolutionary Perspectives," in *Mediterranean-Type Ecosystems: Origin and Structure,* ed. F. di Castri and H.A. Mooney, vol. 7 of *Ecological Studies: Analysis and Synthesis* (Berlin: Springer-Verlag, 1973), 373–389.

81. P. L. Fall, C. A. Lindquist, and S. E. Falconer, "Fossil Hyrax Middens from the Middle East: A record of Paleovegetation and Human Disturbance," in *Packrat Middens: The Last Forty Thousand Years of Change*, ed. J. L. Betancourt et al. (Tucson: University of Arizona Press, 1990), 408–427.

82. V. G. Childe, *Man Makes Himself* (London: Watts and Co., 1936).

83. P. L. Fall, S. E. Falconer, and L. Lines, "Agricultural Intensification and the Secondary Products Revolution Along the Jordan Rift," *Human Ecology* 30 (2002): 445–482.

84. U. Baruch, "Palynological Evidence of Human Impact on the Vegetation as Recorded in Late Holocene Lake Sediments in Israel," in *Man's Role in the Shaping of the Eastern Mediterranean Landscape*, ed. S. Bottema, G. Entjes-Nieborg, and W. van Zeist (Rotterdam: Balkema, 1990), 283–293.

85. U. Baruch and S. Bottema, "Palynological Evidence for Climate Changes in the Levant 17,000–9,000 BP," in *The Natufian Culture in the Levant*, ed. O. Bar-Yosef and F. Valla, International Monographs in Prehistory (Ann Arbor, Mich., 1991), 11–20.

86. A. G. Sherratt, "Plough and Pastoralism: Aspects of the Secondary Products Revolution," in *Pattern of the Past: Studies in Memory of David Clarke*, ed. I. Hodder, G. Isaac, and N. Hammond (Cambridge: Cambridge University Press, 1980), 261–305; A. G. Sherratt, "The Secondary Exploitation of Animals in the Old World," *World Archaeology* 15 (1983): 90–103.

87. D. Zohary and M. Hopf, *Domestication of Plants in the Old World* (Oxford: Clarendon, 1988); D. Zohary and P. Spiegel-Roy, "The Beginnings of Fruits Growing in the Old World," *Science* 187 (1975): 319–327.

88. M. E. Kislev, A. Hartmann, and O. Bar-Yosef, "Early Domesticated Fig in the Jordan Valley," *Science* 312 (2006): 1372–1374.

89. S. E. Falconer and P. L. Fall, "Human Impacts on the Environment During the Rise and Collapse of Civilization in the Eastern Mediterranean," in *Late Quaternary Environments and Deep History: A Tribute to Paul S. Martin*, ed. J. L. Mead and D. Steadman (Hot Springs, Ark.: The Mammoth Site, 1995). K. Kenyon, *Excavations at Jericho*, vol. 1 (London: British School of Archaeology at Jerusalem, 1960). O. Bar-Yosef, "The Walls of Jericho: An Alternative Interpretation," *Current Anthropology* 27 (1986): 157–162. G. Rollefson, A. Simmons, and Z. Kaffafi, "Neolithic Cultures at 'Ain Ghazal, Jordan," *Journal of Field Archaeology* 19 (1992): 443–470. H. Mahasneh, "A PPNB Settlement at Es-Safiya in Wadi el-Mujib," in *Studies in the History and Archaeology of Jordan*, vol. 6 (Amman: Department of Antiquities of Jordan, 1997), 227–234. A. Simmons and M. Najjar, "Current Investigations at Ghwair I, a Neolithic Settlement in Southern Jordan," *Neolithics* 98 (1999): 5–7.

90. I. Kohler-Rollefson, "A Model for the Development of Nomadic Pastoralism on the Trans-Jordanian Plateau," in *Pastoralism in the Southern Levant: Archaeological Materials in Anthropological Perspectives*, ed. O. Bar-Yosef and A. Khazarov, Monographs in World Archaeology 10 (Madison, Wis.: Prehistory Press, 1992), 11–18. I. Kohler-Rollefson and G. O. Rollefson, "The Impact of Neolithic Subsistence Strategies on the Environment: The Case of 'Ain Ghazal, Jordan," in *Man's Role in the*

Shaping of the Eastern Mediterranean Landscape, ed. S. Bottema, G. Entjes-Nieborg, and W. van Zeist (Rotterdam: Balkema, 1990), 3–14.

91. W. M. Denevan, "The Pristine Myth: The Landscape of the Americas in 1492," *Annals of the Association of American Geographers* 82 (1992): 369–385; T. M. Whitmore and B. L. Turner II, "Landscapes of Cultivation in Mesoamerica on the Eve of the Conquest," *Annals of the Association of American Geographers* 82 (1992): 402–425.

92. J. E. Bergmann, "The Distribution of Cacao Cultivation in Pre-Columbian America," *Annals of the Association of American Geographers* 59 (1969): 85–96; B. L. Turner II and C. H. Miksicek, "Economic Plant Species Associated with Prehistoric Agriculture in the Maya Lowlands," *Economic Botany* 38 (1984): 179–193.

93. N. D. Burrows et al., "Evidence of Altered Fire Regimes in the Western Desert Region of Australia," *Conservation Science Western Australia* 5 (2006): 272–284; N. D. Burrows and P. E. Christensen, "A Survey of Aboriginal Fire Patterns in the Western Desert of Australia," in *Fire and the Environment: Ecological and Cultural Perspectives*, ed. S. C. Nodvin and T. E. Waldrop, U.S. Dept. of Agriculture Forest Service, General Technical Report SE-69 (Asheville, N.C., 1990), 20–24.

94. R. A. Bradstock, J. E. Williams, and A. M. Gill, *Flammable Australia: The Fire Regimes and Biodiversity of a Continent* (Cambridge: Cambridge University Press, 2002); M. Letnic et al., "The Responses of Small Mammals and Lizards to Post-Fire Succession and Rainfall in Arid Australia," *Journal of Arid Environments* 59 (2004): 85–114; R. W. Braithwaite, "Biodiversity and Fire in the Savanna Landscape," in *Biodiversity and Savanna Ecosystem Processes*, ed. O. Solbrig, E. Medina, and J. F. Silva (Berlin: Springer, 1996), 121–140.

95. D. Bowman, A. Walsh, and L. D. Prior, "Landscape Analysis of Aboriginal Fire Management in Central Arnhem Land, North Australia," *Journal of Biogeography* 31 (2004): 207–223.

96. R. Bliege Bird et al., "The 'Fire Stick Farming' Hypothesis: Australian Aboriginal Foraging Strategies, Biodiversity, and Anthropogenic Fire Mosaics," *Proceedings of the National Academy of Sciences* 105 (2008): 14796–14801.

97. D. Bird et al., "Aboriginal Burning Regimes and Hunting Strategies in Australia's Western Desert," *Human Ecology* 33 (2005): 443–464.

98. Bliege Bird et al., "The 'Fire Stick Farming' Hypothesis."

99. Ibid.

100. Crosby is credited with coining this very evocative term. See A. W. Crosby, *Ecological Imperialism: The Biological Expansion of Europe, 900–1900* (Cambridge: Cambridge University Press, 1986).

101. Many of the plants in this "kit" would not have survived the frosts of temperate New Zealand. This is likely to have enforced a heightened dependency on the native flora and fauna for the Maori of New Zealand.

102. See P. A. Cox and S. A. Banack, eds., *Islands, Plants, and Polynesians: An Introduction to Polynesian Ethnobotany* (Portland, Ore.: Dioscorides, 1991). This volume contains several chapters outlining both the use and the transportation of plants by Polynesians and is a useful recent reference summarizing this topic.

103. P. V. Kirch, "Changing Landscapes and Sociopolitical Evolution in Mangaia, Central Polynesia," in *Historical Ecology in the Pacific Islands*, ed. P. V. Kirch and T. L. Hunt (New Haven, Conn.: Yale University Press, 1995), 147–165.

104. D. W. Steadman, "Prehistoric Extinctions of Pacific Island Birds: Biodiversity Meets Zooarcheology," *Science* 267 (1995): 1123–1131.

105. Thomas Jefferson, *Notes on the State of Virginia*, anonymously published in Paris. http://web.archive.org/web/20080914030942/http://etext.lib.virginia.edu /toc/modeng/public/JefVirg.html.

106. In a paper read before the Connecticut Academy of Sciences in 1799. The text of the address can be found in N. Webster, *A Collection of Papers on Political, Literary, and Moral Subjects* (New York: B. Franklin, 1843).

107. H. H. Shugart and F. I. Woodward, *Global Change and the Terrestrial Biosphere: Achievements and Challenges* (Oxford: Wiley-Blackwell, 2011).

108. Webster, *A Collection of Papers on Political, Literary, and Moral Subjects*

109. G. B. Bonan, "Frost Followed the Plow: Impacts of Deforestation on the Climate of the United States," *Ecological Applications* 9 (1999): 1305–1315.

110. B. Keen, *The Life of Admiral Christopher Columbus* (New Brunswick, N.J.: Rutgers University Press, 1959); R. A. Anthes, "Enhancement of Convective Precipitation by Mesoscale Variations in Vegetative Covering in Semiarid Regions," *Journal of Climate and Applied Meteorology* 23 (1984): 540–553.

111. B. P. Hayden, "Ecosystem Feedbacks on Climate at the Landscape Scale," *Philosophical Transactions of the Royal Society of London Series B.* 353 (1998): 5–18.

4. Freeing the Onager: Feral and Introduced Animals

1. J. Clutton-Brock, *Horse Power: A History of the Horse and Donkey in Human Societies* (Cambridge, Mass.: Harvard University Press, 1993).

2. P. D. Moehlman et al., "*Equus africanus*," in IUCN Red List of Threatened Species, version 2010.4. http://www.iucnredlist.org.

3. H. P. Uerpmann, "*Equus africanus* in Arabia," in *Equids in the Ancient World*, vol. 2, ed. R. H. Meadows and H. P. Uerpmann (Weisbaden: Dr Reichert, 1986); M. I. Grinder, P. R. Krausman, and R. S. Hofmann, *Equus asinus. Mammalian Species* (n.p.: American Society of Mammalogists, 2006), 794:1–9.

4. Clutton-Brock, *Horse Power*.

5. P. D. Moehlman, N. Shah, and C. Feh, "*Equus hemionus*," in IUCN Red List of Threatened Species, version 2010.4. http://www.iucnredlist.org.

6. Weight of 290 kilograms (640 pounds) and head-body length of 2.1 meters (6.9 feet).

7. M. S. Dimand, "A Sumerian Sculpture of the Third Millennium BC," *Metropolitan Museum of Art Bulletin* n.s. 3 (1945): 253–256.

8. M. Hilzheimer, *Animal Remains from Tell Asmar*, Studies in Ancient Oriental Civilization 20 (Chicago: University of Chicago Press, 1941).

9. F. H. Zeuner, *A History of Domesticated Animals* (London: Hutchinson, 1963).

10. H. Epstein, "Ass, Mule, and Onager," in *Evolution of Domesticated Mammals*, ed. I. L. Mason (London: Longman, 1984), 174–184.

11. Ibid.

12. Ibid.; Zeuner, *A History of Domesticated Animals*.

13. J. Clutton-Brock, *A Natural History of Domesticated Animals* (Austin: University of Texas Press, 1989); Clutton-Brock, *Horse Power*.

14. L. Woolley, *Excavations at Ur: A Record of Twelve Years Work* (London: Ernest Benn, 1955).

15. Hilzheimer, *Animal Remains from Tell Asmar*.

16. Clutton-Brock, *Horse Power*.

17. Zeuner, *A History of Domesticated Animals*.

18. M. A. Levine, "Botai and the Origins of Horse Domestication," *Journal of Anthropological Archaeology* 18 (1999): 29–78.

19. A. K. Outram et al., "The Earliest Horse Harnessing and Milking," *Science* 323 (2009): 1332–1335.

20. Ibid.

21. G. Lindgren et al., "Limited Number of Patrilines in Horse Domestication," *Nature Genetics* 36 (2004): 335–336.

22. C. Vilà et al., "Widespread Origins of Domestic Horse Lineages," *Science* 291 (2001): 474–477.

23. T. Kavar and P. Dovč, "Domestication of the Horse: Genetic Relationships Between Domestic and Wild Horses," *Livestock Science* 116 (2008): 1–14.

24. Ibid.

25. Clutton-Brock, *Horse Power*.

26. J. N. Postgate, "The Equids of Sumer, Again," in *Equids in the Ancient World*, ed. R. H. Meadows and H. P. Uerpmann (Weisbaden: Dr Reichert, 1986), 194–206.

27. Ibid.

28. J. Clutton-Brock, *A Natural History of Domesticated Mammals* (Cambridge: Cambridge University Press, 1999).

29. Postgate, "The Equids of Sumer, Again."

30. J. Zarins, "The Domesticated Equidae of Third Millennium BC Mesopotamia," *Journal of Cuneiform Studies* 30 (1978): 3–17.

31. Moehlman, Shah, and Feh, "*Equus hemionus.*"

32. L. Boyd, W. Zimmermann, and S. R. B. King, "*Equus ferus,*" in IUCN Red List of Threatened Species, version 2010.4. http://www.iucnredlist.org.

33. P. Grubb, "Order Perissodactyla," in *Mammal Species of the World: A Taxonomic and Geographic Reference*, 3rd ed., ed. D. E. Wilson and D. M. Reeder (Baltimore, Md.: Johns Hopkins University Press, 2005), 629–636.

34. Clutton-Brock, *Horse Power.*

35. Moehlman, Shah, and Feh, "*Equus hemionus.*"

36. R. T. Paine, "Food Web Complexity and Species Diversity," *American Naturalist* 100 (1966): 65–75.

37. Ibid.

38. R. N. Owen-Smith, *Megaherbivores: The Influence of Very Large Body Size on Ecology*, Cambridge Studies in Ecology (Cambridge: Cambridge University Press, 1988).

39. N. S. Ribeiro et al., "Aboveground Biomass and Leaf Area Index (LAI) Mapping for Niassa Reserve, Northern Mozambique," *Journal of Geophysical Research-Biogeosciences* 113 (2008), G02S02, doi:10.1029/207JG000550; N. S. Ribeiro, H. H. Shugart, and R. A. Washington-Allen, "Ecological Dynamics of Miombo Woodlands Within Niassa Reserve, Northern Mozambique," *Forest Ecology and Management* 255 (2008): 1626–1636.

40. L. van der Pijl, *Principles of Dispersal in Higher Plants* (Berlin: Springer-Verlag, 1972).

41. R. A. Hynes and A. K. Chase, "Plants, Sites, and Domiculture: Aboriginal Influence Upon Plant Communities in Cape York Peninsula," *Archeology in Oceania* 17 (1982): 38–50.

42. Ibid.

43. D. W. McCreery, "Flotation of the Bab edh-Dhra and Numeria Plant Remains," *Annuals of the American School of Oriental Research* 46 (1979): 165–169.

44. J. P. Grime, *Plant Strategies, Vegetation Processes, and Ecosystem Properties*, 2nd ed. (Chichester: John Wiley and Sons, 2002).

45. G. Ladizinsky, *Plant Evolution Under Domestication* (Dordrecht: Kluwer, 1998).

46. Ibid.

47. Ibid.

48. A. Knight and L. Ruddock, eds., *Advanced Research Methods in the Built Environment* (Oxford: Wiley-Blackwell, 1998).

49. For the rat, the likely origin is northern China: H. N. Southern, *The Handbook of the British Mammals* (Oxford: Blackwell Scientific, 1964); T. H. Yoshida,

Cytogenetics of the Black Rat: Karyotype Evolution and Species Differentiation (Tokyo: University of Tokyo Press, 1980). For the mouse, the likely origin is northern India: P. Boursot et al., "Origin and Radiation of the House Mouse: Mitochondrial DNA Phylogeny," *Journal of Evolutionary Biology* 9 (1996): 391–415.

50. E. P. Russell, *War and Nature* (Cambridge: Cambridge University Press, 2001).

51. Pliny the Elder, *Natural History. Book 19, Chapter 58: The Proper Remedies for These Maladies. How Ants Are Best Destroyed. The Best Remedies Against Caterpillars and Flies* [c. 77 CE], trans. J. Bostock and H. T. Riley (London: Taylor and Francis). Perseus Digital Library, Tufts University: http://www.perseus.tufts.edu/hopper/.

52. E. E. Holton, "Insecticides and Fungicides," *Industrial and Engineering Chemistry* 18 (1926): 931–933.

53. J. Powell, "Pyrethrum 1917–1931," *Soap and Sanitary Chemicals* 7 (1931): 93.

54. T. R. Dunlap, *DDT: Scientists, Citizens, and Public Policy* (Princeton, N.J.: Princeton University Press, 1981).

55. Russell, *War and Nature*.

56. International Programme on Chemical Safety, *DDT and Its Derivatives* (Geneva: World Health Organization, 1979); U.S. Department of Health and Human Services, *Toxicological Profile for DDT, DDE, and DDD* (Atlanta, Ga.: Agency for Toxic Substances and Disease Registry, Division of Toxicology/Toxicology Information Branch, 2002).

57. R. Carson, *Silent Spring* (New York: Houghton Mifflin, 2002).

58. S. A. Gauthreaux Jr. and H. H. Shugart, "Changing Seasons: The Nesting Season, 1969: Pesticides, Population Changes, and Unusual Records," *Audubon Field Notes* 23 (1969): 632–636; S. A. Gauthreaux Jr. and H. H. Shugart, "The Nesting Season, 1970: More Pesticide Problems, Range Extensions, and Unusual Records," *Audubon Field Notes* 24 (1970): 655–659.

59. M. T. Kaufman, "Obituary: Robert K. Merton, Versatile Sociologist and Father of the Focus Group, Dies at Ninety-Two," *New York Times*, February 24, 2003.

60. J. M. May, *Ecology of Human Disease* (New York: MD Publications, 1958).

61. L. M. Camargo et al., "Unstable Hypoendemic Malaria in Rondonia (Western Amazon Region, Brazil): Epidemic Outbreaks and Work-Associated Incidence in an Agroindustrial Rural Settlement," *American Journal of Tropical Medical Hygiene* 51 (1994): 16–25.

62. N. Nogueira and J. R. Coura, "American Trypanosomiasis (Chagas' disease)," in *Tropical and Geographical Medicine*, 2nd ed., ed. K. S. Warren and A. A. F. Mahmoud (New York: McGraw-Hill, 1990).

63. X. Pourrut et al., "Spatial and Temporal Patterns of Zaire Ebolavirus Antibody Prevalence in the Possible Reservoir Bat Species," *Journal of Infectious Diseases* 196 suppl. 2 (November 2007): S176–S183.

64. P. J. Crutzen, "Geology of Mankind," *Nature* 415 (2002): 23, doi:10.1038/415023a; J. Zalasiewicz et al., "The Anthropocene: A New Epoch of Geological Time?" *Philosophical Transactions of the Royal Society Series A* 369 (2011): 835–841.

65. E. Haeckel, *Generelle Morphologie der Organismen* (Berlin: G. Reimer, 1866).

66. C. Linnaeus, *Systema naturae, sive regna tria naturae systematice proposita per classes, ordines, genera, & species* (Lugduni Batavorum: Haak, 1735).

67. A. Jönsson, "*Odium Botanicorum*: The Polemics Between Carl Linneaus and Johann Georg Siegesbeck," in *Språkets speglingar. Festskirft till Birger Bergh* (Skäneförlaget, Sweden, 2000), 555–566.

68. Ibid.

69. C. Linnaeus, *Systema naturae per regna tria naturae, secundum classes, ordines, genera, species, cum characteribus, differentiis, synonymis, locis. Tomus I. Editio decima, reformata* (Stockholm: Salvius, 1758); C. Linnaeus, *Systema naturae per regna tria naturae, secundum classes, ordines, genera, species, cum characteribus, differentiis, synonymis, locis. Tomus II. Editio decima, reformata* (Stockholm: Salvius, 1759).

70. C. Linnaeus, *Species plantarum: exhibentes plantas rite cognitas, ad genera relatas, cum differentiis specificis, nominibus trivialibus, synonymis selectis, locis natalibus, secundum systema sexuale digestas. Tomus I.* (Stockholm: Impensis Laurentii Salvii, 1753); C. Linnaeus, *Species plantarum: exhibentes plantas rite cognitas, ad genera relatas, cum differentiis specificis, nominibus trivialibus, synonymis selectis, locis natalibus, secundum systema sexuale digestas. Tomus II* (Stockholm: Impensis Laurentii Salvii, 1753).

71. Linnaeus, *Systema naturae . . . Tomus I.*

72. T. Frängsmyr et al., *Linnaeus, The Man and His Work* (Berkeley: University of California Press, 1983).

73. P. E. Zimansky, *Ancient Ararat: A Handbook of Urartian Studies* (Delmar, N.Y.: Caravan, 1998).

74. C. de Buffon, *Histoire naturelle, gènèrale et particulière*, vol. 9 (Paris: Imprimerie Royale, 1761).

75. A. von Humboldt, "Sur les lois que l'on observe dans la distribution des formes vègètales," *Annals of Chim Physica*, series 2 (1816): 234; A. von Humboldt, "Nouvelles enquêtes sur les lois que l'on observe dans la distribution des formes vègètales," *Dict Science Nat.* 18 (1820): 422; C. B. Cox, "The Biogeographical Regions Reconsidered," *Journal of Biogeography* 28 (2001): 51–523.

76. P. M. Vitousek et al., "Introduced Species: A Significant Component of Human-Caused Global Change," *New Zealand Journal of Ecology* 21 (1997): 1–16.

77. Ibid.

78. H. H. Shugart, *How the Earthquake Bird Got Its Name and Other Tales of an Unbalanced Nature* (New Haven, Conn.: Yale University Press, 2004).

79. Andrew Grey, commissary to the colony, listed five rabbits in "An Account of Livestock in the Settlement" on May 1, 1788; cited in C. Lever, *Naturalized Mammals of the World* (London: Longman, 1985).

80. Reported by James Calder, surveyor-general for Tasmania in 1869; cited in Lever, *Naturalized Mammals of the World*.

81. E. C. Rolls, *They All Ran Wild: The Story of Pests on the Land in Australia* (Sydney: Angus and Robertson, 1969).

82. Shugart, *How the Earthquake Bird Got Its Name*.

83. Ibid.

84. G. C. Caughley, *Analysis of Vertebrate Populations* (New York: John Wiley, 1977).

85. Pliny the Elder, *Natural History. Book 8, Chapter 81: The Different Species of Hares* [c. 77 CE], trans. J. Bostock and H. T. Riley (London: Taylor and Francis). Perseus Digital Library, Tufts University: http://www.perseus.tufts.edu/hopper/.

86. See Shugart, *How the Earthquake Bird Got Its Name*, for a review of the European rabbit as an invasive species.

87. T. T. Veblen and G. H. Stewart, "The Effects of Introduced Wild Animals on New Zealand," *Annals of the Association of American Geographers* 72 (1982): 372–397.

88. R. B. Allen and W. G. Lee, eds., *Biological Invasions in New Zealand* (Berlin: Springer-Verlag, 2006).

89. G. M. Thomson, *The Naturalization of Animals and Plants in New Zealand* (Cambridge: Cambridge University Press, 1922).

90. Veblen and Stewart, "The Effects of Introduced Wild Animals on New Zealand."

91. I. G. Saint-Hilaire, *Domestication et naturalisation des animaux utiles, rapport général à M. le ministre de l'agriculture* (Paris: Dusacq, 1854).

92. S. Mirsky, "Call of the Reviled," *Scientific American* 298 (2008): 44.

93. Shugart, *How the Earthquake Bird Got Its Name*.

94. C. Elton, *Ecology of Invasions by Animals and Plants* (London: Methuen, 1958).

95. W. Nentweg, "Biological Invasions: Why It Matters," in *Biological Invasions*, Ecological Studies 183, ed. W. Nentweg (Berlin: Springer-Verlag, 2007), 1–8.

96. H. A. Mooney and R. J. Hobbs, *Invasive Species in a Changing World* (Washington, D.C.: Island, 2001).

97. R. Rejmánek et al., "Ecology of Invasive Plants: State of the Art," in *Invasive Alien Species*, ed. H. A. Mooney et al. (Washington, D.C.: Island, 2005), 104–161.

98. T. Hariot, *A Briefe and True Report of the New Found Land of Virginia*, the 1590 Theodor de Bry Latin Edition (facsimile edition accompanied by the modernized

English text) (Charlottesville: Published for the Library at the Mariners' Museum, University of Virginia Press).

99. F. Fenner et al., *Smallpox and Its Eradication* (Geneva: World Health Organization, 1988).

100. Ibid.

101. E. A. Foster, trans. and ed., *Motolinia's History of the Indians of New Spain* (Berkeley, Calif.: Cortés Society, 1950; repr. Westport, Conn.: Greenwood, 1973).

102. Y. Furuse, A. Suzuki, and H. Oshitani, "Origin of Measles Virus: Divergence from Rinderpest Virus Between the Eleventh and Twelfth Centuries," *Virology Journal* 7 (2010): 52; A. J. Hay et al., "The Evolution of Human Influenza Viruses," *Philosophical Transactions of the Royal Society Series B* 356 (2001): 1861–1870.

103. C. Peerings, H. Mooney, and M. Williamson, *Bioinvasions and Globalization: Ecology, Economics, Management, and Policy* (Oxford: Oxford University Press, 2010).

104. W. Nentwig, ed., *Biological Invasions* (Berlin: Springer-Verlag, 2007).

105. Vitousek et al., "Introduced Species."

106. One can argue that truly undisturbed ecosystems are an idealized but not a realized concept given the omnipresent changes in climate and other factors such as the migration and local extinctions. See H. H. Shugart and F. I. Woodward, *Global Change and the Terrestrial Biosphere: Achievements and Challenges* (Oxford: Wiley-Blackwell, 2011).

107. P. Hochedez et al., "Chikungunya Infection in Travelers," *Emerging Infectious Diseases* 12 (2006): 1565–1567.

5. BOUNDING THE SEAS, FREEZING THE FACE OF THE DEEP: WHEN THE SEA IS LOOSED FROM ITS BONDS

1. K. Schifferdecker, *Out of the Whirlwind: Creation Theology in the Book of Job*, Harvard Theological Studies 61 (Cambridge, Mass.: Harvard University Press, 2008), 64.

2. Ibid.

3. C. A. Newson, *The Book of Job: A Contest of Moral Imagination* (Oxford: Oxford University Press, 2003), 244.

4. Schifferdecker, *Out of the Whirlwind*.

5. C. T. O'Reilly, R. Solvason, and C. Solomon, "Resolving the World's Highest Tides," in *The Changing Bay of Fundy—Beyond Four Hundred Years. Proceedings of the Sixth Bay of Fundy Workshop, Corwallis, Nova Scotia, September 29–October 2, 2004*, ed. J. A. Percy et al., Environment Canada–Atlantic Region, Occasional Report 23 (Dartmouth, Nova Scotia/Sackville, New Brunswick: Environment Canada, 2005), 153–157.

6. Pytheus' work has not survived, but he is quoted in the works of later scholars, such as Strabo's *Geographica* and Pliny's *Natural History*.

7. B. L. Van der Waerden, "The Heliocentric System in Greek, Persian, and Hindu Astronomy," *Annals of the New York Academy of Sciences* 500 (1987): 525–545.

8. From Pauly-Wissowa's *Realencyclopädie der Classischen Altertumswissenschaft* [1931], supplementband V, Agamemnon-Statilius: cols. 962–963; cited in Van der Waerden, "The Heliocentric System."

9. N. M. Swerdlow and O. Neugebauer, *Mathematical Astronomy in Copernicus's De Revolutionibus, Volume 1* (Berlin: Springer-Verlag, 1984).

10. *Oxford English Dictionary* (Oxford: Oxford University Press, 2012).

11. From the Old English *nēpflōd* (neap flood). Ibid.

12. D. W. Olson, "Perigean Spring Tides and Apogean Neap Tides in History," published abstract: American Astronomical Society meeting no. 219, contribution 115.03 (2012).

13. M. Manzaloui, "Chaucer and Science," in *Geoffrey Chaucer*, ed. D. Brewer (London: Bell, 1974), 230; E. S. Laird and D. W. Olson, "Boethius, Boece, and Böotes: A Note on the Chronology of Chaucer's *Astronomical Learning*," *Modern Philology* 88 (1990): 147–149.

14. From L. D. Benson, ed., *The Riverside Chaucer* (Oxford: Oxford University Press, 2012).

15. D. W. Olson et al., "Perfect Tide, Ideal Moon: An Unappreciated Aspect of Wolfe's Generalship at Québec, 1759," *William and Mary Quarterly* 59 (2002): 957–974.

16. D. H. Fischer, *Paul Revere's Ride* (New York: Oxford University Press, 1994), 104–105, 312–313. The "Portsmouth Alarm" occurred in December 1774 and should not be confused with Revere's more famous midnight ride, which warned of British troop movements on April 18, 1775, toward Concord, Massachusetts.

17. D. W. Olson and Russell L. Doescher, "The Boston Tea Party," *Sky and Telescope* 86 (1993): 83–86.

18. D. Turner, *Last Dawn: The Royal Oak Tragedy at Scapa Flow*, 3rd ed. (Glendaruel: Argyll, 2011).

19. J. H. Alexander, *Utmost Savagery: The Three Days at Tarawa* (New York: Ivy, 1996).

20. B. Parker, "The Tide Predictions for D-Day," *Physics Today* 64 (2011): 35–40.

21. W. Thomson, "The Tide Gauge, Tidal Harmonic Analyser, and Tide Predicter," *Proceedings of the Institution of Civil Engineers* 65 (1881): 3–24. The machine was subsequently upgraded to ten components.

22. W. Ferrel, "A Maxima and Minima Tide-Predicting Machine," in *U S Coast Survey (1883), Appendix 10* (1883), 253–272. Also see "The Maxima and Minima Tide-Predicting Machine," *Science* 3 (1884): 408–410.

23. E. Roberts, "A New Tide-Predicter," *Proceedings of the Royal Society* 29 (1879): 198–201.

24. E. Roberts, *Description of the U.S. Coast and Geodetic Survey Tide-Predicting Machine No. 2*, Special Publication 32 (Washington, D.C.: U.S. Department of Commerce, 1915).

25. http://www.deutsches-museum.de/en/collections/transport/maritime-exhibition /tide-predicter/.

26. Parker, "The Tide Predictions for D-Day."

27. A. T. Doodson, "The Harmonic Development of the Tide-Generating Potential," *Proceedings of the Royal Society of London Series A* 100 (1921): 305–329.

28. Parker, "The Tide Predictions for D-Day."

29. Ibid.

30. B. Fergusson, *The Watery Maze* (New York: Holt, Rinehart & Winston, 1961).

31. Parker, "The Tide Predictions for D-Day."

32. D. D. Eisenhower, *Crusade in Europe* (Garden City, N.Y.: Doubleday, 1948).

33. Parker, "The Tide Predictions for D-Day"; D. Irving, *The Trail of the Fox— the Search for the True Field Marshal Rommel* (London: Weidenfeld and Nicolson, 1977).

34. Olson, "Perigean Spring Tides and Apogean Neap Tides in History."

35. R. D. Müller et al., "Long-Term Sea-Level Fluctuations Driven by Ocean Basin Dynamics," *Science* 319 (2008): 1357–1362.

36. Ibid.

37. K. Lambeck, T. M. Esat, and E. K. Potter, "Links Between Climate and Sea Levels for the Past Three Million Years," *Nature* 419 (2007): 199–206.

38. J. Lean, J. Beer, and R. Bradley, "Reconstruction of Solar Irradiance Since 1610: Implications for Climate Change," *Geophysical Research Letters* 22 (1995): 3195–3198.

39. M. R. Chapman, N. J. Shackleton, and J.-C. Duplessey, "Sea Surface Temperature Variability During the Last Glacial-Interglacial Cycle Assessing the Magnitude and Pattern of Climate Change in the North Atlantic," *Palaeogeography, Palaeoclimatology, Palaeoecology* 157 (2000): 1–25.

40. A. C. Mix and W. F. Ruddiman, "Oxygen Isotope Analyses and Pleistocene Ice Volumes," *Quaternary Research* 21 (1984): 1–20.

41. G. H. Haug and R. Tiedeman, "Effect of the Formation of the Isthmus of Panama on Atlantic Ocean Thermohaline Circulation," *Nature* 393 (1998): 673–676.

42. M. E. Raymo, "Global Climate Change: A Three-Million Year Perspective," in *Start of a Glacial, Proceedings of the Mallorca NATO ARW, NATO ASI Series I*, ed. G. J. Kukla and E. Went (Heidelberg: Springer Verlag, 1992), 3:207–233.

43. A. Berger, "Milankovitch Theory and Climate," *Reviews of Geophysics* 26 (1988): 624–657; C. Emiliani, "Pleistocene Temperatures," *Journal of Geology* 63 (1955): 538–578.

44. N. J. Shackleton, "Oxygen Isotope Calibration of the Onset of Ice-Rafting and History of Glaciation in the North Atlantic Region," *Nature* 307 (1984): 620–623.

45. W. W. Hay, "The Cause of the Late Cenozoic Northern Hemisphere Glaciations: A Climate Change Enigma," *Terra Nova* 4 (1992): 305–311; Haug and Tiedeman, "Effect of the Formation of the Isthmus of Panama."

46. Lambeck, Esat, and Potter, "Links Between Climate and Sea Levels." But see M. F. Raymo, "The Initiation of North Hemisphere Circulation," *Annual Review of Earth and Planetary Science* 22 (1994): 353–383, for a review of alternate theories.

47. Raymo, "The Initiation of North Hemisphere Circulation."

48. G. H. Haug et al., "North Pacific Seasonality and the Glaciation of North America 2.7 Million Years Ago," *Nature* 433 (2005): 821–825.

49. W. S. Broecker and G. H. Denton, "The Role of Ocean-Atmosphere Reorganizations in Glacial Cycles," *Geochimica et Cosmochimica Acta* 53 (1989): 2465–2501; J. F. McManus, D. W. Oppo, and J. L. Cullen, "A 0.5-Million-Year Record of Millennial-Scale Climate Variability in the North Atlantic," *Science* 283 (1999): 971–975; D. R. MacAyeal, "Binge/Purge Oscillations of the Laurentide Ice Sheets as a Cause of the North Atlantic's Heinrich Events," *Paleoceanography* 8 (1993): 775–784; R. B. Alley and P. U. Clark, "The Deglaciation of the Northern Hemisphere: A Global Perspective," *Annual Reviews of Earth and Planetary Science* 27 (1999): 149–182.

50. Lambeck, Esat, and Potter, "Links Between Climate and Sea Levels."

51. K. Kawamura et al., "Northern Hemisphere Forcing of Climatic Cycles in Antarctica Over the Past 360,000 Years," *Nature* 448 (2007): 912–916.

52. A. M. Tushingham and W. R. Peltier, "ICE-3G: A New Global Model of Late Pleistocene Deglaciation Based Upon Geophysical Predictions of Postglacial Sea-Level Change," *Journal of Geophysical Research* 96 (1991): 4497–4523; K. Lambeck, C. Smither, and P. Johnston, "Sea-Level Change, Glacial Rebound, and Mantle Viscosity for Northern Europe," *Geophysical Journal International* 134 (1998): 647–651.

53. R. G. Fairbanks, "A 17,000-Year Glacio-Eustatic Sea Level Record: Influence of Glacial Melting Rates on the Younger Dryas Event and Deep-Ocean Circulation," *Nature* 342 (1989): 637–642; P. Blanchon and J. Shaw, "Reef Drowning During the Last Deglaciation: Evidence for Catastrophic Sea-Level Rise and Ice-Sheet Collapse," *Geology* 23 (1995): 4–8; J. D. Stanford et al., "Timing of Meltwater Pulse 1a and Climate Responses to Meltwater Injections," *Paleoceanography* 21 (2006): PA4103.

54. A. E. Carlson et al., "Rapid Early Holocene Deglaciation of the Laurentide Ice Sheet," *Nature Geoscience* 1 (2008): 620–624.

55. G. A. Milne et al., "Identifying the Causes of Sea-Level Change," *Nature Geoscience* 2 (2009): 471–478.

56. K. M. Cuffey and S. J. Marshall, "Substantial Contribution to Sea-Level Rise During the Last Interglacial from the Greenland Ice Sheet," *Nature* 404 (2000):

591–594; B. L. Otto-Bliesner et al., "Simulating Arctic Climate Warmth and Icefield Retreat in the Last Interglaciation," *Science* 311 (2006): 1751–1753; R. P. Scherer et al., "Pleistocene Collapse of the West Antarctic Ice Sheet," *Science* 281 (1998): 82–85.

57. J. T. Overpeck et al., "Paleoclimatic Evidence for Future Ice-Sheet Instability and Rapid Sea-Level Rise," *Science* 311 (2006): 1747–1750; E. J. Rohling et al., "High Rates of Sea-Level Rise During the Last Interglacial Period," *Nature Geoscience* 1 (2008): 38–42; D. Dahl-Jensen et al., "Past Temperatures Directly from the Greenland Ice Sheet," *Science* 282 (1998): 268–271.

58. S. Solomon et al., eds., *Climate Change 2007: The Physical Science Basis. Contribution of Working Group I to the Fourth Assessment Report of the Intergovernmental Panel on Climate Change* (Cambridge: Cambridge University Press, 2007).

59. S. Rahmstorf, "A Semi-Empirical Approach to Projecting Future Sea-Level Rise," *Science* 315 (2007): 368–370.

60. S. Solomon et al., eds., *Climate Change 2007*.

61. M. R. Raupach et al., "Global and Regional Drivers of Accelerating CO_2 Emissions," *Proceedings of the National Academy of Sciences* 104 (2007): 10288–10293.

62. http://www.ipcc.ch/publications_and_data/publications_ipcc_fourth _assessment_report_wg1_report_the_physical_science_basis.htm.

63. Overpeck et al., "Paleoclimatic Evidence"; Rohling et al., "High Rates of Sea-Level Rise."

64. N. L. Bindoff et al., "Observations: Oceanic Climate Change and Sea Level," in *Climate Change 2007: The Physical Science Basis. Contribution of Working Group I to the Fourth Assessment Report of the Intergovernmental Panel on Climate Change*, ed. S. Solomon et al. (Cambridge: Cambridge University Press, 2007).

65. The AIFI scenario. See IPCC, "Summary for Policymakers," in *Climate Change 2007: The Physical Science Basis. Contribution of Working Group I to the Fourth Assessment Report of the Intergovernmental Panel on Climate Change*, ed. S. Solomon et al. (Cambridge: Cambridge University Press, 2007).

66. S. Rahmstorf, "Recent Climate Observations Compared to Projections," *Science* 316 (2007): 709.

67. Bindoff et al., "Observations: Oceanic Climate Change and Sea Level."

68. G. A. Meehl et al., "2007: Global Climate Projections," in *Climate Change 2007: The Physical Science Basis. Contribution of Working Group I to the Fourth Assessment Report of the Intergovernmental Panel on Climate Change*, ed. S. Solomon et al. (Cambridge: Cambridge University Press, 2007).

69. S. Rahmstorf, "Sea-Level Rise: Towards Understanding Local Vulnerability," *Environmental Research Letters* 7 (2012): 021001.

70. Ibid.

71. Arctic Monitoring and Assessment Programme, *Snow, Water, Ice, and Permafrost in the Arctic* (Oslo: Arctic Monitoring and Assessment Programme, 2011);

Scientific Committee on Antarctic Research, *Antarctic Climate Change and the Environment* (Cambridge: Scott Polar Research Institute, 2009); U.S. Army Corps of Engineers, *Sea-Level Change Considerations for Civil Works Programs* (Washington, D.C.: Department of the Army, 2011); P. Vellinga et al., *Exploring High-End Climate Change Scenarios for Flood Protection of the Netherlands: International Scientific Assessment Carried Out at the Request of the Delta Committee* (De Bilt: Koninklijk Nederlands Meteorologisch Instituut, 2009).

72. G. McGranahan, D. Balk, and B. Anderson, "The Rising Tide: Assessing the Risks of Climate Change and Human Settlements in Low-Elevation Coastal Zones," *Environment and Urbanization* 19 (2007): 17.

73. B. Strauss et al., "Tidally Adjusted Estimates of Topographic Vulnerability to Sea Level Rise and Flooding for the Contiguous United States," *Environmental Research Letters* 7 (2012): 014033.

74. C. Tebaldi, B. Strauss, and C. Zervas, "Modelling Sea Level Rise Impacts on Storm Surges Along U.S. Coasts," *Environmental Research Letters* 7 (2012): 014032.

75. Ibid.

76. R. J. Nicholls et al., "Coastal Systems and Low-Lying Areas," in *Climate Change 2007: Impacts, Adaptation, and Vulnerability. Contribution of Working Group II to the Fourth Assessment Report of the Intergovernmental Panel on Climate Change*, ed. M. L. Parry et al. (Cambridge: Cambridge University Press, 2007), 315–356.

77. *Living with Coastal Erosion in Europe: Sediment and Space for Sustainability*, part 1, *Major Findings and Policy Recommendations of the EUROSION Project. Guidelines for Implementing Local Information Systems Dedicated to Coastal Erosion Management*, service contract B4-3301/2001/329175/MAR/B3, "Coastal Erosion—Evaluation of the Need for Action" (Directorate General Environment, European Commission, 2004).

78. W. S. Ng and R. Mendelsohn. "The Impact of Sea-Level Rise on Singapore," *Environmental Development and Economics* 10 (2005): 201–215.

79. E. Ohno, "Economic Evaluation of Impact of Land Loss Due to Sea-Level Rise in Thailand," in *Proceedings of the APN/SURVAS/LOICZ Joint Conference on Coastal Impacts of Climate Change and Adaptation in the Asia—Pacific Region, 14–16th November 2000, Kobe, Japan* (Asia Pacific Network for Global Change Research), 231–235.

80. M. El-Raey, "Vulnerability Assessment of the Coastal Zone of the Nile Delta of Egypt, to the Impacts of Sea-Level Rise," *Ocean Coastal Management* 37 (1997): 29–40.

81. G. Yohe, "Assessing the Role of Adaptation in Evaluating Vulnerability to Climate Change," *Climatic Change* 46 (2000): 371–390; G. Yohe and R. S. J. Tol, "Indicators for Social and Economic Coping Capacity—Moving Toward a Working Definition of Adaptive Capacity," *Global Environmental Change* 12 (2002): 25–40.

82. J. B. C. Jackson, "The Future of Oceans Past," *Philosophical Transactions of the Royal Society Series B* 365 (2010): 3765–3778.

6. The Ordinances of the Heavens and Their Rule
on Earth: Adaptation and the Cycles of Life

1. C. Mondragón, "Of Winds, Worms, and *Mana*: The Traditional Calendar of the Torres Islands, Vanuatu," *Oceania* 74 (2004): 289–308.

2. J. del Hoyo, A. Elliott, and J. Sargatal, eds., *Handbook of the Birds of the World*, vol. 3, *Hoatzin to Auks* (Barcelona: Lynx, 1996).

3. H. Caspers, "Spawning Periodicity and Habitat of the Palolo Worm *Eunice viridis* (Polychaeta: Eunicidae) in the Samoan Islands," *Marine Biology* 79 (1984): 229–236.

4. Mondragón, "Of Winds, Worms, and *Mana*."

5. W. J. Durrad, "Notes on the Torres Islands," *Oceania* 11 (1940): 186–201.

6. J. De Smedt and H. De Cruz, "The Role of Material Culture in Human Time Representation: Calendrical Systems as Extensions of Mental Time Travel," *Adaptive Behavior* 19 (2011): 63–76.

7. M. Bassi, "On the Borana Calendrical System: A Preliminary Field Report," *Current Anthropology* 29 (1988): 619–624.

8. The calculation is actually a bit more complex because of conditions and definitions defined by religious convention. For example, the vernal equinox is set to March 21 for Easter calculations. It occurs around this date but not always on it. See the U.S. Naval Observatory website: http://aa.usno.navy.mil/faq/docs/easter.php.

9. K. Schmidt, "Investigations in the Upper Mesopotamian Early Neolithic: Göbekli Tepe and Gürcütepe," *Neo-Lithics* 2 (1995): 9–10; K. Schmidt, "Göbekli Tepe, Southeastern Turkey. A Preliminary Report on the 1995–1999 Excavations," *Paléorient* 26 (2001): 45–54. T. Tedlock and B. Tedlock, "The Sun, Moon, and Venus Among the Stars: Methods for Mapping Mayan Sidereal Space," *Archaeoastronomy* 17 (2002–2003): 6–21; S. Iwaniszewski, "Glyphs D and E of the Lunar Series at Yaxchilán and Piedras Negras," *Archaeoastronomy* 18 (2004): 67–80. M. J. Zawaski and J. M. Malville, "An Archaeoastronomical Survey of Major Inca Sites in Peru," *Archaeoastronomy* 21 (2007–2008): 21–38. A. Thom, "The Solar Observatories of Megalithic Man," *Journal of the British Astronomical Association* 64 (1954): 397; A. Thom, "A Statistical Examination of the Megalithic Sites in Britain," *Journal of the Royal Statistical Society* 118 (1955): 275–295. J. Cox, "The Orientations of Prehistoric Temples in Malta and Gozo," *Archaeoastronomy* 16 (2001): 24–37.

10. E. C. Baity (with comments by A. F. Aveni et al.), "Archaeoastronomy and Ethnoastronomy So Far [and Comments and Reply]," *Current Anthropology* 14 (1973): 389–449.

11. Sponsored by ISAAC, the International Society for Archaeoastronomy and Astronomy in Culture. Other archaeoastronomy societies include SEAC (La Société Européenne pour l'Astronomie dans la Culture) and SIAC (La Sociedad Interamericana de Astronomía en la Cultura).

12. C. Ruggles and M. Cotte, *Heritage Sites of Astronomy and Archaeoastronomy in the Context of the UNESCO World Heritage Convention: A Thematic Study* (Paris: International Secretariat of the International Council on Monuments, 2010).

13. C.-S. Holdermann, H. Müller-Beck, and U. Simon, *Eiszeitkunst im süddeutschschweizerischen Jura: Anfänge der Kunst* (Stuttgart: Karl Theiss Verlag, 2001).

14. A. Marshack, "The Taï Plaque and Calendrical Notation in the Upper Palaeolithic," *Cambridge Archaeological Journal* 1 (1991): 25–61.

15. C. Ruggles, "Astronomy and Stonehenge," in *Science and Stonehenge*, ed. B. Cunliffe and C. Renfrew, Proceedings of the British Academy 92 (Oxford: Oxford University Press, 1997), 203–229; C. Ruggles, "Interpreting Solstitial Alignments in Late Neolithic Wessex," *Archaeoastronomy* 20 (2007): 1–27.

16. J. Druett, *Tupaia: Captain Cook's Polynesian Navigator* (New York: Praeger, 2011).

17. *Transcription of Banks's Endeavour Journal* (South Seas, 2004), 1:299. http://nla.gov.au/nla.cs-ss-jrnl-banks-17690712.

18. D. Lewis, *We, the Navigators: The Ancient Art of Landfinding in the Pacific.* 2nd ed. (Honolulu: University of Hawaii Press, 1994).

19. A. Di Piazza and E. Pearthree, "A New Reading of Tupaia's Chart," *Journal of the Polynesian Society* 116 (2007): 321–340.

20. R. Langdon, "The European Ships on Tupaia's Chart: An Essay in Identification," *Journal of Pacific History* 15 (1980): 225–232.

21. The locations of star rise and star set change seasonally when one is away from the equator.

22. T. Gladwin, *The East Is a Big Bird: Navigation and Logic on Puluwat Atoll* (Cambridge, Mass.: Harvard University Press, 1970).

23. M. Ascher, "Models and Maps from the Marshall Islands: A Case of Ethnomathematics," *Historia Mathematica* 22 (1995): 347–3370, lists sixty-nine such objects and their museum repositories.

24. C. Winkler, "On Sea Charts Formally Used in the Marshall Islands, with Notices on the Navigation of These Islanders in General," in *Annual Report of the Smithsonian Institution for the Year Ending June 30, 1899* (Washington, D.C., 1899), 487–508.

25. Ascher, "Models and Maps from the Marshall Islands."

26. Ibid.

27. C. E. Herdendorf, "Captain James Cook and the Transits of Mercury and Venus," *Journal of Pacific History* 21 (1986): 39–55; D. Sobel, *Longitude: The True Story*

of a Lone Genius Who Solved the Greatest Scientific Problem of His Time (New York: Penguin, 1995).

28. Pliny the Elder, *Natural History. Book 8, Chapter 42: Prognostics of Danger Derived from Animals*, trans. J. Bostock and H. T. Riley (London: Taylor and Francis). http://www.perseus.tufts.edu/hopper/.

29. F. C. James and H. H. Shugart, "The Phenology of the Nesting Season of the Robin (Turdus migratorius) in the United States," *Condor* 76 (1974): 159–168.

30. M. H. Drummond, preface to *The Great Fight: Poems and Sketches by William Henry Drummond, MD*, ed. M. H. Drummond (New York: G. P. Putnam's Sons, 1908).

31. T. Harrisson, "Men and Birds in Borneo," in *The Birds of Borneo*, ed. B. E. Smythies (Edinburgh: Oliver and Boyd, 1960), 20–61; E. Jensen, *The Iban and Their Religion*, Oxford Monographs on Social Anthropology (Oxford: Oxford University Press, 1974); A. Richards, "Iban Augury," *Sarawak Museum Journal* 20 (1972): 63–81. M. R. Dove, "Education, Uncertainty, Humility, and Adaptation in the Tropical Forest: The Agricultural Augury of the Kantu,' " *Ethnology* 32 (1993): 145–167; M. R. Dove, "Process Versus Product in Bornean Augury: A Traditional Knowledge System's Solution to the Problem of Knowing," in *Redefining Nature: Ecology, Culture, and Domestication*, ed. R. Ellen and K. Fukui (Oxford: Berg, 1996), 557–596.

32. According to Dove, "Education, Uncertainty, Humility, and Adaptation," and Richards, "Iban Augury," the birds vary in their "authority." In ascending order they are the *Nenak* (white-rumped shama), *Ketupong* (rufous piculet), *Beragai* (scarlet-rumped trogon), *Papau* (Diard's trogon), *Memuas* (banded kingfisher), *Kutok* (maroon woodpecker), and *Bejampong* (crested jay).

33. N. Nicholls, "ENSO, Drought, and Flooding Rain in South-East Asia," in *South-East Asia's Environmental Future: The Search for Sustainability*, ed. H. Brookfield and Y. Bryon (Tokyo: United Nations University Press/Oxford Press, 1993), 154–175.

34. O. K. Moore, "Divination: A New Perspective," *American Anthropologist* 59 (1957): 69–74.

35. R. Lawless, "Effects of Population Growth and Environment Changes on Divination Practices in Northern Luzon," *Journal of Anthropological Research* 31 (1975): 18–33.

36. R. Marsham, "Indications of Spring, Observed by Robert Marsham, Esquire, F.R.S. of Stratton in Norfolk. Latitude 52°45'," *Philosophical Transactions of the Royal Society of London* 79 (1789): 154–156.

37. W. Markwick, "A Comparative View of the Naturalist's Calendar, as Kept at Selbourne, in Hampshire, by the Late Gilbert White, M.A.; and at Catsfield, near Battle, in Sussex, by William Markwick from the Year 1768 to the Year 1793," in *The Natural History of Selbourne and the Naturalist's Calendar*, ed. G. C. Davies (London: Frederick Warne and Co., 1825), 447–458.

38. G. White, *The Natural History of Selbourne* (London: Frederick Warne and Co., 1789). Currently available as G. White, *The Illustrated Natural History of Selbourne* (London: Thames and Hudson, 2007).

39. R. Mabey, *Gilbert White: A Biography of the Author of the Natural History of Selbourne* (Charlottesville: University of Virginia Press, 2007).

40. K. A. Brinkman and V. Vankus, "Cornaceae—Dogwood family, *Cornus* L., dogwood," in *The Woody Plant Seed Manual,* Agriculture Handbook 727 (Washington, D.C.: Agriculture Dept., Forest Service, 2008), 428–432.

41. M. J. Yanovsky and S. A. Kay, "Living by the Calendar: How Plants Know When to Flower," *Nature Reviews Molecular Cell Biology* 4 (2003): 265–275.

42. Marsham, "Indications of Spring."

43. T. H. Sparks and P. D. Carey, "The Responses of Species to Climate Over Two Centuries: An Analysis of the Marsham Phenological Record, 1736–1947," *Journal of Ecology* 83 (1995): 321–329.

44. C. D. Keeling, "The Concentration and Isotopic Abundance of Carbon Dioxide in the Atmosphere," *Tellus* 12 (1960): 200–203.

45. Sparks and Carey, "The Responses of Species to Climate Over Two Centuries."

46. A. Menzel et al., "European Phenological Response to Climate Change Matches the Warming Pattern," *Global Change Biology* 12 (2006): 1969–1976.

47. M. D. Schwartz and T. M. Crawford, "Detecting Energy-Balance Modifications at the Onset of Spring," *Physical Geography* 21 (2001): 394–409; M. D. Schwartz and B. E. Reiter, "Changes in North American Spring," *International Journal of Climatology* 20 (2000): 929–932; M. D. Schwartz, R. Ahas, and A. Aasa, "Onset of Spring Starting Earlier Across the Northern Hemisphere," *Global Change Biology* 12 (2006): 343–351; Menzel et al., "European Phenological Response to Climate Change."

48. Schwartz, Ahas, and Aasa, "Onset of Spring Starting Earlier."

49. G. R. Walther, "Plants in a Warmer World," *Perspectives in Plant Ecology Evolution and Systematics* 6 (2003): 169–185.

50. D. R. Easterling, "Recent Changes in Frost Days and the Frost-Free Season in the United States," *Bulletin of the American Meteorological Society* 83 (2002): 1327–1332.

51. A. H. Fitter and R. S. R. Fitter, "Rapid Changes in Flowering Time in British Plants," *Science* 296 (2002): 1689–1691.

52. R. B. Myneni et al., "Increased Plant Growth in the Northern High Latitudes from 1981 to 1991," *Nature* 386 (1997): 698–701.

53. *Climate Change 2007: The Physical Science Basis. Contribution of Working Group I to the Fourth Assessment Report of the Intergovernmental Panel on Climate Change,* ed. S. Solomon et al. (Cambridge: Cambridge University Press, 2007).

54. S. Arrhenius, "On the Influence of Carbonic Acid in the Air Upon the Temperature of the Ground," *Philosophical Magazine and Journal of Science, Series 5* 41 (1896): 237–276.

55. J. E. Kutzbach, "Steps in the Evolution of Climatology: From Descriptive to Analytic," in *Historical Essays on Meteorology, 1919–1995*, ed. J. R. Fleming (Boston: American Meteorological Society, 1996), 353–377.

56. L. Partridge and P. H. Harvey, "The Ecological Context of Life History Evolution," *Science* 241 (1998): 1449–1455.

57. M. H. Hastings and B. K. Follett, "Toward a Molecular Biological Calendar," *Journal of Biological Rhythms* 16 (2001): 424–430.

58. R. C. Babcock et al., "Synchronous Spawnings of 105 Scleractian Coral Species on the Great Barrier Reef," *Marine Biology* 90 (1986): 379–394; E. Naylor, "Marine Animal Behaviour in Relation to Lunar Phase," *Earth Moon and Planets* 85–86 (2001): 291–302.

59. Caspers, "Spawning Periodicity and Habitat of the Palolo Worm."

60. R. G. Foster and T. Roenneburg, "Human Responses to the Geophysical Daily, Annual, and Lunar Cycles," *Current Biology* 18 (2008): R784–R794.

61. V. M. Bell-Pedersen et al., "Circadian Rhythms from Multiple Oscillators: Lessons from Diverse Organisms," *Nature Reviews Genetics* 6 (2005): 544–556.

62. M. J. Paul, I. Zucker, and W. J. Schwartz, "Tracking the Seasons: The Internal Calendars of Vertebrates," *Philosophical Transactions of the Royal Society of London Series B* 363 (2008): 341–361; D. S. Farner, "Annual Rhythms," *Annual Review of Physiology* 47 (1985): 65–82.

63. Farner, "Annual Rhythms."

64. See K. P. Able, "The Scope and Evolution of Migration," in *Gatherings of Angels: Migrating Birds and Their Ecology*, ed. K. P. Able (Ithaca, N.Y.: Comstock, 1999), 1–11.

65. I. C. T. Nisbet et al., "Transoceanic Migration of the Blackpoll Warbler: Summary of Scientific Evidence and Response to Criticisms by Murray," *Journal of Field Ornithology* 66 (1995): 612–622.

66. J. D. Cherry, D. H. Doherty, and K. D. Powers, "An Offshore Nocturnal Observation of Migrating Blackpoll Warblers," *Condor* 87 (1985): 548–549; C. P. McClintock, T. C. Williams, and J. M. Teal, "Autumnal Migration Observed from Ships in the Western North Atlantic Ocean," *Bird-Banding* 49 (1978): 262–275.

67. Able, "The Scope and Evolution of Migration."

68. The same result can be obtained by placing the birds in cages in rooms with controlled lighting conditions and constant temperatures.

69. *Zeitgebers* (German for "time givers"), in the standard literature of avian physiology.

70. H. G. Wallraff, "Does Pigeon Homing Depend on Stimuli Perceived During Displacement? I. Experiments in Germany," *Journal of Comparative Physiology* A139 (1980): 193–201.

71. K. P. Able, ed., *Gatherings of Angels: Migrating Birds and Their Ecology* (Ithaca, N.Y.: Comstock, 1999).

72. R. Wiltschko and W. Wiltschko, *Magnetic Orientation in Animals* (Berlin: Springer-Verlag, 1995).

73. A very readable account about this remarkable capability can be found in K. P. Able, "A Sense of Magnetism," *Birding* 30 (1998): 314–321.

74. A. Le Floch et al., "The Polarization Sense in Human Vision," *Vision Research* 50 (2010): 2048–2054.

75. K. P. Able, "How Birds Migrate," in *Gatherings of Angels: Migrating Birds and Their Ecology*, ed. K. P. Able (Ithaca, N.Y.: Comstock, 1999), 11–26.

76. K. Schmidt-Koenig and H.-J. Schlichte, "Homing in Pigeons with Reduced Vision," *Proceedings of the National Academy of Sciences USA* 69 (1972): 2446–2447.

77. M. M. Walker et al., "Structure and Function of the Vertebrate Magnetic Sense," *Nature* 390 (1997): 371–376; C. E. Duebel et al., "Magnetite Defines a Vertebrate Magnetoreceptor," *Nature* 406 (2000): 299–302.

78. F. Papi, "Pigeons Use Olfactory Clues to Navigate," *Ethology, Ecology, and Evolution* 1 (1989): 219–231; K. P. Able, "The Debate Over Olfactory Homing in Pigeons," *Journal of Experimental Biology* 199 (1996): 121–124.

79. W. L. Donn and B. Naini, "The Sea-Wave Origin of Microbaroms and Microseisms," *Journal of Geophysical Research* 78 (1973): 4428–4488; J. T. Hagstrum, "Infrasound and the Avian Navigational Map," *Journal of Experimental Biology* 203 (2000): 1103–1111.

80. P. Berthold, *Control of Bird Migration* (London: Chapman and Hall, 1996).

81. Able, ed., *Gatherings of Angels*.

82. Ibid.

83. W. P. Brown, *Seven Pillars of Creation: The Bible, Science, and the Ecology of Wonder* (New York: Oxford University Press, 2010).

84. The Wisdom of Solomon is composed in Greek and is likely from Alexandria, in Egypt, between the first century BCE and the first century CE. It is included as a deuterocanonical book by the Roman Catholic and Eastern Orthodox Churches and as part of the Apocrypha by Protestant churches. It is noncanonical in the Rabbinical Jewish tradition. See notes on the Wisdom of Solomon, NSRV.

85. R. Przeslawski et al., "Beyond Corals and Fish: The Effects of Climate Change on Noncoral Benthic Invertebrates of Tropical Reefs," *Global Change Biology* 14 (2008): 2773–2795.

7. THE DWELLING OF THE LIGHT AND THE PATHS TO ITS HOME: WINDS, OCEAN CURRENTS, AND THE GLOBAL ENERGY BALANCE

1. See National Imagery and Mapping Agency (NIMA), *The American Practical Navigator: An Epitome of Navigation*, originally by N. Bowditch (Bethesda, Md.:

National Imagery and Mapping Agency, 2002), chap. 34; Golden Gate Weather Service, "Names of Winds," http://ggweather.com/winds.html.

2. Siroccos also have a second peak of occurrence in March. They originate in the Sahara and Arabian deserts.

3. C. D. Whiteman, *Mountain Meteorology: Fundamentals and Applications* (Oxford: Oxford University Press, 2000).

4. G. Masselink, "Sea Breeze Activity and Its Effect on Coastal Processes Near Perth, Western Australia," *Journal of the Royal Society of Western Australia* 79 (1996): 199–205.

5. NIMA, *The American Practical Navigator*.

6. R. Swap et al., "Saharan Dust in the Amazon Basin," *Tellus* 44B (1992): 133–149.

7. This particular chinook produced a temperature rise from −48°C to 9°C (−54°F to 49°F) in a twenty-four-hour period on January 15, 1972. A. H. Horvitz et al., *A National Temperature Record at Loma, Montana* (National Weather Service, Cooperative Observer Program, 2002), http://www.nws.noaa.gov/om/coop/standard.html.

8. National Weather Service Weather Forecast Office, Sioux Falls, S.D., http://www.crh.noaa.gov/fsd/?n=fsdtrivia01.

9. North Atlantic Ocean, northeastern Pacific Ocean east of the International Dateline, and South Pacific Ocean east of 160°E. See K. Emanuel, *Divine Wind: The History and Science of Hurricanes* (Oxford: Oxford University Press, 2005).

10. Northwestern Pacific west of the International Dateline. Ibid.

11. Southwestern Pacific Ocean west of 160°E and southeastern Indian Ocean east of 90°E. Ibid.

12. The United States a one-minute average wind to compute the maximum sustained wind; most countries use the World Meteorological Organization standard of using a ten-minute average wind for this calculation. Ibid.

13. H. Piddington, *The Sailors' Horn Book for the Law of Storms in All Parts of the World* (New York: John Wiley, 1848), 8.

14. "May originate," in that another possible origin is from *jufeng*, a Chinese word implying a wind coming in all directions and with a first appearance in a Chinese text written in 470 CE (*Nan Yue Zhi*, or *Book of the Southern Yue Region*). The Cantonese *tai-fung*, which sounds like typhoon, derives from *jufeng*. See Emanuel, *Divine Wind*, chap. 3.

15. Described in the ancient Greek book *Bibliotheca*, by Pseudo-Apollodorus. See A. Diller, "The Text History of the *Bibliotheca* of Pseudo-Apollodorus," *Transactions and Proceedings of the American Philological Association* 66 (1935): 296–313.

16. Emanuel, *Divine Wind*.

17. P. D. Wilson, "Wragge, Clement Lindley (1852–1922), Meteorologist," in *Australian Dictionary of Biography*, vol. 12, ed. J. Ritchie (Melbourne: Melbourne University Press, 1990). http://www.adb.online.anu.edu.au/biogs/A120646b.htm.

18. R. A. Anthes, *Tropical Cyclones: Their Evolution, Structure, and Effects.* Meteorological Monographs 41 (American Meteorological Society, 1982).

19. Emanuel, *Divine Wind.*

20. NOAA, *Hurricanes: Releasing Nature's Fury, A Preparedness Guide*, NOSS/PA 94050 (National Oceanic and Atmospheric Administration, National Weather Service, 2001).

21. C. M. Graney, "Coriolis Effect, Two Centuries Before Coriolis," *Physics Today* 64 (2011): 8.

22. In classical mechanics, the outward acceleration from centrifugal force increases with the velocity squared and as the inverse of the radius.

23. Emanuel, *Divine Wind.*

24. See R. H. Thurston, trans., *Reflections on the Motive Power of Heat* (New York: John Wiley and Sons, 1897).

25. Ibid., 37–38.

26. Emanuel, *Divine Wind.*

27. G. Cartwright, "Dan Rather Retorting," *Texas Monthly* (March 2005): 136.

28. E. S. Blake et al., *The Deadliest, Costliest, and Most Intense United States Tropical Cyclones from 1851 to 2006 (and Other Frequently Requested Hurricane Facts)*, NOAA Technical Memorandum NWS TPC-5 (Miami, Fla.: National Weather Service, National Hurricane Center, 2007).

29. N. L. Frank and S. A. Husain, "The Deadliest Tropical Cyclone in History," *Bulletin of the American Meteorological Society* 32 (1980): 438–444.

30. Ibid.

31. G. M. Dunnavan and J. W. Dierks, "An Analysis of Super Typhoon Tip (October 1979)," *Monthly Weather Review* 108 (1980): 1915–1923.

32. J. M. Wilmshurst et al., "High-Precision Radiocarbon Dating Shows Recent and Rapid Initial Human Colonization of East Polynesia," *Proceedings of the National Academy of Science* 108 (2011): 1815–1820.

33. S. Bedford, C. Sand, and S. P. Connaughton, eds., *Oceanic Explorations: Lapita and Western Pacific Settlement* (Canberra: Australian National University Press, 2007).

34. E. Halley, "An Historical Account of the Trade Winds, and Monsoons, Observable in the Seas Between and Near the Tropics, with an Attempt to Assign the Physical Cause of Said Winds," *Philosophical Transactions of the Royal Society* 16 (1686): 153–186.

35. G. Hadley, "On the Cause of the General Trade Winds," *Philosophical Transactions of the Royal Society* 34 (1735): 58–62; reprinted in C. Abbe, *The Mechanics of the Earth's Atmosphere*, Smithsonian Misc. Collections 51, no. 1 (1910); D. B. Shaw, *Meteorology Over the Tropical Oceans* (Royal Meteorological Society, 1979).

36. Ibid.

37. G.-G. Coriolis, "Sur les équations du mouvement relatif des systèmes de corps," *Journal de l'Ecole Royale Polytechnique* 15 (1835): 144–154.

38. E. Chambers, *Cyclopaedia; or, a Universal Dictionary of Arts and Sciences* (London: James and John Knapton, 1728), http://digital.library.wisc.edu/1711.dl/HistSciTech .Cyclopaedia01.

39. A. O. Persson, "Hadley's Principle: Understanding and Misunderstanding the Trade Winds," *History of Meteorology* 3 (2006): 17–42.

40. The Dalton publications of 1793 and 1834 mentioned in his letter to the editor of the *Philosophical Magazine* are Dalton's first monograph and its second edition: J. Dalton, *Meteorological Observations and Essays*, 2nd ed. (London: Baldwin and Cradock, 1834).

41. Persson, "Hadley's Principle."

42. M. G. Gardner, *The Annotated Ancient Mariner* (New York: Clarkson Potter, 1965).

43. M. Garstang, *Climate and the Mfecane* (draft manuscript, 2012).

44. See R. W. Katz, "Sir Gilbert Walker and the Connection Between El Niño and Statistics," *Statistical Sciences* 1 (2002): 97–112, for a scientifically annotated biography. See also G. T. Walker, "Correlation in Seasonal Variations of Weather. II," *Memoirs of the Indian Meteorological Department* 21 (1910): 22–45; G. T. Walker, "Correlation in Seasonal Variations of Weather. III. On the Criterion for the Reality of Relationships or Periodicities," *Memoirs of the Indian Meteorological Department* 21 (1914): 13–15; G. T. Walker, "Correlation in Seasonal Variations of Weather," *Quarterly Journal of the Royal Meteorological Society* 44 (1918): 223–224; G. T. Walker, "Correlation in Seasonal Variations of Weather. VIII. A Preliminary Study of World-Weather," *Memoirs of the Indian Meteorological Department* 24 (1923): 75–131; G. T. Walker, "Correlation in Seasonal Variations of Weather. IX. A Further Study of World Weather," *Memoirs of the Indian Meteorological Department* 24 (1924): 275–332.

45. Garstang, *Climate and the Mfecane.*

46. J. Bjerknes, "Atmospheric Teleconnections from the Equatorial Pacific," *Monthly Weather Review* 97 (1969): 163–172.

47. K. E. Trenberth, "General Characteristics of El Niño–Southern Oscillation," in *Teleconnections Linking Worldwide Climate Anomalies: Scientific Basis and Societal Impact*, ed. M. H. Glantz, R.W. Katz, and N. Nicholls (Cambridge: Cambridge University Press, 1991), 13–42.

48. K. E. Trenberth, "Signal Versus Noise in the Southern Oscillation," *Monthly Weather Review* 112 (1984): 326–332.

49. Walker, "Correlation in Seasonal Variations of Weather. IX"

50. Walker, "Correlation in Seasonal Variations of Weather. VII."

51. P. J. Lamb and R. A. Peppler, "North Atlantic Oscillation: Concept and an Application," *Bulletin of the American Meteorological Society* 68 (1987): 1218–1225;

J. W. Hurrell, "Decadal Trends in the North Atlantic Oscillation: Regional Temperatures and Precipitation," *Science* 269 (1995): 676–679; J. W. Hurrell, "Influence of Variations in Extratropical Wintertime Teleconnections on Northern Hemisphere Temperature," *Geophysical Research Letters* 23 (1996): 665–668.

52. J. M. Wallace and D. S. Gutzler, "Teleconnections in the Geopotential Height Field During the Northern Hemisphere Winter," *Monthly Weather Review* 109 (1981): 784–812.

53. W. Ferrel, *An Essay on the Winds and Currents of the Oceans* (Nashville, Tenn.: Nashville Journal of Medicine, 1856).

54. Forty-five days, 13 hours, 42 minutes, and 53 seconds, by Loïck Peyron on the trimaran *Banque Populaire V*, completed on January 6, 2012. http://www.sailspeed records.com.

55. P. E. Russell, *Prince Henry "The Navigator": A Life* (New Haven, Conn.: Yale University Press, 2000).

56. N. Mostert, *Frontiers* (New York: Knopf, 1992).

57. M. Mitchell, *Friar Andrés de Urdaneta, O.S.A.* (London: Macdonald and Evans, 1964).

58. C. G. Mann, *1493: Uncovering the New World Columbus Created* (New York: Knopf, 2012).

59. A. W. Sleeswyk, "Carvel-Planking and Carvel Ships in the North of Europe," *Archaeonautica* 14 (1998): 223–228.

60. See E. T. Dankwa, *Development of the Square-Rigged Ship from the Carrack to the Full-Rigger* (1999), http://www.in-arch.net/index.html.

61. C. Wunsch, "What Is the Thermohaline Circulation?" *Science* 298 (2002): 1179–1180.

62. Ibid.

63. W. S. Broecker, "The Great Ocean Conveyor," *Oceanography* 4 (1991): 79–89.

64. W. H. Berger, "The Younger Dryas Cold Spell: A Quest for Causes," *Global and Planetary Change* 3 (1990): 219–237.

65. R. B. Alley et al., "Abrupt Increase in Greenland Snow Accumulation at the End of the Younger Dryas Event," *Nature* 362 (1993): 527–529.

66. W. S. Broecker, "Was the Younger Dryas Triggered by a Flood?" *Science* 312 (2006): 1146–1148.

67. I. Eisenman, C. M. Bitz, and E. Tziperman, "Rain Driven by Receding Ice Sheets as a Cause of Past Climate Change," *Paleoceanography* 24 (2009): PA4209.

68. Berger, "The Younger Dryas Cold Spell."

69. R. B. Firestone et al., "Evidence for an Extraterrestrial Impact 12,900 Years Ago That Contributed to the Megafaunal Extinctions and the Younger Dryas Cooling," *Proceedings of the National Academy of Sciences* 104 (2007): 6016–6021; T. E. Bunch et al., "Very High-Temperature Impact Melt Products as Evidence for Cosmic

Airbursts and Impacts 12,900 Years Ago," *Proceedings of the National Academy of Sciences* 109 (2012): E1903–1912.

70. H. N. Pollack, S. J. Hurter, and J. R. Johnson, "Heat Flow from the Earth's Interior: Analysis of the Global Data Set," *Reviews in Geophysics* 30 (1993): 267–280; J. H. Davies and D. R. Davies, "Earth's Surface Heat Flux," *Solid Earth* 1 (2010) :5–24.

71. J. Kiehl and K. Trenberth, "Earth's Annual Global Mean Energy Budget," *Bulletin of the American Meteorological Society* 78 (1997): 197–206.

72. Ibid.

73. Ibid.

74. H. Le Treut et al., "Historical Overview of Climate Change," in *Climate Change 2007: The Physical Science Basis. Contribution of Working Group I to the Fourth Assessment Report of the Intergovernmental Panel on Climate Change*, ed. S. Solomon, et al. (Cambridge: Cambridge University Press, 2007), 93–127.

75. Ibid.

76. Ibid.

77. J.-B. J. Fourier, "Remarques générales sur les températures du globe terrestre et des espaces planétaires," *Annales de Chiie et de Physique, 2nd Series* 27 (1824): 136–167; translation by Ebeneser Burgess in *American Journal of Science* 32 (1837): 1–20.

78. J. R. Fleming, *Historical Perspectives on Climate Change* (New York: Oxford University Press, 1998).

79. S. Arrhenius, "On the Influence of Carbonic Acid in the Air Upon the Temperature of the Ground," *Philosophical Magazine and Journal of Science, Series 5* 41 (1896): 265.

80. J. Uppenbrink, "Arrhenius and Global Warming," *Science* 272 (1996): 1122.

81. See Arrhenius, "On the Influence of Carbonic Acid"; summary in Fleming, *Historical Perspectives on Climate Change*.

82. S. Solomon et al., eds., *Climate Change 2007: The Physical Science Basis. Contribution of Working Group I to the Fourth Assessment Report of the Intergovernmental Panel on Climate Change* (Cambridge: Cambridge University Press, 2007).

83. J. E. Kutzbach, "Steps in the Evolution of Climatology: From Descriptive to Analytic," in *Historical Essays on Meteorology, 1919–1995*, ed. J. R. Fleming (Boston: American Meteorological Society, 1996), 353–377.

84. Ibid.

8. MAKING THE GROUND PUT FORTH GRASS: THE RELATIONSHIP BETWEEN CLIMATE AND VEGETATION

1. K. Schifferdecker, *Out of the Whirlwind: Creation Theology in the Book of Job*, Harvard Theological Studies 61 (Cambridge, Mass.: Harvard University Press, 2008).

2. G. M. Tucker, "Rain on a Land Where No One Lives: The Hebrew Bible and the Environment," *Journal of Biblical Literature* 116 (1997): 3–17.

3. Ibid.

4. The standard references are J. Grinnell, "The Niche Relations of the California Thrasher," *Auk* 34 (1917): 364–382; J. Grinnell, "Field Tests and Theories Concerning Distributional Control," *American Naturalist* 51 (1917): 115–128. However, many of the ideas are presented in the less frequently cited J. Grinnell, "The Origin and the Distribution of the Chestnut-Backed Chickadee," *Auk* 21 (1904): 364–382. For early uses of the word in an ecological context, see P. M. Gaffney, "The Roots of the Niche Concept," *American Naturalist* 109 (1975): 490; D. L. Cox, "A Note on the Queer History of 'Niche,'" *Bulletin of the Ecological Society of America* 61 (1980): 201–202. The later references illustrate that the word "niche" was in common use to describe where animals were found. Grinnell formalized the niche concept but did not invent its general use.

5. H. W. Grinnell, "Joseph Grinnell: 1877–1939, with Frontispiece and Eleven Other Illustrations," *Condor* 42 (1940): 3–34.

6. C. Elton, *Animal Ecology* (New York: MacMillan and Co., 1927); P. S. Giller, *Community Structure and the Niche* (London: Chapman and Hall, 1984); J. R. Griesemer, "Niche: Historical Perspectives," in *Keywords in Evolutionary Biology*, ed. E. F. Keller and E. A. Lloyd (Cambridge, Mass.: Harvard University Press, 1992), 231–240.

7. H. H. Shugart, *How the Earthquake Bird Got Its Name and Other Tales of an Unbalanced Nature* (New Haven, Conn.: Yale University Press, 2004).

8. W. Tang and G. Eisenbrand, *Chinese Drugs of Plant Origin* (New York: Springer-Verlag, 1992); D. Wujastyk, *The Roots of Ayurveda: Selections from Sanskrit Medical Writings* (London: Penguin, 2003); M. Stuart, ed., *Herbs and Herbalism* (London: Orbis, 1979).

9. G. E. Smith, *The Papyrus Ebers*, trans. C. P. Bryan (London: Garden City, 1930).

10. B. Ebbell, *The Papyrus Ebers: The Greatest Egyptian Medical Document* (Copenhagen: Levin and Munsgaard, 1937).

11. M. Clagett, *Ancient Egyptian Science, A Source Book*, vol. 3: *Ancient Egyptian Mathematics* (Philadelphia: American Philosophical Society, 1999); C. Rossi, *Architecture and Mathematics in Ancient Egypt* (Cambridge: Cambridge University Press, 2007).

12. Stuart, ed., *Herbs and Herbalism*.

13. A. Hort, *Enquiry Into Plants and Minor Works on Odours and Weather Signs. By Theophrastus and Translated by Sir Albert Hort*, 2 vols. (London: Heinemann, 1916); A. G. Morton, *History of Botanical Science* (London: Academic Press, 1981).

14. H. H. Shugart and F. I. Woodward, *Global Change and the Terrestrial Biosphere: Achievements and Challenges* (Oxford: Wiley-Blackwell, 2011).

15. G. Fowden, *The Egyptian Hermes: A Historical Approach to the Late Pagan Mind* (Cambridge: Cambridge University Press, 1986).

16. L. Berggren and A. Jones, *Ptolemy's Geography: An Annotated Translation of the Theoretical Chapters* (Princeton, N.J.: Princeton University Press, 2000).

17. For example, see Yāqūt ibn ʿAbd Allāh al-Ḥamawī, *The Introductory Chapters of Yāqūt's Muʿjam al-Buldān* (Leiden: Brill, 1959).

18. C. Linnaeus, *Systema naturae, sive regna tria naturae systematice proposita per classes, ordines, genera, & species* (Lugduni Batavorum: Haak, 1735).

19. C. Schweizer, "Migrating Objects: The Bohemian National Museum and Its Scientific Collaborations in the Early Nineteenth Century," *Journal of the History of Collections* 18 (2006): 187–199.

20. C. M. von Sternberg, *Versuch einer geognostisch-botansuchen Darstellung der Flora der Vorvelt* (Leipzig: Fleischer, 1820–1838).

21. Shugart and Woodward, *Global Change and the Terrestrial Biosphere*.

22. M. C. Ebach and D. F. Goujet, "The First Biogeographical Map," *Journal of Biogeography* 33 (2006): 761–769

23. T. Rice, *Voyages of Discovery* (London: Allen and Unwin, 2010).

24. L. Schiebinger, "Jeanne Baret: The First Woman to Circumnavigate the Globe," *Endeavour* 27 (2003): 22–25.

25. L. A. Gilbert, "Banks, Sir Joseph (1743–1820)," *Australian Dictionary of Biography* (National Centre of Biography, Australian National University, 2011), http://adb.anu.edu.au/biography/banks-sir-joseph-1737/text1917. Of the four servants, two were white men from the town of Revesby (James Roberts and Peter Briscoe) and two were black men (Thomas Richmond and George Dorlton).

26. G. Blainey, *Sea of Dangers: Captain Cook and His Rivals* (Sydney: Penguin, 2008).

27. Gilbert, "Banks, Sir Joseph (1743–1820)."

28. Rice, *Voyages of Discovery*.

29. C. Darwin, *Journal of the Researches Into the Natural History and Geology of the Countries Visited During the Voyage of the HMS Beagle Round the World Under the Command of Capt. Fitz Roy, R.N.* (London: W. Clowes and Sons, 1860).

30. F. N. Egerton, "A History of the Ecological Sciences, Part 32: Humboldt, Nature's Geographer," *Bulletin of the Ecological Society of America* 90 (2009): 253–282.

31. W. George, "Louis Antoine de Bougainville (1729–1811)," *Dictionary of Scientific Biography* 2 (1970): 342–343; G. Forster, *A Voyage Round the World*, vol. 1, ed. N. Thomas and O. Berghof (Honolulu: University of Hawai'i Press, 2001); Egerton, "A History of the Ecological Sciences, Part 32."

32. Egerton, "A History of the Ecological Sciences, Part 32."

33. F. W. H. A. von Humboldt, *Florae Fribergensis specimen plantas cryptogamicas praesertim subterraneas exhibens* (Berlin: Henr. Augustum Rottmann, 1793). Cited

in R. Hartshorne, "The Concept of Geography as a Science of Space, from Kant and Humboldt to Hettner," *Annals of the Association of American Geographers* 48 (1958): 97–108.

34. H. Beck, "Das Ziel der grossen Reise Alexander von Humboldts," *Erdkunde* 12 (1958): 42–58; reprinted in W. Stearn, ed., *Humboldt, Bonpland, Knuth, and Tropical American Botany* (Lehre: J. Cramer, 1968).

35. F. Fleming, *Off the Map: Tales of Endurance and Exploration* (London: Weidenfeld and Nicolson, 2004); Egerton, "A History of the Ecological Sciences, Part 32."

36. Baudin aux Citoyens Professeurs et Adminitstrateurs du Museum nationale d'histoire naturelle à Paris, 6 Thermidor An VI [24 Juillet 1798], *Archives nationales* (Paris), AJXV 569, 178–179; cited in Madeleine Ly-Tio-Fane, "A Reconnaissance of Tropical Resources During Revolutionary Years: The Role of the Paris Museum d'Histoire Naturelle," *Archives of Natural History* 18 (1991): 333–362.

37. Egerton, "A History of the Ecological Sciences, Part 32."

38. Ibid.

39. E. Knoblich, "Alexander von Humboldt—The Explorer and the Scientist," *Centaurus* 47 (2007): 3–14.

40. D. Botting, *Humboldt and the Cosmos* (New York: Harper and Row, 1973). Cited in Egerton, "A History of the Ecological Sciences, Part 32."

41. Egerton, "A History of the Ecological Sciences, Part 32."

42. C. Limoges, "The Development of the Muséum d'Histoire Naturelle of Paris, c. 1808–1914," in *The Organization of Science and Technology in France, 1808–1914* (Paris: Maison des Sciences de l'Homme, 1980), 211–240.

43. Egerton, "A History of the Ecological Sciences, Part 32."

44. G. Helferich, *Humboldt's Cosmos: Alexander von Humboldt and the Latin American Journey That Changed the Way We See the World* (New York: Gotham, 2004).

45. J. Löwenberg, "Alexander von Humboldt: bibliographische Uebersicht seiner Werke, Schriften und zerstreuten Abhandlungen," in *Alexander von Humboldt: eine wissenschaftliche Biographie* (Leipzig: F. A. Brockhaus, 1872), 2:487–552.

46. F. W. H. A. von Humboldt and A. J. A. Bonpland, *Essai sur la géographie des plantes, accompagné d'un tableau physique des regions équinoxiales, fondé sur des measures executes* (Paris: Fr. Schoell, 1807).

47. Egerton, "A History of the Ecological Sciences, Part 32."

48. A. L. de Lavoisier, *Traité élémentaire de chimie* (Paris: Cuchet, 1789); C. Lyell, *Principles of Geology* (London: John Murray, 1830–1833).

49. Egerton, "A History of the Ecological Sciences, Part 32."

50. K. R. Bierman, "Streiflichter auf geophysikalische Aktivitäten Alexander von Humboldts," *Gerlands Beiträge zur Geophysik* 80 (1971): 277–291.

51. P. J. Bowler, "Climb Chimborazo and See the World," *Science* 208 (2002): 63–63.

52. Correspondence from F. W. H. A. von Humboldt to C. Darwin, September 18, 1839. Darwin Correspondence Database, http://www.darwinproject.ac.uk /entry534. Original text: "Vous me dites dans Votre aimable lettre que, tres jeune, ma maniere d'étudier et de peindre la nature sous la zone torride avoit pu contribuer à exciter en Vous l'ardeur et le désir des voyages lointains. D'après l'importance de Vos travaux, Monsieur, ce seroit là le plus grand succes que mes faibles travaux auroient pu obtenir."

53. Knoblich, "Alexander von Humboldt."

54. A. von Humboldt, *Reise auf dem Rio Magdalena, durch die Anden und Mexico* [Alexander von Humboldt's journals for his travels to the new world from 1799 to 1804], vol. 1: *Texts*, transcribed and ed. Margot Faak (Berlin: Akademie Verlag, 1986). Quotation is from *Reise T. II: Übersetzung*, op. cit., 258.

55. A. G. Tansley, "The Use and Abuse of Vegetational Concepts and Terms, *Ecology* 16 (1935): 284–307.

56. A. H. R. Grisebach, *Die Vegetation der Erde nach ihrer klimatoschen Anordnung*, 2 vols. (Leipzig: Verlag von Wilhelm Engelmann, 1872).

57. C. C. Raunkiaer, *The Life Forms of Plants and Statistical Plant Geography* (Oxford: Clarendon, 1934).

58. J. Grace, "Climatic Tolerance and the Distribution of Plants," *New Phytologist* 106 (1987): 113–130 (suppl.)

59. F. I. Woodward, *Climate and Plant Distribution* (Cambridge: Cambridge University Press, 1987).

60. W. Köppen and R. Geiger, *Handbuch der Climatologie* (Berlin: Teil I D, 1930); C. W. Thornthwaite, "The Climates of North America According to a New Classification," *Geographical Review* 21 (1931): 633–655.

61. Ibid.

62. Shugart and Woodward, *Global Change and the Terrestrial Biosphere*.

63. S. Manabe and R. J. Stouffer, "Sensitivity of a Global Climate to an Increase in CO_2 Concentration in the Atmosphere," J. *Geophy. Res.* 85 (1980): 5529–5554.

64. M. E. Schlesinger and Z.-C. Zhao, *Seasonal Climatic Changes Induced by Doubled CO_2 as Simulated by the OSU Atmospheric GCM/Mixed Layer Ocean Model* (Corvallis, Ore.: Climate Research Institute, Oregon State University, 1988); S. Manabe and R. T. Wetherald, "Large-Scale Changes in Soil Wetness Induced by an Increase in Carbon Dioxide," *Journal of Atmospheric Science* 44 (1987): 1211–1235; J. Hansen et al., "Global Climate Changes as Forecast by the Goddard Institute for Space Studies' Three Dimensional Model," *Journal of Geophysical Research* 93 (1988): 9341–9364; J. F. B. Mitchell, "The Seasonal Response of a General Circulation Model to Changes in CO2 and Sea Temperatures," *Quarterly Journal of the Royal Meteorological Society* 109 (1983): 113–152.

65. A. Henderson-Sellers and K. McGuffie, "Land-Surface Characterization in Greenhouse Climate Simulations," *International Journal of Climatology* 14 (1994): 1065–1094.

66. M. I. Budyko, *Heat Balance of the Earth's Surface* [in Russian] (Leningrad: Gidrometeoizdat, 1956); M. I. Budyko, *Climate and Life* (New York: Academic Press, 1971); M. I. Budyko, *The Evolution of the Biosphere* (Dordrecht: D. Reidel, 1986).

67. N. M. Tchebakova et al., "A Global Vegetation Model Based on the Climatological Approach of Budyko," *Journal of Biogeography* 25 (1993): 59–83; N. M. Tchebakova, R. Monserud, and D. I. Nazimova, "A Siberian Vegetation Model Based on Climatic Parameters," *Canadian Journal of Forest Research* 24 (1994): 1597–1607.

68. K. C. Prentice and I. Y. Fung, "Bioclimatic Simulations Test the Sensitivity of Terrestrial Carbon Storage to Perturbed Climates," *Nature* 346 (1990): 48–51.

69. T. M. Smith and H. H. Shugart, "The Transient Response of Terrestrial Carbon Storage to a Perturbed Climate," *Nature* 361(1993): 523–526.

70. Ibid.

71. E. O. Box, *Macroclimate and Plant Forms: An Introduction to Predictive Modeling in Phytogeography* (The Hague: Dr. W. Junk, 1981).

72. Ibid.

73. Ibid.

74. T. M. Smith, H. H. Shugart, and P. N. Halpin, "Computer Models of Forest and Global Changes in the Environment," in *Responses of Forest Ecosystems to Environmental Change*, ed. A. Teller, P. Mathy, and J. N. R. Jeffers (Essex: Elsevier, 1992), 91–102; T. M. Smith, J. B. Smith, and H. H. Shugart, "Modeling the Response of Tropical Forests to Climate Change: Integrating Regional and Site-Specific Studies," in *Tropical Forests in Transition: Ecology of Natural and Anthropogenic Disturbance Processes in Tropical Forest Biomes*, ed. J. G. Goldammer (Basel: Birkhauser-Verlag, 1992), 253–268.

75. T. M. Smith, H. H. Shugart, and F. I. Woodward, eds., *Plant Functional Types: Their Relevance to Ecosystem Properties and Global Change* (Cambridge: Cambridge University Press, 1997).

76. C. Zabinski and M. B. Davis, "Hard Times Ahead for Great Lakes Forests: A Climate Threshold Model Predicts Responses to CO_2-Induced Climate Change," in *The Potential Effects of Global Climate Change on the United States (EPA-230-05-89-054)*, ed. J. B. Smith and D.A. Tirpak (Washington, D.C.: U.S. Environmental Protection Agency, 1989), 5-1–5-19.

77. Hansen et al., "Global Climate Changes"; Manabe and Stouffer, "Sensitivity of a Global Climate to an Increase in CO_2."

78. B. Huntley, "Plant Species' Response to Climate Change: Implications for the Conservation of European Birds," *Ibis* 137 (1994): S127–S138.

79. Mitchell, "The Seasonal Response of a General Circulation Model"; Schlesinger and Zhao, *Seasonal Climatic Changes Induced by Doubled CO2.*

80. I. Newton, *Finches* (London: Collins, 1972).

81. E. Dexter and A. Chapman, "Climate Change and Endangered Species," *Erinyes* (Newsletter of the Environmental Resources Information Network, Department of the Environment, Sport and Territories, Canberra, Australia) 23 (1995): 1, 6.

82. T. Webb III, "The Appearance and Disappearance of Major Vegetational Assemblages: Long-Term Vegetation Dynamics in Eastern North America," *Vegetatio* 69 (1987): 177–187; J. C. Bernabo and T. Webb III, "Changing Patterns in the Holocene Pollen Record of Northeastern North America: A Mapped Summary," *Quaternary Research* 8 (1977): 64–96; I. C. Prentice et al., "Reconstructing Biomes from Palaeoecological Data: A General Method and Its Application to European Pollen Data at 0 and 6 ka," *Climate Dynamics* 12 (1996): 185–194; B. Huntley et al., "Modeling Present and Potential Future Ranges of Some European Higher Plants Using Climate Response Surfaces," *Journal of Biogeography* 22 (1996): 967–1001.

83. C. S. Holling, "Principles of Insect Predation," *Annual Review of Entomology* 6 (1961): 163–182; C. S. Holling, "The Analysis of Complex Population Processes," *Canadian Entomologist* 96 (1964): 335–347; F. J. Rohlf and D. Davenport, "Simulation of Simple Models of Animal Behavior with a Digital Computer," *J. Theor. Biol.* 23 (1969): 400–424.

84. M. Huston, D. L. DeAngelis, and W. M. Post, "New Computer Models Unify Ecological Theory," *BioScience* 38 (1988): 682–691.

85. H. H. Shugart and D. C. West, "Forest Succession Models," *BioScience* 30 (1980): 308–313.

86. H. H. Shugart, *Terrestrial Ecosystems in Changing Environments* (Cambridge: Cambridge University Press, 1998).

87. J. K. Shuman, H. H. Shugart, and T. L. O'Halloran, "Sensitivity of Siberian Larch Forests to Climate Change," *Global Change Biology* 17 (2011): 2370–2384.

88. A. M. Solomon et al., "Testing a Simulation Model for Reconstruction of Prehistoric Forest-Stand Dynamics," *Quaternary Research* 14 (1980): 275–293; A. M. Solomon and T. Webb III, Computer-Aided Reconstruction of Late Quaternary Landscape Dynamics," *Annual Reviews of Ecology and Systematics* 16 (1985): 63–84; A. M. Solomon, "Comparison of Taxon Calibrations, Modern Analog Techniques, and Forest-Stand Simulation Models for the Quantitative Reconstruction of Past Vegetation: A Critique," *Earth Surface Processes and Landforms* 11 (1986): 681–685; G. B. Bonan and B. P. Hayden, "Using a Forest-Stand Simulation Model to Examine the Ecological and Climatic Significance of the Late-Quaternary Pine/Spruce Pollen Zone in Eastern Virginia," *Quaternary Research* 33 (1990): 204–218.

89. H. H. Shugart, T. M. Smith, and W. M. Post, "The Potential for Application of Individual-Based Simulation Models for Assessing the Effects of Global Change," *Annual Reviews of Ecology and Systematics* 23 (1992): 15–38.

90. H. H. Shugart and T. M. Smith, "A Review of Forest-Patch Models and Their Application to Global Change Research," *Climatic Change* 34 (1996): 131–153; H. Bugmann, J. F. Reynolds, and L. F. Pitelka, "How Much Physiology Is Needed in Forest Gap Models for Simulating Long-Term Vegetation Response of Forests?" *Climatic Change* 51 (2001): 249–250.

91. H. H. Shugart and F. I. Woodward, "Global Change and the Terrestrial Biosphere"; A. M. Solomon, "Transient Response of Forests to CO_2-Induced Climate Change: Simulation Experiments in Eastern North America," *Oecologia* 68 (1986): 567–579; G. B. Bonan, "Environmental Factors and Ecological Processes Controlling Vegetation Patterns in Boreal Forests," *Landscape Ecology* 3 (1989): 111–130; T. M. Smith et al., "Modeling the Potential Response of Vegetation to Global Climate Change," *Advances in Ecological Research* 22 (1992): 93–116.

92. H. Bossel, "Modelling Forest Dynamics: Moving from Description to Explanation," *Forest Ecology and Management* 42 (1991): 129–142; H. Bossel, "TREEDYN3 Forest Simulation Model," *Ecological Modelling* 90 (1996): 187–227; H. Bossel and H. Krieger, "Simulation Model of Natural Tropical Forest Dynamics," *Ecological Modelling* 59 (1991): 37–71; H. Bossel and H. Krieger, "Simulation of Multispecies Tropical Forest Dynamics Using a Vertically and Horizontally Structured Model," *Forest Ecology and Management* 69 (1994): 123–144; L. Kammesheidt, P. Köhlert, and A. Huth, "Sustainable Timber Harvesting in Venezuela: A Modelling Approach," *Journal of Applied Ecology* 36 (2001): 756–770; L. Liu and P. S. Ashton, "FORMO-SAIC: An Individual-Based Spatially Explicit Model Simulation Forest Dynamics in Landscape Mosaics," *Ecological Modelling* 106 (1998): 177–200; P. Köhler and A. Huth, "The Effects of Tree Species Grouping in Tropical Forest Modeling: Simulations with the Individual-Based Model FORMIND," *Ecological Modelling* 109 (1998): 301–321; J. Chave, "Study of Structural, Successional, and Spatial Patterns in Tropical Rain Forests Using TROLL, a Spatially Explicit Forest Model," *Ecological Modelling* 124 (1999): 233–254; A. Huth and T. Ditzer, "Long-Term Impacts of Logging in a Tropical Rain Forest—A Simulation Study," *Forest Ecology and Management* 142 (2001): 33–51; S. Gourley-Fleury et al., "Using Models to Predict Recovery and Assess Tree Species Vulnerability in Logged Tropical Forests: A Case Study from French Guiana," *Forest Ecology and Management* 209 (2005): 69–86.

93. M. F. Acevedo, D. L. Urban, and H. H. Shugart, "Models of Forest Dynamics Based on Roles of Tree Species," *Ecological Modelling* 87 (1996): 267–284; G. Shao, H. H. Shugart, and T. M. Smith, "A Role-Type Model (ROPE) and Its Application in Assessing Climate Change Impacts on Forest Landscapes," *Vegetatio* 121 (1996): 135–146.

94. T. Kohyama, "Size-Structured Tree Populations in Gap Dynamic Forests—The Forest Architecture Hypothesis for Stable Coexistence of Species," *Journal of Ecology* 81 (1993): 131–143.

95. P. R. Moorcroft, G. C. Hurtt, and S. W. Pacala, "A Method for Scaling Vegetation Dynamics: The Ecosystem Demography Model (ED)," *Ecological Monographs* 74 (2001): 557–586.

96. J. K. Shuman, H. H. Shugart, and T. L. O'Halloran, "Sensitivity of Siberian Larch Forests to Climate Change," *Global Change Biology* 17 (2011): 2370–2384.

97. Ibid.

98. D. B. Zilversmit, C. Entenmann, and M. C. Fishler, "On the Calculation of Turnover Time and Turnover Rate from Experiments Involving the Use of Labeling Agents," *J. Gen. Physiol.* 26 (1943): 325–331.

99. C. F. Jordan, "Ecological Effects of Nuclear Radiation," in *Ecological Knowledge and Environmental Problem Solving: Concepts and Case Studies*, ed. G. H. Orians (Washington, D.C.: National Academy Press, 1986), 331–344.

100. G. P. Peters et al., "Rapid Growth in CO_2 Emissions After the 2008–2009 Global Financial Crisis," *Nature Climate Change* (2011), doi:10.1038/nclimate1332.

101. Y. Pan et al., "A Large and Persistent Carbon Sink in the World's Forests," *Science* 333 (2011): 988–993.

102. F. I. Woodward, *Climate and Plant Distribution* (Cambridge: Cambridge University Press, 1987).

103. H. L. Penman, "Natural Evaporation from Open Water, Soil, and Grass," *Proceedings of the Royal Society of London, Series A* 193 (1948): 120–145; J. L. Monteith, "Evaporation and the Environment," in *The State and Movement of Water in Living Organisms*, ed. C. E. Fogg (Cambridge: Cambridge University Press, 1981), 205–234.

104. T. M. Smith et al., "Modeling the Potential Response of Vegetation to Global Climate Change," *Advances in Ecological Research* 22 (1992): 93–116.

105. T. J. Blasing, *Recent Greenhouse Gas Concentrations* (Oak Ridge, Tenn.: Carbon Dioxide Information Analysis Center, Oak Ridge National Laboratory, 2011), http://cdiac.ornl.gov/pns/current_ghg.htmlDOI: 10.3334/CDIAC/atg.032.

106. D. B. Jiang, et al., "Modeling the Middle Pliocene Climate with a Global Atmospheric General Circulation Model," *Journal of Geophysical Research* 110 (2005), D14107, doi:10.1029/2004JD005639; M. E. Raymo and G. H. Rau, "Plio-Pleistocene Atmospheric CO_2 Levels Inferred from POM $\delta^{13}C$ at DSDP Site 607," *Eos* 73 (1992): 95; M. E. Raymo et al., "Mid-Pliocene Warmth: Stronger Greenhouse and Stronger Conveyor," *Marine Micropaleontology* 27 (1996): 313–326; Z. T. Guo et al., "Late Miocene-Pliocene Development of Asian Aridification as Recorded in the Red-Earth Formation in Northern China," *Global and Planetary Change* 41 (2004): 135–145; N. J. Shackleton, J. C. Hall, and D. Pate, "Pliocene Stable Isotope Stratigraphy of ODP Site 846," in *Proceedings of the Ocean Drilling Program, Scientific Results*, vol. 138 (College Station, Texas: Ocean Drilling Program, 1995), 337–356.

107. *Climate Change 2007: The Physical Science Basis. Contribution of Working Group I to the Fourth Assessment Report of the Intergovernmental Panel on Climate Change*, ed. S. Solomon et al. (Cambridge: Cambridge University Press, 2007).

108. F. I. Woodward, "Stomatal Numbers Are Sensitive to Increases in CO_2 from Preindustrial Levels," *Nature* 327 (1987): 617–618.

109. F. I. Woodward, "Plant Responses to Past Concentrations of CO_2," *Vegetatio* 104/105 (1993): 145–155.

110. R. J. Norby et al., "Forest Response to Elevated CO_2 Is Conserved Across a Broad Range of Productivity," *Proceedings of the National Academy of Sciences of the United States of America* 102 (2005): 18052–18056.

111. A. D. Friend, "Terrestrial Plant Production and Climate Change," *Journal of Experimental Botany* 61 (2011): 1293–1309.

112. R. J. Norby et al., "CO_2 Enhancement of Forest Productivity Constrained by Limited Nitrogen Availability," *Proceedings of the National Academy of Sciences* 104 (2011): 19368–19373.

113. C. Körner, "Plant CO_2 Responses: An Issue of Definition, Time, and Resource Supply," *New Phytologist* 172 (2006): 393–411.

114. Ibid.

115. F. I. Woodward and W. L. Steffen, *Natural Disturbances and Human Land Use in Dynamic Global Vegetation Models*. IGBP Report 38 (Stockholm, 1996).

116. W. Cramer et al., "Global Response of Terrestrial Ecosystem Structure and Function to CO_2 and Climate Change: Results from Six Dynamic Global Vegetation Models," *Global Change Biology* 7 (2001): 357–373.

117. Friend, "Terrestrial Plant Production and Climate Change"; Shugart and Woodward, *Global Change and the Terrestrial Biosphere*.

118. A. Voinov and H. Shugart, "'Integronsters' and the Special Role of Data," in *International Environmental Modelling and Software Society 2010 International Congress on Environmental Modelling and Software Modelling for Environment's Sake*, ed. D. A. Swayne et al. Fifth Biennial Meeting, Ottawa, Canada, 2011.

119. R. A. Betts et al., "Contrasting Physiological and Structural Vegetation Feedbacks in Climate Change Simulations," *Nature* 387 (1997): 796–799.

120. P. M. Cox et al., "Acceleration of Global Warming Due to Carbon-Cycle Feedbacks in a Coupled Climate Model," *Nature* 408 (2000): 184–187.

121. P. Friedlingstein, et al., "Climate-Carbon Cycle Feedback Analysis: Results from the C4MIP Model Intercomparison," *Journal of Climate* 19 (2006): 3337–3353.

122. Cox et al., "Acceleration of Global Warming."

123. Shuman, Shugart, and O'Halloran, "Sensitivity of Siberian Larch Forests to Climate Change."

124. Smith and Shugart, "The Transient Response of Terrestrial Carbon Storage to a Perturbed Climate."

125. Cox et al., "Acceleration of Global Warming."

126. P. Cox and C. Jones, "Illuminating the Modern Dance of Climate and CO_2," *Science* 321 (2008): 1642–1644.

127. Shugart and Woodward, *Global Change and the Terrestrial Biosphere.*

128. R. J. Norby et al., "Forest Response to Elevated CO_2 Is Conserved Across a Broad Range of Productivity," *Proceedings of the National Academy of Sciences of the United States of America* 102 (2005): 18052–18056.

129. Tucker, "Rain on a Land Where No One Lives"; Schifferdecker, *Out of the Whirlwind*; W. P. Brown, *Seven Pillars of Creation: The Bible, Science, and the Ecology of Wonder* (New York: Oxford University Press, 2010).

130. Brown, *Seven Pillars of Creation*, 13.

9. Feeding the Lions: The Conservation of Biological Diversity on a Changing Planet

1. S. J. O'Brien et al., "Biochemical Genetic Variation in Geographic Isolates of African and Asiatic Lions," *National Geographic Research* 3 (1987): 114.

2. U. Breitenmoser et al., "*Panthera leo* ssp. *persica*," in IUCN Red List of Threatened Species, version 2010.4. http://www.iucnredlist.org.

3. H. Bauer, K. Nowell, and C. Packer, "*Panthera leo*," in IUCN Red List of Threatened Species, version 2010.4. http://www.iucnredlist.org.

4. K. Nowell and P. Jackson, *Wild Cats: Status Survey and Conservation Action Plan* (Gland: IUCN/SSC Cat Specialist Group, 1996).

5. P. Jackson, "Nearly One-Tenth of Last Asiatic Lions Died This Year," *Cat News* 47 (2008): 36–37.

6. M. Tsevat, "The Meaning of the Book of Job," *Hebrew Union College Annual* 37 (1966): 73–106. Before his death on March 13, 2010, Matitiahu Tsevat was Professor Emeritus of the Bible at Hebrew Union College–Jewish Institute of Religion in Cincinnati, Ohio.

7. T. P. Burns, "Lindeman's Contradiction and the Trophic Structure of Ecosystems," *Ecology* 70 (1989): 1355–1362.

8. F. B. Golley, "Energy Dynamics of a Food Chain of an Old Field Community," *Ecological Monographs* 30 (1960): 187–206.

9. In this case, the units would be in calories per unit of area or kilojoules per unit of area.

10. W. F. Humphreys, "Production and Respiration in Animal Populations," *Journal of Animal Ecology* 48 (1979): 427–453.

11. Ibid.

12. C. S. Elton, *Animal Ecology* (London: Sidgwick and Jackson, 1927).

13. C. Carbone and J. L. Gittleman, "A Common Rule for the Scaling of Carnivore Density," *Science* 295 (2002): 2273–2276; J. L. Gittleman, "Carnivore Body Size: Ecological and Taxonomic Correlates," *Oecologia* 67 (1985): 540–554.

14. See M. E. Power, "Top-Down and Bottom-Up Forces in Food Webs: Do Plants Have Primacy?" *Ecology* 73 (1992): 733–746; Y. Ayal, "Productivity, Organism Size, and the Trophic Structure of the Major Terrestrial Biomes," *Theoretical Ecology* 4 (2011): 1–11; D. M. Post, M. L. Pace, and N. G. Hairston Jr., "Ecosystem Size Determines Food-Chain Length in Lakes," *Nature* 405 (2001): 1047–1049.

15. Elton, *Animal Ecology*.

16. J. R. Kercher and H. H. Shugart, "Trophic Structure, Effective Trophic Position, and Connectivity in Food Webs," *American Naturalist* 109 (1975): 191–206.

17. J. H. Brown et al., "Toward a Metabolic Theory of Ecology," *Ecology* 85 (2004): 1771–1789.

18. N. G. Hairston, F. E. Smith, and L. B. Slobodkin, "Community Structure, Population Control, and Competition," *American Naturalist* 94 (1960): 421–425; N. G. Hairston Jr. and N. G. Hairston Sr., "Cause-Effect Relationships in Energy Flow, Trophic Structure, and Interspecific Interactions," *American Naturalist* 142 (1993): 379–411.

19. Power, "Top-Down and Bottom-Up Forces in Food Webs."

20. T. R. C. White, "The Importance of Relative Shortage of Food in Animal Ecology," *Oecologia* 33 (1978): 1–86.

21. Ibid.

22. The four studied savanna ecosystems are all protected areas: Hwange National Park, Zimbabwe (19.18S, 26.68E); Kruger National Park, South Africa (24.08S, 31.58E); Serengeti National Park, Tanzania (2.58S, 34.68E); Ngorongoro Crater National Park, Tanzania (3.18S, 35.48E).

23. H. Fritz et al., "A Food Web Perspective on Large Herbivore Community Limitation," *Ecography* 34 (2011): 196–202.

24. N. Owen-Smith and M. L. G. Mills, "Predator–Prey Size Relationships in an African Large-Mammal Food Web," *Journal of Animal Ecology* 77 (2008): 173–183; H. Joubert, "Hunting Behavior of Lions (*Pantera leo*) on Elephants (*Loxodonta africana*) in the Chobe National Park, Botswana," *African Journal of Ecology* 44 (2006): 279–281.

25. R. N. Owen-Smith, *Megaherbivores: The Influence of Very Large Body Size on Ecology* (Cambridge: Cambridge University Press, 1988).

26. W. J. Ripple and B. Van Valkenburgh, "Linking Top-Down Forces to Pleistocene Megafaunal Extinctions," *BioScience* 60 (2010): 516–526.

27. C. Carbone et al., "Energetic Constraints on the Diet of Terrestrial Carnivores," *Nature* 402 (1999): 286–288.

28. C. Carbone, A. Teacher, and J. M. Rowcliffe, "The Costs of Carnivory," *PLOS Biology* 5, no. 2 (1999): e22.

29. Ibid.

30. C. Carbone, N. Pettorelli, and P. A. Stephens, "The Bigger They Come, the Harder They Fall: Body Size and Prey Abundance Influence Predator–Prey Ratios," *Biology Letters* 7 (2011): 312–315.

31. K. G. van Orsdol, J. P. Hanby, and J. D. Bygott, "Ecological Correlates of Lion Social Organization (*Panthera leo*)," *Journal of Zoology* (2006): 97–112; J. T. Du Toit and N. Owen-Smith, "Body Size, Population Metabolism, and Habitat Specialization Among Large African Herbivores," *American Naturalist* 133 (1989): 736–740; C. A. Carbone, A. Teacher, and J. M. Rowcliffe, "The Costs of Carnivory," *PLOS Biology* 5 (2007): e22; F. G. T. Radloff and J. T. Du Toit, "Large Predators and Their Prey in a Southern African Savanna: A Predator's Size Determines Its Prey Size Range," *Journal of Animal Ecology* 73 (2004): 410–423; Owen-Smith and Mills, "Predator–Prey Size Relationships."

32. Carbone, Teacher, and Rowcliffe, "The Costs of Carnivory."

33. B. D. Patterson et al., "Livestock Predation by Lions (*Panthera leo*) and Other Carnivores on Ranches Neighboring Tsavo National Parks, Kenya," *Biological Conservation* 119 (2004): 507–516.

34. H. Bauer, *Lion Conservation in West and Central Africa: Integrating Social and Natural Science for Wildlife Conflict Resolution Around Waza National Park, Cameroon* (Institute for Environmental Sciences, Leiden University, 2003).

35. UNICEF data for GNI per capita see: http://www.unicef.org/index.php.

36. G. B. Schaller, *The Serengeti Lion: A Study of Predator-Prey Relations* (Chicago: University of Chicago Press, 1972).

37. W. C. Allee et al., *Principles of Animal Ecology* (Philadelphia: W. B. Saunders, 1949); P. A. Stephens, W. J. Sutherland, and R. P. Freckleton, "What Is the Allee Effect?" *Oikos* 87 (1999): 185–190.

38. M. Bjorklund, "The Risk of Inbreeding Due to Habitat Loss in the Lion (*Panthera leo*)," *Conservation Genetics* 4 (2003): 515–523.

39. H. Bauer, K. Nowell, and C. Packer, "Panthera leo."

40. Ibid.

41. D. M. Raup and J. J. Sepkoski, "Periodicity of Extinctions in the Geologic Past," *Proc. Nat. Acad. Sci. USA Phys. Sci.* 81 (1984): 801–805; D. Jablonski, "Background and Mass Extinctions: The Alternation of Macroevolutionary Regimes," *Science* 231 (1986): 129–133. Also see review in J. B. Harrington, "Climatic Change: A Review of Causes," *Canadian Journal of Forest Research* 17 (1987): 1313–1339.

42. D. H. Erwin, J. W. Valentine, and J. J. Sepkoski, "A Comparative Study of Diversification Events: The Early Paleozoic Versus the Mesozoic," *Evolution* 41 (1987): 365–389; M. Jenkins, "Species Extinctions," in *Global Biodiversity: Status of the Earth's Living Resources*, ed. B. Groombridge (London: Chapman and Hall, 1992), 192–205.

43. W. W. Alvarez, F. Asaro, and H. V. Michel, "Extraterrestrial Cause for the Cretaceous-Tertiary Extinction," *Science* 208 (1980): 1095–1108; R. Ganapathy, "A Major Meteorite Impact on the Earth 65 Million Years Ago: Evidence from the Cretaceous-Tertiary Boundary Clay," *Science* 209 (1980): 921–923.

44. R. Lewin, "Mass Extinctions Select Different Victims," *Science* 231 (1986): 219–220.

45. J. B. Pollack et al., "Environmental Effects of an Impact-Generated Dust Cloud: Implications for the Cretaceous-Tertiary Extinctions," *Science* 219 (1983): 287–289.

46. P. S. Martin and H. E. Wright Jr., eds., *Pleistocene Extinctions: The Search for a Cause* (New Haven, Conn.: Yale University Press, 1967); P. S. Martin and R. G. Klein, eds., *Quaternary Extinctions: A Prehistoric Revolution* (Tucson: University of Arizona Press, 1984).

47. D. S. Webb, "Ten Million Years of Mammal Extinctions in North America," in *Quaternary Extinctions: A Prehistoric Revolution*, ed. P. S. Martin and R. G. Klein (Tucson: University of Arizona Press, 1984), 189–210.

48. T. Nilsson, *The Pleistocene* (Stuttgart: Ferdinand Enke Verlag, 1983); E. Anderson, "Who's Who in the Pleistocene: A Mammalian Bestiary," in *Quaternary Extinctions: A Prehistoric Revolution*, ed. P. S. Martin and R. G. Klein (Tucson: University of Arizona Press, 1984), 5–39.

49. Ibid.

50. R. N. Owen-Smith, *Megaherbivores: The Influence of Very Large Body Size on Ecology* (Cambridge: Cambridge University Press, 1988).

51. See Martin and Klein, *Quaternary Extinctions*, for a collection of different views and three different summarizations of the then available data.

52. Owen-Smith, *Megaherbivores*.

53. P. S. Martin and D. W. Steadman, "Prehistoric Extinctions on Islands and Continents," in *Extinctions in Near Time: Advances in Vertebrate Paleobiology Series*, ed. R. D. E. McPhee (New York: Kleuwer Academic/Plenum Publishers, 1999), 17–55.

54. Ibid.

55. R. D. E. MacPhee and P. A. Marx, "The 40,000 Year Plague: Humans, Hyperdisease, and First-Contact Extinctions," in *Natural Change and Human Impact in Madagascar*, ed. S. Goodman and B. Patterson (Washington, D.C.: Smithsonian Institution Press, 1997), 169–217.

56. Martin and Klein, *Quaternary Extinctions*.

57. J. E. Morrow and C. Gnecco, eds., *Paleoindian Archaeology: A Hemispheric Perspective* (Gainesville: University Press of Florida, 2006).

58. M. R. Waters and T. W. Stafford Jr., "Redefining the Age of Clovis: Implications for the Peopling of the Americas," *Science* 315 (2007): 1122–1126.

59. G. Haynes et al., "Comment on 'Redefining the Age of Clovis: Implications for the Peopling of the Americas,'" *Science* 317 (2007): 320; M. R. Waters and T. W. Stafford Jr., "Response to "Comment on 'Redefining the Age of Clovis: Implications for the Peopling of the Americas,'" *Science* 317 (2007): 320.

60. T. Goebel, M. R. Waters, and D. H. O'Rourke, "The Late Pleistocene Dispersal of Modern Humans in the Americas," *Science* 319 (2008): 1497–1502.

61. Martin and Wright, *Pleistocene Extinctions*; P. S. Martin, "Prehistoric Overkill: The Global Model," in *Quaternary Extinctions: A Prehistoric Revolution*, ed. P. S. Martin and R. G. Klein (Tucson: University of Arizona Press, 1984), 354–403; P. S. Martin, *Twilight of the Mammoths: Ice Age Extinctions and the Rewilding of North America* (Berkeley: University of California Press, 2005); Martin and Steadman, "Prehistoric Extinctions on Islands and Continents."

62. J. E. Mosimann and P. S. Martin, "Simulating Overkill by Paleoindians," *American Scientist* 63 (1975): 304–313.

63. O. K. Davis, "Spores of the Dung Fungus *Sporomiella*: Increased Abundance in Historic Sediments and Before Pleistocene Megafaunal Extinction," *Quaternary Research* 28 (1987): 290–294; R. A. Kerr, "Megafauna Died from Big Kill, Not Big Chill," *Science* 300 (2003): 885.

64. Mosimann and Martin, "Simulating Overkill by Paleoindians."

65. M. I. Budyko, "On the Causes of the Extinction of Some Animals at the End of the Pleistocene," *Soviet Geography Review and Translation* 8 (1967): 783–793.

66. P. S. Martin, "The Discovery of America," *Science* 179 (1973): 969–974.

67. Goebel, Waters, and O'Rourke, "The Late Pleistocene Dispersal of Modern Humans."

68. J. J. Clague, R. W. Mathewes, and T. A. Ager, "Environments of Northwestern North America Before the Last Glacial Maxima," in *Entering America: Northeast Asia and Beringia Before the Last Glacial Maximum*, ed. D. B. Madsen (Salt Lake City: University of Utah Press, 2004), 63–94.

69. D. Jackson et al., "Initial Occupation of the Pacific Coast of Chile During Late Pleistocene Times," *Current Anthropology* 48: 725–731.

70. J. M. Adovasio and D. R. Pedler, "Monte Verde and the Antiquity of Humankind in the Americas," *Antiquity* 71 (1997): 573–580; T. D. Dillehay, *Monte Verde: A Late Pleistocene Settlement in Chile*, vol. 2: *The Archaeological Context and Interpretation* (Washington, D.C.: Smithsonian Institution Press, 1997); D. J. Meltzer et al., "On the Pleistocene Antiquity of Monte Verde, Southern Chile," *American Antiquity* 62 (1997): 659–663.

71. Goebel, Waters, and O'Rourke, "The Late Pleistocene Dispersal of Modern Humans."

72. M. R. Waters et al., "The Buttermilk Creek Complex and the Origins of Clovis at the Debra L. Friedkin Site, Texas," *Science* 331 (2011): 1599–1603.

73. M. B. Collins and B. B. Bradley, "Evidence for Pre-Clovis Occupation at the Gault Site (41BL323), Central Texas," *Current Research in the Pleistocene* 25 (2008): 70–72.

74. D. J. Joyce, "Chronology and New Research on the Schaefer Mammoth (*Mammuthus primigenius*) site, Kenosha County, Wisconsin, USA," *Quaternary International* 142 (2006): 44–57.

75. J. M. Adovasio and D. R. Pedler, "Pre-Clovis Sites and Their Implications for Human Occupation Before the Last Glacial Maximum," in *Entering America: Northeast Asia and Beringia Before the Last Glacial Maximum*, ed. D. M. Madsen (Salt Lake City: University of Utah Press, 2004), 139–158.

76. M. T. P. Gilbert et al., "DNA from Pre-Clovis Human Coprolites in Oregon, North America," *Science* 320 (2008): 786–789.

77. S. D. Webb, ed., *First Floridians and Last Mastodons: The Page-Ladson Site in the Aucilla River* (Dordrecht: Springer, 2005).

78. J. M. McAvoy and L. D. McAvoy, eds., *Archaeological Investigations of Site 44SX202, Cactus Hill, Sussex County, Virginia*, Research Report Series 8 (Richmond: Virginia Department of Historic Resources, 1997); D. L. Lowery et al., "Late Pleistocene Upland Stratigraphy of the Western Delmarva Peninsula, USA," *Quaternary Science Reviews* 29 (2010): 1472–1480.

79. J. M. Erlandson et al., "The Kelp Highway Hypothesis: Marine Ecology, the Coastal Migration Theory, and the Peopling of the Americas," *Journal of Island and Coastal Archaeology* 2 (2007): 161–174; Dillehay, *Monte Verde*.

80. B. Bradley and D. Stanford, "The North Atlantic Ice-Edge Corridor: A Possible Palaeolithic Route to the New World," *World Archaeology* 36 (2004): 459–478.

81. J. R. Johnson et al., "American Geophysical Union Joint Assembly, Acapulco, 22 to 25 May 2007," *Eos* 88, no. 23, Joint Assembly Supplement Abstract PP42A-03; Goebel, Waters, and O'Rourke, "The Late Pleistocene Dispersal of Modern Humans."

82. Bradley and Stanford, "The North Atlantic Ice-Edge Corridor"; A. J. Jelinek, "Early Man in the New World: A Technological Perspective," *Arctic Anthropology* 8 (1971): 15–21.

83. D. J. Stanford and B. A. Bradley, *Across Atlantic Ice: The Origin of America's Clovis Culture* (Berkeley: University of California Press, 2012).

84. Bradley and Stanford, "The North Atlantic Ice-Edge Corridor."

85. Ibid.; McAvoy and McAvoy, *Archaeological Investigations of Site 44SX202*; Lowery et al., "Late Pleistocene Upland Stratigraphy of the Western Delmarva Peninsula."

86. W. A. Neves et al., "Early Holocene Human Skeletal Remains from Cerca Grande, Lagoa Santa, Central Brazil, and the Origins of the First Americans," *World Archaeology* 36 (2004): 479–501.

87. W. W. Howells, *Cranial Variation in Man: A Study by Multivariate Analysis of Patterns of Difference Among Recent Human Populations*, Papers of the Peabody

Museum of Archaeology and Ethnology 67 (Cambridge, Mass.: Harvard University, 1973); W. W. Howells, *Skull Shapes and the Map: Craniometric Analyses in the Dispersion of Modern* Homo, Papers of the Peabody Museum of Archaeology and Ethnology 79 (Cambridge, Mass.: Harvard University, 1989).

88. J. F. Powell, *The First Americans: Race, Evolution, and the Origin of Native Americans* (Cambridge: Cambridge University Press, 2005).

89. Goebel, Waters, and O'Rourke, "The Late Pleistocene Dispersal of Modern Humans."

90. T. Goebel, "Pleistocene Human Colonization of Siberia and Peopling of the Americas: An Ecological Approach," *Evolutionary Anthropology* 8 (1999): 208–277.

91. Ibid.

92. Ibid.

93. N. Pinter, S. Fiedel, and J. E. Keeley, "Fire and Vegetation Shifts in the Americas at the Vanguard of Paleoindian Migration," *Quaternary Science Reviews* 30 (2011): 269–272.

94. J. T. Faith, "Late Pleistocene Climate Change, Nutrient Cycling, and the Megafaunal Extinctions in North America," *Quaternary Science Reviews* 30 (2011): 1675–1680.

95. MacPhee and Marx, "The 40,000-Year Plague." Taking an opposite point of view: S. K. Lyons et al., "Was a 'Hyperdisease' Responsible for the Late Pleistocene Megafaunal Extinctions?" *Ecology Letters* 7 (2004): 859–868.

96. Lyons et al., "Was a 'Hyperdisease' Responsible?"

97. MacPhee and Marx, "The 40,000-Year Plague."

98. T. Surovell et al., "Global Archaeological Evidence for Proboscidean Overkill," *Proceedings of the National Academy of Sciences of the United States of America* 102 (2005): 6231–6236.

99. N. Owen-Smith, "Pleistocene Extinctions: The Pivotal Role of Megaherbivores," *Paleobiology* 13 (1987): 351–362.

100. D. H. Janzen, "The Pleistocene Hunters Had Help," *American Naturalist* 121 (1983): 598–599; Ripple and Van Valkenburgh, "Linking Top-Down Forces to Pleistocene Megafaunal Extinctions."

101. J. A. Estes et al., "Trophic Downgrading of Planet Earth," *Science* 333 (2011): 301–306.

102. D. J. Kennett et al., "Nanodiamonds in the Younger Dryas Boundary Sediment Layer," *Science* 323 (2009): 94.

103. V. T. Holliday and D. J. Meltzer, "The 12.9-ka ET Impact Hypothesis and North American Paleoindians," *Current Anthropology* 51 (2011): 575–606.

104. R. W. Graham and E. L. Lundelius Jr., "Coevolutionary Disequilibrium and Pleistocene Extinctions," in *Quaternary Extinctions: A Prehistoric Revolution*, ed. P. S. Martin and R. G. Klein (Tucson: University of Arizona Press, 1984), 223–249;

J. E. King and J. A. Saunders, "Environmental Insularity and the Extinction of the American Mastodon," in *Quaternary Extinctions: A Prehistoric Revolution*, ed. P. S. Martin and R. G. Klein (Tucson: University of Arizona Press, 1984), 315–339.

105. D. R. Guthrie, "Rapid Body Size Decline in Alaskan Pleistocene Horses Before Extinction," *Nature* 426 (2003): 169–171.

106. B. Shapiro et al., "Rise and Fall of the Beringian Steppe Bison," *Science* 306 (2003): 1561–1565; M. C. Wilson, "Late Quaternary Vertebrates and the Opening of the Ice-Free Corridor, with Special Reference to Bison," *Quaternary International* 32 (1996): 97–105.

107. D. W. Steadman et al., "Asynchronous Extinction of Late Quaternary Sloths on Continents and Islands," *Proceedings of the National Academy of Sciences* 102 (2005): 11763–11768.

108. S. Wroe and J. Field, "A Review of the Evidence for a Human Role in the Extinction of Australian Megafauna and an Alternative Interpretation," *Quaternary Science Reviews* 25 (2006): 2692–2703.

109. A. D. Barnosky et al., "Assessing the Causes of Late Pleistocene Extinctions on the Continents," *Science* 306 (2004): 70–75.

110. R. L. Beschtii and W. J. Ripple, "Large Predators and Trophic Cascades in Terrestrial Ecosystems of the Western United States," *Biological Conservation* 142 (2009): 2401–2414; Ripple and Van Valkenburgh, "Linking Top-Down Forces to Pleistocene Megafaunal Extinctions."

111. D. F. Morey and M. D. Wiant, "Early Holocene Domestic Dog Burials from the North American Midwest," *Current Anthropology* 33 (1992): 224–229.

112. B. W. Brook and A. D. Barnosky, "Quaternary Extinctions and Their Link to Climate Change," in *Saving a Million Species: Extinction Risk from Climate Change*, ed. L. Hannah (Washington D.C.: Island, 2011), 179–198.

113. G. H. Miller et al., "Ecosystem Collapse in Pleistocene Australia and a Human Role in Megafauna Extinction," *Science* 309 (2005): 287–290.

114. S. Wroe et al., "Estimating the Weight of the Pleistocene Marsupial Lion, *Thylacoleo carnifex* (Thylacoleonidae: Marsupialia): Implications for the Ecomorphology of a Marsupial Superpredator and Hypotheses of Impoverishment of Australian Marsupial Carnivore Faunas," *Australian Journal of Zoology* 47 (1999): 489–498.

115. S. Wroe, "Cranial Mechanics Compared in Extinct Marsupial and Extant African Lions Using a Finite Element Technique," *Journal of Zoology* 274 (2008): 332–339.

116. S. Wroe, "A Review of Terrestrial Mammalian and Reptilian Carnivore Ecology in Australian Fossil Faunas, and Factors Influencing Their Diversity: The Myth of Reptilian Domination and Its Broader Ramifications," *Australian Journal of Zoology* 50 (2002): 1–24.

117. R. E. Molnar, *Dragons in the Dust: The Paleobiology of the Giant Monitor Lizard Megalania* (Bloomington: Indiana University Press, 2004).

118. B. G. Fry et al., "A Central Role for Venom in Predation by *Varanus komodoensis* (Komodo Dragon) and the Extinct Giant *Varanus (Megalania) priscus,*" *Proceedings of the National Academy of Sciences* 106 (2009): 8969–8974.

119. P. M. S. Willis and B. Mackness, "*Quinkana babarra,* a New Species of Ziphodont Mekosuchine Crocodile from the Early Pliocene Bluff Downs Local Fauna, Northern Australia, with a Revision of the Genus," *Proceedings and Journal of the Linnean Society of New South Wales* 116 (1996): 143–151.

120. D. A. Burney and T. F. Flannery, "Fifty Millennia of Catastrophic Extinctions After Human Contact," *Trends in Ecology and Evolution* 20 (2005): 395–401.

121. D. A. Burney and T. F. Flannery, "Reply to Response to Wroe et al.: Island Extinctions Versus Continental Extinctions," *Trends in Ecology and Evolution* 21 (2006): 63–64.

122. J. M. Bowler et al., "New Ages for Human Occupation and Climatic Change at Lake Mungo, Australia," *Nature* 421 (2003): 837–840.

123. R. G. Bednarik, "The Earliest Evidence of Ocean Navigation," *International Journal of Nautical Archaeology* 26 (1997): 183–191.

124. J. F. O'Connell and J. Allen, "Dating the Colonization of Sahul (Pleistocene Australia–New Guinea): A Review of Recent Research," *Journal of Archaeological Science* 31 (2004): 835–853.

125. T. F. Flannery, "Pleistocene Faunal Loss: Implications of the Aftershock for Australia's Past and Future," *Archaeology in Oceania* 25 (1990): 47–67.

126. R. G. Roberts et al., "New Ages for the Last Australian Megafauna: Continent-wide Extinction About 46,000 Years Ago," *Science* 292 (2001): 1888–1892; C. S. M. Turney et al., "Redating the Onset of Burning at Lynch's Crater (North Queensland): Implications for Human Settlement in Australia," *Journal of Quaternary Science* 16 (2001): 767–771; M. I. Bird et al., "Radiocarbon Analysis of the Early Archaeological site of Nauwalabila I, Arnhem Land, Australia: Implications for Sample Suitability and Stratigraphic Integrity," *Quaternary Science Reviews* 21 (2002): 1061–1075.

127. Miller et al., "Ecosystem Collapse in Pleistocene Australia."

128. See discussion in Wroe and Field, "A Review of the Evidence for a Human Role."

129. R. G. Roberts and B. W. Brook, "Turning Back the Clock on the Extinction of Megafauna in Australia," *Quaternary Science Reviews* 29 (2010): 593–595.

130. The title of this section is taken from a New Zealand song. See M. M. Trotter and B. McCulloch, "Moas, Men, and Middens," in *Quaternary Extinctions: A Prehistoric Revolution,* ed. P. S. Martin and R. G. Klein (Tucson: University of Arizona Press, 1984), 709–728.

131. R. Taylor, "An Account of the First Discovery of Moa Remains," *Transactions of the New Zealand Institute* 5 (1873): 97–101.

132. C. J. Burrows, "Diet of the New Zealand Dinornithoformes," *Naturwissenschaffen* 67 (1980): S.151; Trotter and McCulloch, "Moas, Men, and Middens."

133. R. Duff, *The Moa-Hunter Period of Maori Culture* (Wellington: E. C. Keating [government printer]), 1873.

134. Trotter and McCulloch, "Moas, Men, and Middens."

135. A. J. Anderson, "The Extinction of Moa in Southern New Zealand," in *Quaternary Extinctions: A Prehistoric Revolution*, ed. P. S. Martin and R. G. Klein (Tucson: University of Arizona Press, 1984), 728–740.

136. Duff (*The Moa-Hunter Period of Maori Culture*) reports Moa remains on *top* of tussock grass that dates 23060 BP using the ^{14}C-isotope dating technique. There are other possible post-European records discussed in Anderson, "The Extinction of Moa in Southern New Zealand," and Trotter and McCulloch, "Moas, Men, and Middens." There is scientific debate regarding these records. It seems that the moa came tantalizingly close to surviving until European contact. It is highly unlikely that any moas exist today.

137. M. S. McGlone, "Polynesians and the Late Holocene Deforestation of New Zealand," in *Environment and People in Australasia: Abstracts of the Fifty-Second Congress of the Australian and New Zealand Association for the Advancement of Science*, ed. A. Ross (Macquarie University, 1982).

138. The location was Shag Mouth on the South Island. See H. D. Skinner, "Archeology of Canterbury. II. Monck's Cave," *Records of the Canterbury Museum* 2 (1924): 151–162; Trotter and McCulloch, "Moas, Men, and Middens," cite several other such examples.

139. A. J. Anderson, "A Review of Economic Patterns During the Archaic Phase in Southern New Zealand," *New Zealand Journal of Archeology* 3 (1982): 15–20.

140. E. Skokstad, "Divining Diet and Disease from DNA," *Science* 289 (2000): 530–531.

141. Anderson in 1984 (Anderson, "The Extinction of Moa in Southern New Zealand") produced a back-of-the-envelope but conservative estimate of 100,000 to 500,000 Moas in the 117 known Moa-hunting sites based on the size of the sites and the density of Moa bones found. He further notes that the largest sites found are on a coastline that has been receding at a rate of 0.5 to 1 m per year so that some major butchering sites may not be included in these calculations. Anderson feels that the size of the Moa population was likely in the tens of thousands. In 1995, Anderson notes that the count on sites has risen to 300 and reiterates the importance of overhunting as the factor in eliminating these species. A. J. Anderson, "Prehistoric Polynesian Impact on the New Zealand Environment: Te Whenua Hou," in *Historical Ecology in the Pacific Islands*, ed. P. V. Kirch and T. L. Hunt (New Haven, Conn.:

Yale University Press, 1995), 271–283. Also see A. J. Anderson, "Mechanics of Overkill in the Extinction of New Zealand Moas," *Journal of Archeological Science* 16 (1989): 137–151; A. J. Anderson, *Prodigious Birds: Moas and Moa-Hunting in Prehistoric New Zealand* (Cambridge: Cambridge University Press, 1989).

142. R. N. Holdaway and C. Jacomb, "Rapid Extinction of the Moas (Aves: Diornithiformes): Model, Test, and Implications," *Science* 287 (2000): 2250–2254.

143. D. K. Grayson, "The Archaeological Record of Human Impacts on Animal Populations," *Journal of World Prehistory* 15 (2001): 1–68.

144. D. W. Steadman, "Prehistoric Extinctions of Pacific Island Birds: Biodiversity Meets Zooarchaeology," *Science* 267 (1995): 1123–1131.

145. D. A. Burney, "Rates, Patterns, and Processes of Landscape Transformation and Extinction in Madagascar," in *Extinctions in Near Time: Advances in Vertebrate Paleobiology Series*, ed. R. D. E. McPhee (New York: Kleuwer Academic/Plenum Publishers, 1999), 145–164; D. A. Burney et al., "*Sporormiella* and the Late Holocene Extinctions in Madagascar," *Proceedings of the National Academy of Sciences* 100 (2003): 10800–10805.

146. Grayson, "The Archaeological Record of Human Impacts."

147. L. R. Godfrey, "The Tale of the *Tsy-aomby-aomby*," *The Sciences* (1986): 49–51; D. A. Burney et al., "The *Kilopilopitsofy*, *Kidoky*, and *Bokyboky*: Accounts of Strange Animals from Belo-sur-Mer, Madagascar, and the Megafaunal Extinction Window," *American Anthropology* 100 (1998): 957966.

148. N. Owen-Smith, "The Interaction of Humans, Megaherbivores, and Habitats in the Late Pleistocene Extinction Event," in *Extinctions in Near Time*, R. MacPhee (New York: Kleuwer Academic/Plenum, 1999), 57–69; N. Owen-Smith and M. G. Mills, "Shifting Prey Selection Generates Contrasting Herbivore Dynamics Within a Large-Mammal Predator-Prey Web," *Ecology* 89 (2008): 1120–1133; C. N. Johnson, "Determinants of Loss of Mammal Species During the Late Quaternary 'Megafauna' Extinctions: Life History and Ecology, but Not Body Size," *Proceedings of the Royal Academy of Sciences London Series B* 269 (2002): 2221–2227.

149. F. Galton, "The First Steps Toward the Domestication of Animals," *Transactions of the Ethnological Society, London* n.s. 3 (1865): 122–138.

150. A. Purvis et al., "Nonrandom Extinction and the Loss of Evolutionary History," *Science* 288 (2000): 328–330; A. Purvis et al., "Predicting Extinction Risk in Declining Species," *Proceedings of the Royal Academy of Sciences London, Series B* 267 (2000): 1947–1952.

151. R. K. Pachauri and A. Reisinger, eds., *Climate Change 2007: Synthesis Report. Core Writing Team* (Geneva: IPCC, 2007).

152. C. D. Thomas et al., "Extinction Risk from Climate Change," *Nature* 427 (2004): 145–148; C. D. Thomas et al., "Biodiversity Conservation: Uncertainty in Predictions of Extinction Risk/Effects of Changes in Climate and Land Use/Climate Change and Extinction Risk," *Nature* 430 (2004).

153. C. D. Thomas, "First Estimates of Extinction Risk from Climate Change," in *Saving a Million Species: Extinction Risk from Climate Change*, ed. L. Hannah (Washington, D.C.: Island, 2011), 11–28.

154. L. Hannah, ed., *Saving a Million Species: Extinction Risk from Climate Change* (Washington, D.C.: Island, 2011).

155. L. B. Buckley and J. Roughgarden, "Biodiversity Conservation—Effects of Changes in Climate and Land Use," *Nature* 430 (2004); J. Harte et al., "Biodiversity Conservation— Climate Change and Extinction Risk," *Nature* 430 (2004); W. Thuiller, "Biodiversity Conservation—Uncertainty in Predictions of Extinction Risk," *Nature* 430 (2004); H. R. Akçakaya et al., "Use and Misuse of the IUCN Red List Criteria in Projecting Climate Change Impacts on Biodiversity," *Global Change Biology* 12 (2006): 2037–2043.

156. Thomas, "First Estimates of Extinction Risk from Climate Change."

157. E. Jansen et al., "Palaeoclimate," in *Climate Change 2007: The Physical Science Basis. Contribution of Working Group I to the Fourth Assessment Report of the Intergovernmental Panel on Climate Change*, ed. S. Solomon et al. (Cambridge: Cambridge University Press, 2007).

158. T. Lovejoy and L. Hannah, eds., *Climate Change and Biodiversity* (New Haven, Conn.: Yale University Press, 2005).

159. C. D. Thomas et al., "Extinction Risk from Climate Change," *Nature* 427 (2004): 145–148; L. Hannah et al., "Conservation of Biodiversity in a Changing Climate," *Conservation Biology* 16 (2002): 11–17.

10. MAKING WEATHER AND INFLUENCING CLIMATE: HUMAN ENGINEERING OF THE EARTH

1. "If you will only obey the Lord your God, by diligently observing all his commandments that I am commanding you today, the Lord your God will set you high above all the nations of the earth; all these blessings shall come upon you and overtake you, if you obey the Lord your God: . . . The Lord will open for you his rich storehouse, the heavens, to give the rain of your land in its season and to bless all your undertakings." Deuteronomy 28:1–2 . . . 12 (NRSV).

2. G. S. Hopkins, "Language Dissertation no. 12: Indo-European *deiu̯os and Related Words," *Language* 8 (1932): 5–83.

3. According to J. K. Kowalski, *The House of the Governor: A Maya Palace at Uxmal, Yucatán, México* (Norman: University of Oklahoma Press, 1987). For an additional opinion, see L. Schele, "The Iconography of Maya Architectural Façades During the Late Classic Period," in *Function and Meaning in Classic Maya Architecture*, ed. S. D. Houston (Washington, D.C.: Dumbarton Oaks Research Library and Collection, 1994), 479–518.

4. C. Isendahl, "The Weight of Water: A New Look at Pre-Hispanic Puuc Maya Water Reservoirs," *Ancient Mesoamerica* 22 (2011): 185–197.

5. M. Medina-Elizalde et al., "High-Resolution Stalagmite Climate Record from the Yucatán Peninsula Spanning the Maya Terminal Classic Period," *Earth and Planetary Science Letters* 298 (2010): 255–262.

6. T. P. Culbert and D. S. Rice, eds., *Pre-Columbian Population History in the Maya Lowlands* (Albuquerque: University of New Mexico Press, 1990).

7. K. Carmean, N. Dunning, and J. K. Kowalski, "High Times in the Hill Country: A Perspective from the Terminal Classic Puuc Region," in *The Terminal Classic in the Maya Lowlands: Collapse, Transition, and Transformation*, ed. A. A. Demarest, P. M. Rice, and D. S. Rice (Boulder: University Press of Colorado, 2004), 424–449.

8. J. M. Ingham, "Human Sacrifice at Tenochtitlan," *Comparative Studies in Society and History* 26 (1984): 379–400.

9. M. D. Therrell, D. W. Stahle, and R. A. Soto, "Aztec Drought and the 'Curse of One Rabbit,'" *Bulletin of the American Meteorological Society* 85 (2004): 1263–1272.

10. There is a thirteen-year missing segment in the tree ring chronology. Ibid.

11. Ibid.

12. Ibid.

13. B. J. Meggers, "The Transpacific Origin of Mesoamerican Civilization: A Preliminary Review of the Evidence and Its Theoretical Implications," *American Anthropologist* 77 (1975): 1–27; T. A. Dowson, "Rain in Bushman Belief, Politics, and History: The Rock-Art of Rainmaking in the Southeastern Mountains, Southern Africa," in *The Archeology of Rock-Art*, ed. Christopher Chippendale and P. S. C. Taçon (Cambridge: Cambridge University Press, 1998), 73–89; J. W. Schofer, "Theology and Cosmology in Rabbinic Ethics: The Pedagogical Significance of Rainmaking Narratives," *Jewish Studies Quarterly* 12 (2005): 227–259.

14. Hopkins, "Language Dissertation no. 12."

15. C. Bailey, *The Religion of Ancient Rome* (London: Archibald Constable, 1907).

16. "The word for 'god' which is common to the greatest number of Indo-European languages is *deiu̯os*. It is the Indo-Iranian word, Sanskrit *devas* 'god,' Avestan *daēvō* 'demon,' In the Italic branch Latin, *deus* and *divus*, Oscan *deívaí* 'divae,' Volscian *deue* 'divo, divae' represent *deiu̯os* itself, and in Umbrian occurs the derivative *deueia* 'divina.' The same word for 'god' appears throughout the Celtic languages, Old Irish *dia*, Old Welsh *duiu-*, Modern Welsh *duw*, Old Cornish *duy*, Breton *doué*. Old Icelandic plural tīvar 'gods' is the only trace in Germanic of the original meaning for *deiu̯os*; or the singular has become the name of a particular god, Old Icelandic *Tȳr*, Old High German *Zīo*, Anglo-Saxon *Tīg*. The Baltic languages show *deiu̯os* in Lithuanian dièvas, Lettic dìevs, Old Prussian deiw(a)s, all meaning 'god.'" Hopkins, "Language Dissertation no. 12."

17. L. J. LeCount, "Like Water for Chocolate: Feasting and Political Ritual Among the Late Classic Maya at Xunantunich, Belize," *American Anthropologist* 103 (2001): 935–953.

18. S. Paliga, "Influenţe romane şi preromane în limbile slave de sud," http://egg .mnir.ro/pdf/Paliga_InflRomane.pdf.

19. T. W. Patterson, "Hatfield the Rainmaker," *San Diego Historical Society Quarterly* 16, no. 4 (1970).

20. D. Wallechinsky and I. Wallace, *The People's Almanac* (New York: Doubleday, 1975).

21. Patterson, "Hatfield the Rainmaker."

22. M. Sharaf, *Fury on Earth: A Biography of Wilhelm Reich* (Cambridge, Mass.: Da Capo, 1994).

23. *Biography of Wilhelm Reich*. http://www.wilhelmreichtrust.org/.

24. Ibid.

25. Ibid.

26. M. Bradley, "The Strange Case of Wilhelm Reich," *New Republic* (May 26, 1947), http://www.joanbrady.co.uk/pages/content/index.asp?PageID=98.

27. Sharaf, *Fury on Earth*.

28. "Wilhelm Reich," http://www.britannica.com/EBchecked/topic/496351 /Wilhelm-Reich.

29. J. R. Fleming, *Meteorology in America, 1800–1870* (Baltimore, Md.: Johns Hopkins University Press, 1990).

30. J. R. Fleming, "The Pathological History of Weather and Climate Modification: Three Cycles of Promise and Hype," *Historical Studies in the Physical Sciences* 37 (2006): 3–25.

31. Fleming, *Meteorology in America*.

32. Ibid.

33. E. Powers, *War and the Weather; or, The Artificial Production of Rain* (Chicago: S. C. Griggs, 1871). The plates for the 1871 book were destroyed in the 1871 Chicago Fire. Powers later reissued the 1871 book in as E. Powers, *War and the Weather*, rev. ed. (Chicago: Knight & Leonard, 1890).

34. Fleming, "The Pathological History of Weather and Climate Modification."

35. C. C. Spence, *The Rainmakers: American "Pluviculture" to World War II* (Lincoln: University of Nebraska Press, 1980).

36. Fleming, "The Pathological History of Weather and Climate Modification."

37. "Government Rainmaking," *The Nation* 53 (October 22, 1891).

38. D. W. Keith, "Geoengineering the Climate: History and Prospect," *Annual Reviews of Energy and the Environment* 25 (2000): 245–284.

39. V. J. Schaefer, "The Production of Ice Crystals in a Cloud of Supercooled Water Droplets," *Science* 104 (1946): 457–459; I. Langmuir, "The Growth of Particles in Smokes and Clouds and the Production of Snow from Supercooled Clouds," *Proceedings of the American Philosophical Society* 92 (1948): 167–185.

40. B. Vonnegut, "The Nucleation of Ice Formation by Silver Iodide," *Journal of Applied Physics* 18 (1947): 593–595.

41. Fleming, "The Pathological History of Weather and Climate Modification."

42. *New York Times*, November 15 1946, 24; cited in ibid.

43. R. D. Elliott, "Experience of the Private Sector," in *Weather and Climate Modification*, ed. W. N. Hess (New York: John Wiley and Sons, 1974), 45–89.

44. A. M. Choudhury, *The Present Status of Artificial Rain Making*. International Centre for Theoretical Physics Internal Report IC/86/239 (Trieste, Italy: 1987).

45. A. Korolev, "Limitations of the Wegener-Bergeron-Findeisen Mechanism in the Evolution of Mixed-Phase Clouds," *Journal of Atmospheric Physics* 64 (2006): 3372–3375.

46. R. A. Kerr, "Cloud Seeding: One Success in 35 Years," *Science* 217 (2006): 519–521.

47. A. Gagin and J. Neuman, "The Second Israeli Randomized Cloud Seeding Experiment—Evaluation of Results," *Journal of Applied Meteorology* 20 (1981): 1301–1311.

48. B. F. Ryan and W. D. King, "A Critical Review of the Australian Experience in Cloud Seeding," *Bulletin of the American Meteorological Society* 78 (1997): 239–254.

49. R. List et al., "The Rain Enhancement Experiment in Puglia, Italy: Statistical Evaluation," *Journal of Applied Meteorology* 38 (1999): 281–289.

50. R. T. Bruintjes, "A Review of Cloud Seeding Experiments to Enhance Precipitation and Some New Prospects," *Bulletin of the American Meteorological Society* 80 (1997): 805–830.

51. S. A. Tessendorf et al., "The Queensland Cloud Seeding Research Program," *Bulletin of the American Meteorological Association* 93 (2012): 75–90.

52. J. L. Lovasich et al., "Further Studies of the Whitetop Cloud-Seeding Experiment," *Proceedings of the National Academy of Sciences* 68 (1971): 147–151.

53. U.S. National Research Council, *Critical Issues in Weather Modification Research* (Washington, D.C.: U.S. Government Printing Office, 2003).

54. http://www.tyndall.ac.uk.

55. M. Garstang et al., "Weather Modification: Finding Common Ground," *Bulletin of the American Meteorological Society* 85 (2005): 647–655.

56. *New York Times* (June 15, 1947): 46, 1; cited in Fleming, "The Pathological History of Weather and Climate Modification."

57. A. Krock, "An Inexpensive Start at Controlling the Weather," *New York Times* (March 23, 1961): 32; cited in Fleming, "The Pathological History of Weather and Climate Modification."

58. *Gravel v. United States*, 408 U.S. 606 (1972).

59. S. M. Hersch, "Rainmaking Is Used as Weapon by U.S.; Cloud-Seeding in Indochina Is Said to Be Aimed at Hindering Troop Movements and Suppressing Antiaircraft Fire Rainmaking Used for Military Purposes by the U.S. in Indochina Since '63," *New York Times*, July 3, 1972, 1.

60. Subcommittee on Oceans and International Environment (Claiborne Pell, Chairman), *Hearings Before the Subcommittee on Oceans and International Environment of the Committee on Foreign Relations United States Senate, 93rd Congress, 2nd Session on the Need for an International Agreement Prohibiting the Use of Environmental and Geophysical Modification as Weapons of War and Briefing on the Defense Weather Modification Activity*, January 25 and March 20, 1974 (top secret hearing held on March 20, 1974; made public on May 19, 1974).

61. Ibid.

62. Ibid.

63. http://www.un-documents.net/enmod.htm.

64. Fleming, "The Pathological History of Weather and Climate Modification."

65. S. Solomon et al., "Irreversible Climate Change Due to Carbon Dioxide Emissions," *Proceedings of the National Academy of Sciences* 109 (2009): 1704–1709.

66. For examples, *Energy* 18, no. 5 (1993); *Philosophical Transactions of the Royal Society* (2008); *Atmospheric Science Letters* (2011), theme issue: *Geoscale Engineering to Avert Dangerous Climate Change*.

67. Keith, "Geoengineering the Climate."

68. A. E. Burke, "Influence of Man Upon Nature—The Russian View: A Case Study," in *Man's Role in Changing the Face of the Earth*, ed. W. L. Thomas Jr. (Chicago: University of Chicago Press, 1956), 1035–1051.

69. J. O. Fletcher, *Changing Climate*, Rand Publ. 3933 (Santa Monica, Calif.: RAND, 1968).

70. N. T. Zikeev and G. A. Doumani, *Weather Modification in the Soviet Union, 1946–1966; A Selected Annotated Bibliography* (Washington, D.C.: Library of Congress, Science and Technology Division, 1967).

71. N. P. Rusin and L. A. Flit, *Man Versus Climate* (Moscow: Peace, 1960).

72. M. Borisov, "Can We Control the Arctic Climate?" *Bulletin of the Atomic Scientists* 25 (1969): 43–48.

73. Ibid.

74. M. I. Budyko, *Climatic Changes* [1974], trans. American Geophysical Union (Washington, D.C., 1977).

75. P. N. Edwards, *A Vast Machine* (Cambridge: Mass.: MIT Press, 2010).

76. President's Science Advisory Council, *Restoring the Quality of Our Environment, Report of the Environmental Pollution Panel* (Washington, D.C.: U.S. Government Printing Office, 1965).

77. P. J. Crutzen, "Albedo Enhancement by Stratospheric Sulfur Injections: A Contribution to Resolve a Policy Dilemma?" *Climate Change* 77 (2006): 211–220. The same concept was put forward in the same year in T. M. L. Wigley, "A Combined Mitigation/Geoengineering Approach to Climate Stabilization," *Science* 314 (2006): 452–454.

78. Budyko, *Climatic Changes*.

79. J. Hansen et al., "Potential Climate Impact of Mount Pinatubo Eruption," *Geophysical Research Letters* 19 (1992): 215–218; P. Minnis et al., "Radiative Climate Forcing by the Mt. Pinatubo Eruption," *Science* 259 (1993): 1411–1415.

80. Keith, "Geoengineering the Climate."

81. Zikeev and Doumani, *Weather Modification in the Soviet Union*.

82. Committee on Satellite Power Systems, *Electric Power from Orbit: A Critique of a Satellite Power System* (Washington, D.C.: National Academy Press, 1981).

83. Keith, "Geoengineering the Climate."

84. R. Angel, "Feasibility of Cooling the Earth with a Cloud of Small Spacecraft Near the Inner Lagrange point (L1)," *Proceedings of the National Academy of Sciences* 103 (2006): 17184–17189.

85. J. T. Early, "Space-Based Solar Shield to Offset Greenhouse Effect," *Journal of the British Interplanetary Society* 42 (1989): 567–569; W. Seifritz, "Mirrors to Halt Global Warming?" *Nature* 340 (1989): 603.

86. E. Teller, L. Wood, and R. Hyde, *Global Warming and Ice Ages: I. Prospects for Physics Based Modulation of Global Change*, UCRL-JC-128157 (Livermore, Calif.: Lawrence Livermore National Laboratory, 1997).

87. B. Bewick, J. P. Sanchez, and C. R. McInnes, "The Feasibility of Using L1 Positioned Dust Cloud as a Method of Space-Based Engineering," *Advances in Space Research* 49 (2012): 1212–1228.

88. P. J. Rasch et al., "An Overview of Geoengineering of Climate Using Stratospheric Sulfate Aerosols," *Philosophical Transactions of the Royal Society Series A* 366 (2008): 4007–4037; P. J. Rasch, P. J. Crutzen, and D. B. Coleman, "Exploring the Geoengineering of Climate Using Stratospheric Sulfate Aerosols: The Role of Particle Size," *Geophysical Research Letters* 35 (2008), L02809.

89. Ibid.

90. B. Kravitz et al., "Sensitivity of Stratospheric Geoengineering with Black Carbon to Aerosol Size and Altitude of Injection," *Journal of Geophysical Research* 117 (2012), http://dx.doi.org/10.1029/2011JD017341.

91. A.-I. Partanen et al., "Direct and Indirect Effects of Sea Spray Geoengineering and the Role of Injected Particle Size," *Journal of Geophysical Research* 117 (2012): D02203.

92. D. Cressey, "Cancelled Project Spurs Debate Over Geoengineering Patents: SPICE Research Consortium Decides Not to Field-Test Its Technology to Reflect the Sun's Rays," *Nature* 485 (2012): 429.

93. J. K. Shuman, H. H. Shugart, and T. L. O'Halloran, "Sensitivity of Siberian Larch Forests to Climate Change," *Global Change Biology* 17 (2011): 2370–2384; G. B. Bonan, D. Pollard, and S. L. Thompson, "Effects of Boreal Forest Vegetation on Global Climate," *Nature* 359 (1992): 716–718.

94. K. E. Trenbreth and A. Dai, "Effects of Mount Pinatubo Volcanic Eruption on the Hydrological Cycle," *Geophysical Research Letters* 34 (2007): L15702.

95. Rusin and Flit, *Man Versus Climate*.

96. Keith, "Geoengineering the Climate."

97. J. P. Bruce, H. Lee, and E. F. Haites, eds., *Climate Change 1995: Economic and Social Dimensions of Climate Change* (Cambridge: Cambridge University Press, 1996); R. T. Watson, M. C. Zinyowera, and R. H. Moss, eds., *Climate Change 1995: Impacts, Adaptations, and Mitigation of Climate Change: Scientific-Technical Analyses* (Cambridge: Cambridge University Press, 1996); N. J. Rosenberg, R. C. Izaurralde, and E. L. Malone, "Carbon Sequestration in Soils: Science, Monitoring, and Beyond," in *Proceedings of the St. Michaels Workshop* (Columbus, Ohio: Batelle Laboratory, 1998).

98. Panel on Policy Implications of Greenhouse Warming, *Policy Implications of Greenhouse Warming: Mitigation, Adaptation, and the Science Base* (Washington, D.C.: National Academy Press, 1992).

99. A. J. Watson, "Volcanic Iron, CO_2, Ocean Productivity, and Climate," *Nature* 385 (1997): 587–588; M. J. Behrenfeld and Z. S. Kolber, "Widespread Iron Limitation of Phytoplankton in the South Pacific Ocean," *Science* 283 (1999): 840–843.

100. K. H. Coale et al., "IronEx-I, an in situ Iron-Enrichment Experiment: Experimental Design, Implementation, and Results," *Deep-Sea Research* 45 (1998): 919–945; J. H. Martin et al., "Testing the Iron Hypothesis in Ecosystems of the Equatorial Pacific Ocean," *Nature* 371 (1994): 123–129; R. Monastersky, "Iron Versus the Greenhouse," *Science News* 148 (1995): 220–222.

101. T. H. Peng and W. S. Broecker, "Dynamic Limitations on the Antarctic Iron Fertilization Strategy," *Nature* 349 (1991): 227–229.

102. H. S. Kheshgi, "Sequestering Atmospheric CO_2 by Increasing Ocean Alkalinity," *Energy* 20 (1995): 915–922; E. A. Parson and D. W. Keith, "Fossil Fuels Without CO_2 Emissions," *Science* 282 (1998): 1053–1054.

103. Budyko, *Climatic Changes*.

104. J. C. Stroeve et al., "The Arctic's Rapidly Shrinking Sea Ice Cover: A Research Synthesis," *Climatic Change* 110 (2012): 1005–1037.

105. W. F. Ruddiman, *Plows, Plagues, and Petroleum: How Humans Took Control of Climate* (Princeton, N.J.: Princeton University Press, 2005).

106. *Climate Change 2007: The Physical Science Basis. Contribution of Working Group I to the Fourth Assessment Report of the Intergovernmental Panel on Climate Change*, ed. S. Solomon et al. (Cambridge: Cambridge University Press, 2007).

107. J. J. Blackstock and J. C. S. Long, "Climate Change: The Politics of Geoengineering," *Science* 327 (2010): 527.

108. J. Lovelock, "A Geophysiologist's Thoughts on Geoengineering," *Philosophical Transactions of the Royal Society Series A* 366 (2008): 3883–3890.

109. A. Robock, "It Is 5 Minutes to Midnight," *Bulletin of the Atomic Scientists* 64 (2004): 7.

110. N. E. A. Vaugham and T. M. Lenton, "A Review of Climate Geoengineering Proposals," *Climatic Change* 109 (2011): 745–790.

111. R. J. Cicerone, "Geoengineering: Encouraging Research and Overseeing Implementation," *Climatic Change* 77 (2006): 221–226.

11. Conclusion: Comprehending the Earth

1. W. P. Brown, *Seven Pillars of Creation: The Bible, Science, and the Ecology of Wonder* (New York: Oxford University Press, 2010).

2. Letter from Isaac Newton to Robert Hooke, February 5, 1676, as transcribed in J.-P. Maury, *Newton: Understanding the Cosmos* (London: Thames and Hudson, 1992). Hooke and Newton were in the midst of an argument involving optics. Hooke was short and hunchbacked, giving the quote a potential double entendre.

3. D. Brewster, *Memoirs of the Life, Writings, and Discoveries of Sir Isaac Newton*, vol. 2 (Edinburgh: Thomas Constable and Company, 1855), chap. 27.

Index

Aberdeen Bestiary, 37
acclimatization societies, 93
Africa, 168–69, 229
agriculture and agricultural development:
animal labor in, 67; auguries in, 134–
35; chemicals used in, 83–84; climate
change and, 67–69, 304n106; crop,
50, 53; fires in, 63; land impact from,
64–69; life cycles for, 124; movement
and, 64–66, 303n101; Neolithic Revolution and cultures in, 59–60, 64–66,
124–25; pest animals and, 82; rain and
engineering, 259; Secondary Products
Revolution in, 61; soil erosion in,
65–66; swidden, 65; weeds and, 80–81.
See also plant domestication
Alaska, *102*
American robin (*Turdus migratorius*), 134
Americas, 232–42. *See also* North America
animal domestication: artificial selection
in, 46–48, 90; of Australian dingo,
44–45, 298n27; Book of Job on, 35–36,
41, 296n1; breeding in, 42, 89–90; in
Çatal Hüyük, 52, 53; of cattle, 54–55;
conditions for, 41–42, 251–52; of
dogs, 42–43, 44–45, 48–50; Fertile
Crescent and, 50–51; genetic traits
in, 46–47; of horses, 74–75; human
activity impacted by, 36, 42; land cover
impact of, 58–62; movement in, 90; of
onagers, 72–73, 75, 100; predators and,
228; reproductive activity in, 89–90;
size and, 51; tame animals compared
to, 45–46; time and timing of, 47–48,
52; of unicorn, 38, 39; of wild ox, 41,
50–53; of wolf, 43–48; women nursing
in, 43–44, 45
animal preservation: in conservation
ecology, 222; food chain patterns and,
223–26, 228–29; landscapes required in,
222; of large predators, 223, 228–29, 251;
population in, 228–29; in "whirlwind
speech," 254–55; "whirlwind speech"
on, 222, 251, 254–55. *See also* food, food
chain, and food web patterns
animals: agriculture and labor of, 67; alien
species of, 95; biological timing in, 134,
140–42, 150; chorological classification of, 88; disease in, 85, 97–98; in
Earth system, 11; horse as ridden, 74;
human-dominated ecosystems and
pest, 82–84; humans impacted by
disease in, 85; migration and classifying, 142–43; moas and moa extinction,
249; onager as draft, 73, 74; religion
on movement of, 54–55; "whirlwind
speech" on reproduction of, 140.
See also megafauna and megafaunal
extinctions; taming and tame animals;
wildlife; *specific animals*

Anthropocene epoch, 86–87
anthropogenic landscapes, 62–66
antiquity, 8–10
Ararat, 88
archaeoastronomy, 127–30, 317n11
archaeology, 60–62
archaeons, 27–28, 295n61
Archean eon, 30–31
architecture, 127, 128–30
arid and semiarid environments, 183–85
Aristotle, 188
Arrhenius, Svante, 179–80, 284–85
artificial selection, 46–48, 90
Ashurnanipal, 75
asteroid impacts, 27, 29, 31, 177, 230, 240, 287
astrology, 133–34
astronomy, 128–30
asynchronous coupling, 215, 217
atmosphere: carbon dioxide in, 211–13; gases in, 30–31; geoengineering through aerosols in, 276–77; geography and, 185; life-supported, 27–31, 295n70; wind circulation and global, 163–71
auguries: agricultural, 134–35; astrology and, 133–34; in biological and ecological timing, 132–35; bird, 123–24, 134–35, 318n32; geography and location for, 134; historical use of, 133, 134; for Kantu' people, 134–35; Pliny the Elder on, 133
auroch or wild ox: animal domestication of, 41, 50–53; in Book of Job, 35–36; cave paintings of, 39–40; crop agriculture and, 50; description of, 39–40, 297n13; as domestic cattle source, 41; geographical location of, 39–40, 41; hunting, 40–41; in Ishtar Gate, 40; origins of, 39; as power symbol, 41; taming, 50, 51; uses of, 53–54, 56. See also cattle, domestic

Australia: dingo in, 44–45, 298n27, 298nn29–30; exploration of, 191; human arrival, 244–48; humans by boat to, 245; megafaunal extinction in, 242–47; rabbit as feral species in, 90–91; rainmaking attempts in, 269, 270
Aztecs, 259–60

Babylon and Babylonian captivity, 9, 14, 40, 70, 73, 103–4
Baity, Elizabeth Chesley, 127
Banks, Joseph, 130–31, 190–91, 285, 328n25
banteng (Bali cattle), 56
Becquerel, Antoine-César, 69
Behemoth, 5–6
belief, 5, 32
Belyaev, Dmitry K., 46–48
Beringia, 235, 238–39
Bible and biblical text: on agriculture and climate change, 68; on bird migration, 142; creation accounts in, 31–32; environmental implications from, x; Genesis in, 6–7, 9, 13–14; meaning of, ix–x; on planetary engineering and weather, 257–59, 279, 347n1; rain in, 183–84, 185, 186; science and natural world in, 149–50, 183–84, 185, 218, 254–55, 321n84; "whirlwind speech" and translations of, 10–11. See also Book of Job; creation and creation myth
biological and ecological timing: in animals, 134, 140–42, 150; auguries in, 132–35; circadian and circannual rhythm for, 141; climate and climate change impact on, 150–51; clocks for, 140–42, 144–45; in migration, 142–49; phenology and, 135–40; timers, 141. See also auguries; phenology
biology and biologists, 192. See also Humboldt, Friedrich W. H. Alexander von; specific biologists

birds and bird migration: of American robin, 134; auguries involving, 123–24, 134–35, 318n32; Bible on, 142; biological timing in, 142–49; clocks in, 142, 144–45, 320nn68–69; compasses in, 145–47; from food shortage, 143; maps in, 147–48; obligate, 143–44; of Rufous hummingbirds, 148–49; of wandering tattler, 123–24, 142, 143

blackpoll warbler, 143–44

Blake, William, *xiv*

boats, early use of, 237–38, 245

Bonpland, Aimé J. A., 194–95

Book of Job: on animal domestication, 35–36, 41, 296n1; antiquity of, 8–10; on Asiatic lion, 221, 254–55; astrology in, 133–34; Babylonian captivity and, 40; on bird migration, 142; creation in, 6–9, 14–15, 103–4, 283–84; dating, 8–9, 290n8; Earth's systems and, ix–xi, 6–8, 13; Israelites in, 9; on life cycles and heavenly ordinances, 123, 124; oceans in, 103, 120–21, 180–81; onager in, 71–72, 100; religious fusion in, 9–10; Ugaritic legends and, 9–10, 291n26; Uz as homeland in, 2; "whirlwind speech" in, x–xi, *xiv*, 1–2, 3–8. *See also* "whirlwind speech" and questions

Borana people, 126

Botai people, 74

Box, Elgene, 204–5

breeding, 42, 55–56, 74–76, 89–90, 100

Brown, Mike, 18

Caesar, Julius, 40

calendars, 125–26, 128–30, 150, 260, 316n8

canopies, leafy, 211–14

Canterbury Tales (Chaucer), 107–8

carbon and carbon dioxide: in asynchronously coupled models, 215, 217; in atmosphere, 211–13; FACE, 213–14; forests and transfer of, 210; geoengineering and, 278–79; human emission of, 212; leafy plant canopies and, 211–14; plant growth from, 212–14, 217

carnivores, 243–44. *See also* predators, large

Carnot, Nicolas Léonard Sadi, 161

Carson, Rachel, 84

Çatal Hüyük, 52–53

cattle, domestic: auroch as source of, 41; banteng, 56; breeding, 55–56; common features of, 58; in crop agriculture, 53; in culture and society, 54; DNA for determining origin of, 54; domestication of, 54–55; genes and genetic traits of, 54–56; herd management, 58–59; mithan, 56; movement of, 54–56; taurine, 54–55; water buffalo, 57–58; yak, 56–57; zebu, 54, 55

cave paintings, 39–40

Central America, 62

centrifugal force, 159–60, 323n22

Ceres, 18

Chagas' disease, 85–86

Chaldeans, 2, 8

change. *See* climate and climate change; phenology; *specific topics*

Chaucer, Geoffrey, 107–8

chemicals, 83–84, 286

Chichen Itza, 129–30

China, 49

chinook wind, 154–55, 322n7

chorology and chorological classification, 87–89

Christian calendar, 126

circadian and circannual rhythm, 141

climate and climate change: agricultural impact on, 67–69, 304n106; Aristotle on, 188; biological and ecological timing impacted by, 150–51; ecological models of, 201–3, 204–18; ecosystem impact from, 199–200, 225; exploration and, 189–92; GCMs on, 180, 215;

climate and climate change (*continued*)
geoengineering of, 273–79; geography
or location and, 187–88; human-
induced, 138–39, 216–18; IPCC report
on, 117–18; Kaya identity and, 117;
land cover impacting, 67–69; maps,
198–99; megafaunal extinctions and,
240–41; ocean gyres and, 176–77;
ocean levels and, 115–16, 117–19;
phenology and global, 138–40; in
Pleistocene epoch, 240–41; species loss
from, 252–53; vegetation relationship with, 11, 183, 187–219, 225. *See
also* atmosphere; geoengineering or
planetary engineering; greenhouse
gases; plants and vegetation, climate
change impact; rain and rain storms;
temperature; weather

clocks, biological, 140–42, 144–45,
320nn68–69

clouds and cloud seeding, 267–72

Clovis, 233–37, 239

Clutton-Brock, Juliet, 41–42

coats of arms and heraldry, 38–40, 297n11

Columbus, Christopher, 68, 173, 174

comets, 26–28, 294n45

comfort, 42

compasses, 145–47

computers and computer models, 201–3,
204–8, 234, 250

conservation and conservation ecology,
222, 251–54

continents, 98, 230–31

convergence, form, *182*, 184–85

Cook, James, 130, 131, 132

cooling, 28

Coriolis force and effect, 159, 166, 169, 172

creation and creation myth: biblical
accounts of, 31–32; in Book of Job,
6–9, 14–15, 103–4, 283–84; in *Enūma
Elish*, 103–4; in Genesis, 6–7, 9, 13–14;
God in, 13–15, 32; humans as focus of,
7; oceans in, 103–4; "priestly" account
of, 6, 13–14; religious fusion of, 14;
science on, 11, 15, 32; of solar system, 11;
water in, 13–14; "Yahwistic account"
within, 6–7. *See also* formation

Ctesias of Cnidus, 36, 296n2

culture and society: acclimatization, 93;
Americas and human, 235–37; archaeology revealing, 60; Botai, 74; of Çatal
Hüyük, 52–53; cattle in, 54; Clovis,
233–35, 237, 239; Neolithic Revolution,
period, and, 59–60, 64–66, 124–25,
127–30, 150; nonlinear development of,
61–62; phenology in, 136; Solutrean,
237; species introduction and acclimatization, 93

cycles, 19, 104–6, 111–14, 123–25, 128–30,
135–40

Dalton, Francis, 41–42

Dalton, John, 165, 324n40

Darwin, Charles, 46, 66, 88, 192, 330n52

date and dating: Archean eon, 30, 31; of
Australian dingo appearance, 44–45,
298n29; Book of Job, 8–9, 290n8;
Hadean eon, 30; humans in Americas,
235, 236–37; humans in Australia,
245–46; of megafauna extinction,
244, 246; of Pleistocene epoch, 39,
230; weeds, 80; of Zircons, 24–25

da Vinci, Leonardo, 37–38

DDT. *See* dichlorodiphenyl
trichloroethane

Decorated Plate of the Geißenklösterle,
128–29

deserts. *See* arid and semiarid environments

DGVMs. *See* Dynamic Global Vegetation
Models

dichlorodiphenyl trichloroethane (DDT),
83–84

disease: animal and plant, 85, 97–98; in
Çatal Hüyük, 53; classification, 85;

evolution, 86; feral species and, 91; in human-dominated ecosystems, 84–87; humans and novel, 96–97; introduced, 96–98; medical advances and, 86; in megafaunal extinction, 239–40; smallpox, 96–97; species movement and, 84–86, 96–98; zoonoses transmission of, 85. *See also specific diseases*

dispersal systems, 79

distribution, species and plant, 87, 88–92

diversity. *See* species; *specific topics*

DNA, 48–49, 54

dogs, 42–45, 48–52, 242, 298n30

doldrums, 166–68

domestication. *See* animal domestication; plant domestication

domiculture, 79–80

donkeys, 75–77

Doodson, Arthur Thompson, 110, 111

Dove, Heinrich W., 165

drought, 259–60

dry ice, 267–68

Dynamic Global Vegetation Models (DGVMs), 214–15, 216

Earth: archaeoastronomy and rotation of, 128; geoengineering in reflectivity change to, 277–78; greenhouse gases and radiation of, 178–80, 181, 278–79; history of, 24, 28; human-dominated ecosystems and, 98–100; humans as keystone species on, 86–87; human understanding of, 287; Moon of, 22–24, *282*; phenology and greenness of, 139–40; solar energy of, 178; "whirlwind speech" as creation and function account of, 6–9, 103–4, 283–84

Earth, formation of: early, 18–20; elemental signature in, 22, 292nn22–23; gravity in, 20–21, 22; heat in, 21; land and oceans, 24–27; physical systems, 24–31; solar system and, 15, 18–24; stel-

lar dust in, 20–21, 22; Theia collision in, 23–24; violence of, 23; water in, 25

Earth, systems of: animals in, 11; atmosphere supporting life in, 27–31, 295n70; Book of Job and, ix–xi, 6–8, 13; change natural to, 11; component interaction, 11, 21–22, 181; formation of physical, 24–31; life changing, 286–87; nature of, ix; origins, 15–17; sciences in understanding, 284–86; "whirlwind speech" questions on, 1–2, 3–5, 283–84. *See also* creation and creation myth; environment; geoengineering or planetary engineering

Ebola hemorrhagic fever, 86

ecology, ecologists, and ecological view: on climate change ecosystem impact, 199–200, 225; conservation, 222, 251–54; exploration and regularity in, 192; leafy plant canopy models, 211–12; of plants and vegetation, 187, 199, 219; vegetation and climate change models, 201–3, 204–18. *See also* biological and ecological timing; plants and vegetation

economy, 120

ecosystems: climate impact on, 199–200, 225; food chains and control of, 224–26, 337n22; humans and services of, 218–19; material circulation in, 209–10; radiation transfer in, 209–10; solar systems and, 98; undisturbed, 98–99, 310n106

ecosystems, human-dominated: disease in, 84–87; Earth and, 98–100; negatives in, 80–87, 99–100; pest animals in, 82–84; positives in, 99; study of, 98–99; weeds in, 80–82

"Eden hypothesis," 28–29

elements, signature of, 22, 292nn22–23

elephants, African, 78

Ellsburg, Daniel, 271

El Niño, 170–71
Elton, Charles, 94
Emanuel, Kerry, 160
energy and energy balance, 178, 204, 276–79, 287
Enûma Elish, 9, 14, 103–4
environment: arid and semiarid, 183–85; biblical text and implications for, x; built, 82; human change to, 1–2, 11; hunter-gatherers impacting, 247; periodicity to, 11; plant phenology and, 137–38; reproduction coordinated with, 140–42; synthesis of, ix. See also ecosystems
equids, wild, 76–77
Eris, 18
Espy, James Pollard, 265–66
Europe and Europeans. See specific topics
evolution, 81–82, 86, 192, 230, 248, 345n136
exploration and expeditions: of Australia, 191; by Banks, 190–91, 328n25; on biological and ecological regularity, 192; climate and vegetation through, 189–92; involving Humboldt, 193–95; navigation and, 122, 130–32, 173–74, 317n21; ships used in, 174; species diversity in, 191–92
extinction, 229–32, 251–53. See also megafauna and megafaunal extinctions

FACE. See Free Air Carbon Dioxide Experiments
Ferrel, William, 171, 284–85
Ferrel cell, 171–72
Fertile Crescent, 50–51, 60
fires and fire-stick hunting, 63–64, 66, 239, 246, 249, 266
föhn wind, 154–55
food, food chain, and food web patterns: animal preservation and, 223–26, 228–29; bird migration and short-

age of, 143; ecosystem control and, 224–26, 337n22; food chain analysis in, 223; large predators in, 224, 226–29; megafaunal extinction impacted by, 240; plant production required for, 223–24, 228–29, 336n9; predator size in, 226–28; warm-blooded compared to cold-blooded, 223
foraminifera, 114–15
forces, 159–60, 163, 166, 169, 171–72, 323n22
forests, 210, 249
formation: of Earth and solar system, 15, 18–24; of Earth's physical systems, 24–31; gravity in planetary, 20–21, 23; of land and oceans, 24–27, 28–29, 294n45; of Moon, 23–24; stellar dust in, 19, 20–21, 22, 292n18; of Sun, 19–20; violence in planetary, 23. See also Earth, formation of
Fourier, Joseph, 179, 284–85
foxes, 46–48
Free Air Carbon Dioxide Experiments (FACE), 213–14
frequencies of variation, 105

Gaia hypothesis, 29–30
Galilei, Galileo, 17
Gallic Wars (Caesar), 40
Galton, Francis, 251–52
game ranches, 93–94
gap models, 207, 208
gases, 30–31. See also greenhouse gases
general circulation models (GCMs), 180, 215
General Electric, 267–68, 271
genes and genetic traits, 46–47, 54–56, 72, 238
Genesis, 6–7, 9, 13–14
geoengineering or planetary engineering: for anthropogenic global warming, 275–79; atmospheric aerosols as,

276–77; Bible on, 257–59, 279, 347n1; carbon and carbon dioxide in, 278–79; of climate, 273–79; complexity and consequences of, 278, 280–81; decision makers in, 278; early ideas in, 273–75; Earth reflectivity change as, 277–78; energy balance incoming change in, 276–78; energy balance outgoing change in, 278–79; glaciers or arctic ice melted as, 274–75; human, 11, 86–87; reflecting materials in orbit as, 276; "rings of Saturn" project in, 274; in Russia, 273–76; for scientists, 280–81; summary of, 279–81; volcanic eruptions involved in, 275–76. *See also* rainmaking

geography and location: atmosphere and, 185; for auguries, 134; for cattle domestication, 54–55; climate and, 187–88; continents and, 98, 230–31; of Fertile Crescent, 50–51; mountains in, 187–88, 194–95; plant, 195–96; vegetation by, 188; wild ox in, 39–40, 41; winds and, 154–55. *See also* land, landscapes, and land cover; movement; *specific continents*; *specific geography*

glaciers, glaciations, and ice ages: in Alaska's Inner Passage, *102*; causes, 114–15; cycles for, 113, 114; European, 177; extinction from, 231, 233; future ocean levels and, 116, 118–19; geoengineering in melting, 274–75; Laurentide ice sheet, 116–17; Little, 216; in North America, 233; past ocean levels and, 112–17; in Pleistocene epoch, 230–31; Sun's summer radiation and, 114

global warming, 215–18, 275–79

God, 2–5, 8, 13–15, 31–32, 158–59, 348n16

gods, 32, *152*, 154, 259–61

gravity, 20–21, 23

greenhouse gases: Earth's radiation, heat, and, 178–80, 181, 278–79; ocean levels and, 117; in photosynthesis, 208–9, 213–14, 216; science and scientists on, 179–80, 253–54; vegetation impact from, 202–3, 208–9, 213–14, 216. *See also* carbon and carbon dioxide

greenness, 139–40

Grinnell, Joseph, 184

Grisebach, A. H. R., 197–98

"Guiana Indians," 43–44

habitat, alteration, 231–32

Hadean eon, 24–25, 27

Hadley, George, 164–66, 284–85

Hadley circulation, 166–69

Halley, Edmond, 164–65

hardiness, animal, 42, 251

Hariot, Thomas, 96

Hatfield, Charles Mallory, 262–63

heat and warming: early life and, 27–29; in Earth's formation, 21; Earth's radiation, greenhouse gases, and, 178–80, 181, 278–79; global, 215–18, 275–79; during Hadean eon, 27; hurricanes as engines of, 160–61; matter and, 21; in ocean formation, 25–26; ocean gyres and, 177, 178; plant phenology and, 137; winds and, 164, 166, 169–70, 178

heavenly ordinances. *See* stars or heavenly ordinances

heraldry. *See* coats of arms and heraldry

herbivores, 78

herd and herd management, 58–59

heritage sites, 128–30

history. *See specific topics*

HMS *Cancaeux*, 108, 311n16

HMS *Endeavour*, 130–31, 132, 190

HMS *Royal Oak*, 108

Holdridge-model, 202–3

honeyeaters, 78

Hooke, Robert, 284, 354n2

horses, 74–77, 297n12

house finches (*Carpodacus cassinii*), 148

house mouse, 82–83

humans and humanity: Americas and
arrival of, 235–39; Americas and
culture of, 235–37; animal and
wildlife disease impacting, 85; animal
domestication and regard for, 42;
animal domestication impact on,
36, 42; Anthropocene epoch and
impact of, 86–87; appearance and
genes of early, 238; auguries in history
of, 133, 134; Australia and arrival
of, 244–48; boats used by early,
237–38, 245; carbon dioxide emission by, 212; chorological classification of, 88; climate change induced
by, 138–39, 216–18; as creation story
focus, 7; DDT risk to, 84; disease and
movement of, 84–86, 96–98; Earth
understanding of, 287; ecosystems
and radioactive transfer to, 209–10;
ecosystem services to, 218–19; energy
use of, 287; environmental change
from, 1–2, 11; extinctions caused by,
230, 251; as keystone species, 77–79,
86–87; land and landscapes altered by,
62–66, 69; in megafaunal extinctions,
232, 233–35, 239, 241–42, 244–47,
248, 249, 250–51; natural world and
place of, 183–84, 185, 218, 255; novel
diseases and, 96–97; pest animal
cohabitation with, 82–83; planetary
engineering by, 11, 86–87; religion and
"why" questions for, 283–84; species
introduction by, 89–94, 100–101. See
also animal domestication; ecosystems,
human-dominated; geoengineering or
planetary engineering

Humboldt, Friedrich W. H. Alexander
von, 284–86; background of, 192–93;
on climate and vegetation relationship, 193, 194; exploration by, 193–95;
geomagnetism contribution of, 196;

isothermal lines theory from, 196–97;
natural history for, 193; plant geography contribution of, 196; scientific
contribution of, 195–97, 330n52

hunter-gatherers, 63–64, 247

hunting: aurochs, 40–41; by Clovis, 234,
239, 241–42; fires and fire-stick, 63–64,
239; in megafaunal extinctions, 232,
233–35, 239, 241–42, 244–47, 248, 249,
250–51; tradition of, 40–41; unicorns,
37–38

hurricanes, 169; cloud seeding and, 269–
70; components of, 158–60; definition
and nature of, 155–56, 322nn9–11; deri-
vation of, 156; eye in, 157–58, 159–60;
forces involved in, 159–60, 163, 323n22;
as heat engines, 160–61; media on,
161–62; naming, 156–57, 322nn14–15;
rain storms and, 158; record, 161–63;
spiraling morphology, 158; temperature and, 157–58, 160, 161; violence of,
156, 162; water in, 157–58, 161; winds
in, 155–56, 157–58, 162–63, 322n12; as
working mechanism, 157–61

hyraxes, 59

IBM. See individual-based model

ICOMOS. See International Council on
Monuments and Sites

Indica (Ctesias), 36, 296n2

individual-based model (IBM), 206, 207

Intergovernmental Panel on Climate
Change (IPCC), 117–18

International Astronomical Union, 17,
18, 128

International Council on Monuments and
Sites (ICOMOS), 128

Intertropical Convergence Zone (ITCZ),
167, 168–69

IPCC. See Intergovernmental Panel on
Climate Change

Isaac, Rabbi Shelomo Ben "Rashi," ix–x

Ishtar Gate, 40
Islamic calendar, 126
isothermal lines, theory of, 196–97
Israel and Israelites, 9, 269
ITCZ. *See* Intertropical Convergence
    Zone

Jefferson, Thomas, 67, 68
Jewish calendar, 126
Job, 2–4, 185, 289n1, 290n2. *See also* Book
    of Job
Julian year, 125
Jun-Feng Pang, 49

kangaroos, 243
Kantu' people, 134–35
Kaya identity, 117
Keeling, Charles, 138
keystone species, 77–79, 86–87, 254

land, landscapes, and land cover: agri-
    cultural impact on, 64–69; animal
    domestication impact on, 58–62; for
    animal preservation, 222; anthropo-
    genic, 62–66; climate impacted by,
    67–69; fires altering, 63–64, 66; for-
    mation of, 24–27; herd management
    impacting, 58–59; human alteration of,
    62–66, 69; hunter-gatherer alteration
    of, 63–64; Martu altering, 63–64, 69;
    overgrazing impact on, 59; Polynesian
    impact on, 64–66; soil erosion in,
    65–66
Lapita people, 44–45
Late Heavy Meteorite Bombardment, 24
Late Quaternary Extinction, 232
Laurentide ice sheet, 116–17
Leviathan, 5–6
life: atmosphere supporting, 27–31,
    295n70; cycles, 19, 123–25, 128–30,
    136–37; Earth changes from, 286–87;
    "Eden hypothesis" for, 28–29; Gaia

hypothesis for, 29–30; gases and,
    30–31; during Hadean eon, 27; heat
    and early, 27–29; on Mars, 29–30;
    "Noah hypothesis" for, 29; plants by
    form of, 204–5, 207–8; sedentary vil-
    lage, 59–60, 61–62
Linnaeus, 87–88, 188
lions: in Africa, 229; animal preserva-
    tion and care of, 228–29, 251; Asiatic,
    221–22, 254–55; "Dying Lioness," *220*,
    221; ecosystem control by, 226, 337n22;
    food webs for, 224, 226–28; popula-
    tion of, 221–22; size of, 226–28
literature, tides in, 107–8
location. *See* geography and location
Lovelock, James, 29–30
Lyme disease, 85

magnetism and magnetic field, 146, 196
malaria, 83–85
Mangaia, 65–66
Maori, 248–50, 303n101
maps and mapping, 147–48, 197–202,
    205–6
Mars, 29–30
Marsham, Robert, 136, 138
marsupials, 242–44
Martu, 63–64, 69
masks of Chaac, *256*, 259
material circulation, 209–10
matter, 21
Mayans, 259
medicine, 86, 186–87
medieval period, 37–38
megafauna and megafaunal extinctions:
    asteroid impact in, 240; Australian,
    242–47; carnivores, 243–44; causes
    of, 231; classical explanation for North
    American, 233–35; climate change in,
    240–41; computer models for, 234,
    250; dating, 244, 246; definition and
    nature of, 231, 232; disease in, 239–40;

megafauna and megafaunal extinctions
(*continued*)
  fires in, 239, 246, 249; food chain
    effects impacting, 240; habitat altera-
    tion by modern, 231–32; humans and
    hunting in, 232, 233–35, 239, 241–42,
    244–47, 248, 249, 250–51; kangaroos
    in, 243; as Late Quaternary Extinction,
    232; marsupial, 242–44; moas in, 247–
    50, 345n138, 345n141; in New Zealand,
    247–51; North American, 232–42; in
    Pleistocene epoch, 229–47, 251
Mellaart, James, 53
migration and migrants: ability, devel-
    opment and loss, 148–49; animal
    classification and, 142–43; biological
    timing in, 142–49; clocks in, 142,
    144–45, 320nn68–69; compasses in,
    145–47; irruptive, 143, 144; maps in,
    147–48; obligate, 143–44; returning
    home in, 144–48. *See also* birds and
    bird migration
Milankovitch cycles, 113–14, 115
military and military history, 108–11,
    271–72, 279, 311n16
mithan, 51, 56
moas, 247–50, 345n136, 345n138, 345n141
monolatry, 31–32
monotheism, 32
Moon and moons, 22–24, 104–7, 109,
    125–26, 150, *282*
mountains, 187–88, 194–95
movement: agricultural development
    through, 64–66, 303n101; of animals in
    religion, 54–55; of cattle, 54–56; disease
    and species, 84–86, 96–98; of domes-
    ticated animals, 90; of feral species,
    89–92; genes and, 54–56; Noah's ark
    and species, 88; Pleistocene and conti-
    nental, 230–31. *See also* distribution
Müller, P. H., 83, 84

NACA. *See* National Advisory Commit-
    tee for Aeronautics
Narváez, Pánfilo de, 97
NASA. *See* National Aeronautics and
    Space Administration
National Advisory Committee for Aero-
    nautics (NACA), 295n66
National Aeronautics and Space Adminis-
    tration (NASA), 29–30, 295n66
nature and natural world, 183–85, 193, 218,
    254–55, 321n84. *See also* ecosystems;
    specific topics
navigation, *122*, 130–32, 173–74, 317n21
neap tides, 106–9
nebulas, 19–20
Neolithic cultures, Revolution, and
    period, 59–60, 64–66, 124–25, 127–30,
    150
New Revised Standard Version of the
    Bible (NRSV). *See* Bible and biblical
    text
Newton, Sir Isaac, 284, 285, 354n2
New Zealand, 92–93, 247–51, 303n101,
    344n130
niche theory, 184–85, 327n4
"Noah hypothesis," 29
Noah's ark, 88
Normandy invasion, 109–11
North America: Beringia route to, 235,
    238–39; boats and humans arriving
    in, 237–38; glaciations in, 233; human
    appearance and genes in, 238; human
    culture in, 236–37; megafaunal extinc-
    tions in, 232–42
Norway rat, 82–83
NRSV. *See* New Revised Standard Version
    of the Bible

ocean gyres and currents: climate and,
    176–77; Coriolis force and, 172; major,
    172, *175*; summary of, 181; tempera-

ture and, 175–77, 178; thermohaline circulation and, 174–77; thermohaline circulation and conveyer belt of, 176–77; *volta do mar* and, 172–74; water density in, 174–76; winds and, 171–77

ocean levels: cities vulnerable to rising, 120; climate, climate change, and, 115–16, 117–19; economic cost of rising, 120; foraminifera and, 114–15; future rise in, 116–19; glaciers and future, 116, 118–19; glaciers and past, 112–17; greenhouse gas and, 117; impact of rising, 119–20; oxygen and, 114–15; past, 112–16; in Pleistocene epoch, 112–13, 114, 115–16; population and rising, 119; sea basin size in, 112; summary of, 120–21; Sun and, 113–14

oceans: in Alaska's Inner Passage, *102*; in Book of Job, 103, 120–21, 180–81; in creation accounts, 103–4; early, origin of, 25–26; formation of, 24–27, 28–29, 294n45; keystone species in, 77–78; summary of, 120–21; tides of, 104–11. *See also* hurricanes; tides

onager or wild ass: in Book of Job, 71–72, 100; breeding of, 75–76, 100; description of, 72, 300n58; domestication and taming of, 72–73, 75, 100; as draft animal, 73, 74; genetic nature of, 72; in Standard of Ur, 73, 75; Sumerian writings on, 75–76; summary of, 100; wild, 76–77

orbits, regularities in, 16

Origin of the Species (Darwin), 88

origins: of astrology, 133; DNA for determining, 48–49, 54; of dogs, 48–50, 51–52; of early oceans, 25–26; of Earth, 15–17; of wild ox, 39. *See also* formation

Orion nebula, *12*, 15

overgrazing, 59

ox. *See* auroch or wild ox

oxygen, 30–31, 114–15

Paleolithic period, 128–30, 150

Palolo sea worms, 124, 132, 140–41, 150

Parker, Bruce, 110–11

Pentagon Papers, 271–72

perennating tissue, 198

periodicity, 11

pest animals, 82–84

Petra, 59

phenology: and biological cycles, 135–40; biological timing, ecological timing, and, 135–40; in culture and society, 136; definition and nature of, 135–36; Earth's greenness and, 139–40; European network for, 139; global climate change and, 138–40; plant, 136–38; research, 136, 138–40; temperature and, 137, 139

photosynthesis, 208–9, 213–14, 216, 286–87

Physiologus, 37, 296n6

Piazzi, Giuseppe, 18

Piddington, Henry, 156

pigeons, homing, 145–48

planetary engineering. *See* geoengineering or planetary engineering

planets, 16–18, 20–21, 23. *See also* Earth

plant domestication: agricultural development, movement, and, 64–66, 303n101; through domiculture, 79; in Neolithic Revolution, 59–60; pollen grains revealing, 60–61, 62; sedentary villages and, 59–60, 61–62

plants and vegetation: as alien species, 94–95; chorological classification of, 87–88; climate relationship with, 11, 183, 187–219, 225; convergence in arid conditions for, *182*, 184–85; disease in, 97–98; distribution of, 89; ecosystem

plants and vegetation (*continued*)
control by, 225; environmental changes
registered by, 137–38; food chains and
required, 223–24, 228–29, 336n9; as
keystone species, 78; life cycles of,
136–37; phenology involving, 136–38;
water in, 204, 211–13
plants and vegetation, climate change
impact: asynchronous coupling of
models for, 215; carbon dioxide and
growth of, 212–14, 217; DGVMs for,
214–15, 216; ecological models of,
201–3, 204–18; ecological view of, 187,
199, 219; ecosystem radiation transfer
through, 210; energy and water in, 204,
211–13; functional types or life-form,
204–5, 207–8; gap models in, 207,
208; geography and, 195–96; greenhouse gases, 202–3, 208–9, 213–14,
216; human, 216–18; individual and
IBMs, 206–8; isothermal lines theory
for, 196–97; leafy canopies of, 211–14;
maps and mapping, 197–98, 200–202,
205–6; as medicinal, 186–87; perennating tissue for classifying, 198; scale
and classification of, 200–203; science
on, 186–87, 188–89; species mapping,
205–6. See also Humboldt, Friedrich
W. H. Alexander von
Pleistocene epoch: climate change in,
240–41; date and dating, 39, 230; dogs
origins during, 51–52; megafaunal
extinctions in, 229–47, 251; nature of,
230–31; ocean levels in, 112–13, 114,
115–16. See also megafauna and megafaunal extinctions
Pliny the Elder, 83, 91, 133
Pluto, 17, 18
polar cells, 171
polarized light, 146–47
politics and politicization, 5
pollen, grains of, 60–62

Polynesians: agriculture and land alteration of, 64–66; heavenly ordinances
for, 130–32; as keystone species, 78–79;
Maori as, 248–50, 303n101; movement
of, 64–66; navigation by, 130–32,
317n21; trade wind use by, 163
population, 119, 206, 221–22, 228–29, 259
Portugal and Portuguese, 173
Powers, Edward, 266
predators, large: animal domestication
and, 228; animal preservation of, 223,
228–29, 251; ecosystem control by,
224–25, 226, 337n22; in food chains,
224, 226–29; size of, 226–28
pressure, 166, 171
pressure-gradient force, 159, 166
Ptolemy, Claudius, 188
Pytheus of Massalia, 105–6, 311n6

rabbits, 90–92, 95, 309n79
radiation, 114, 178–80, 181, 209–10,
278–79
rain and rain storms: agricultural engineering and, 259; in arid and semiarid
environments, 183, 185; in Bible,
183–84, 185, 186; hurricanes and, 158;
makers, 262–64; population impacted
by drought and, 259; religion and,
258–62; sacrifices for, 259–60; seasonality, 199; as steam engine, 265–66;
trade winds and, 166, 168–69
rainmaking: cloud seeding in, 267–71; dry
ice in, 267–68; by military, 271–72;
post-World War II, 267–71; prominent
purveyors of, 262–64; *Science* on,
268–69; as scientific endeavor, 264–71,
349n33; U.S. and early attempts at,
265–67
Rather, Dan, 161–62
Raunkiaer, Christen Christensen, 198
red deer, 92–93
re'em, 35–36, 38–41, 296n1

Reich, Wilhelm, 263–64
religion: belief in, 5, 32; Book of Job and
    fusion of, 9–10; calendars by, 126,
    316n8; chorological classification
    and, 88; creation myths and fusion
    of, 14; humans and "why" questions,
    283–84; Job and devotion to, 2–3; on
    movement of animals, 54–55; rain
    and, 258–62; Ugaritic, 10; unicorn in,
    38; weather made through, 261–62;
    "whirlwind speech" questions relation
    to, 7–8
reproduction and reproductive activity,
    89–90, 134, 140–42
research. *See specific topics*
rocky intertidal, 77–78
rotations, 16
Roth, W. E., 43
Rufous hummingbirds (*Selasphorus rufus*),
    148–49
Ruggles, Daniel, 266
Russia, 267–71, 273–76

Sabeans, 2, 8
sacrifice, 8, 51, 259–60
Saussure, Horace-Bénédict de, 179
scale, 200–203
Schaefer, Vincent, 267–68, 269
Schimper, A. F. W., 197–98
Science, 268–69
science and scientists: belief in, 5, 32; bibli-
    cal view of natural world and, 149–50,
    183–84, 185, 218, 254–55, 321n84; of
    biology, 192; of chorology, 87–89;
    on climate and vegetation, 186–87,
    188–89; on creation, 11, 15, 32; on
    Earth's origins, 15; Earth understanding
    through, 284–86; geoengineering for,
    280–81; on greenhouse gases, 179–80,
    253–54; Humboldt's contribution to,
    195–97, 330n52; on species distribu-
    tion, 88–89; technology impacting,

32–33, 286; on trade winds, 163–66;
    truth and, 5; weather modification
    or rainmaking through, 264–71,
    349n33; "whirlwind speech" questions
    answered by, 4–5. *See also* ecology,
    ecologists, and ecological view
seas. *See* oceans
seasons and seasonality, 199
Secondary Products Revolution, 61–62
sedentary villages and village life, 59–60,
    61–62
seeds, evolution of, 81
Shakespeare, William, 93, 136–37
ships, in exploration, 174
Siegesbeck, Johann Georg, 87–88
Silent Spring (Carson), 84
smallpox, 96–97
society. *See* culture and society
soil erosion, 65–66
Solander, Daniel, 190–91
solar energy, 178
solar system: creation of, 11; ecosystems
    and, 98; formation of, 15, 18–24; moon
    regularities in, 17; orbit regularities
    in, 16; Orion nebula and, *12*; patterns
    in, 15–17; planetary regularities in, 16,
    17; rotational regularities in, 16; stellar
    dust in formation of, 19, 22, 292n18
Solomon, 149
Solutrean people, 237
South Pacific, 66, 123–24. *See also*
    Polynesians
species: acclimatization societies introduc-
    ing, 93; animal alien, 95; arid environ-
    ments and survival of, 184; climate
    change and loss of, 252–53; continents
    and diversity of, 98; disease and move-
    ment of, 84–86, 96–98; distribution,
    87, 88–92; domiculture, 79; evolution
    for diversity of, 230; exploration and
    diversity of, 191–92; extinction of, 229,
    231, 252–53; feral, 89–92; game ranches

species (*continued*)
 for introduction of, 93–94; human introduction of, 89–94, 100–101; keystone, 77–79, 86–87, 254; magnitude of diversity of alien, 94–95; mapping plant, 205–6; Noah's ark and movement of, 88; photosynthetic, 286–87; plants as alien, 94–95; purposeful introduction of, 92–94; rare, 254; wild equids endangerment as, 76–77. *See also* animals; megafauna and megafaunal extinctions; *specific types*
Species Plantarum (Linnaeus), 88
spectrum, matter, 21
Spence, Clark C., 264–65
spring tides, 106–9
Standard of Ur, *70*, 73, 75
starfish, 77–78
stars or heavenly ordinances, 19, 123–25, 128–32, 146, 150
steam engines, 265–66
stellar dust, 19–22, 292n18
Sternberg, Kaspar Maria Graf von, 188–89
stick navigation charts, *122*, 132
Stonehenge World Heritage Site, 129
stromolites, 28–29
Sumeria and Sumerians, 75–76
Sun: compass, 146; as energy source, 178; glaciers and summer radiation from, 114; nebulas and formation of, 19–20; ocean levels and, 113–14; thermonuclear reaction in, 20; tides and, 104, 105–7, 109; time and calendars from, 125, 150. *See also* geoengineering or planetary engineering
Systema Naturae (Linnaeus), 87
syzygy, 106

taming and tame animals: artificial selection for, 46–48; aurochs, 50, 51; crop agriculture and, 50; domesticated compared to, 45–46; of onagers, 72–73, 75, 100; for ritual sacrifice, 51; salt and, 51. *See also* animal domestication
Taylor, Richard, 165, 247
technology, 32–33, 286
temperature: hurricanes and, 157–58, 160, 161; ocean gyres and, 175–77, 178; phenology and, 137, 139; winds and, 155, 164, 166, 169–70, 178. *See also* heat and warming
Thaïs bone, 129
Theia, 23–24
Theoprastus, 187
thermohaline circulation, 174–77
thermonuclear reaction, 20
tides: cycles of, 104–5, 106, 111–12, 114; extreme, 105; frequencies of variation in, 105; in literature, 107–8; in military history, 108–11, 311n16; Moon and Sun in, 104, 105–7, 109; neap, 106, 107–9; Normandy invasion and, 109–11; ocean, 104–11; prediction machines for, 109–11, 311n21; spring, 106–9; syzygy and, 106
time and timing: in animal domestication, 47–48, 52; biological and ecological, 132–51; calendars for, 125–26, 128–30, 150, 260, 316n8; of dog origins, 48–50; Moon and Sun in, 125–26, 150; phenology and biological, 135–40; tide cycles, 111–12; of wild ox domestication, 52; for years, 125. *See also* biological and ecological timing; date and dating
trees, IBMs for, 207
tropical cyclones. *See* hurricanes
tropical storms, 155–63. *See also* hurricanes
truth, 5
Tupaia, 130–31, 132
typhoons. *See* hurricanes

Ugarit and Ugaritic, 9–10, 291n26
Unicorn in Captivity, The, 34

unicorns, 296n6; ancient scholars on, 36–37; in Book of Job, 35–36, 296n1; in coats of arms and heraldry, 38–39, 297n11; Ctesias' description of, 36; domestication of, 38, 39; horns of, 38, 297n10; hunting, 37–38; in medieval period, 37–38; in religion, 38; young women and, 37–38
United States, 265–71
Urdaneta, Andrés de, 173–74, 285
Uz, 2

vegetation. See plants and vegetation
Viking missions, 29–30, 295n66
volcanoes and volcanic eruptions, 275–76
volta do mar, 172–74
vortices. See whirlwinds or vortices

Walker, Sir Gilbert, 169, 170–71
Walker circulation, 169–71
wandering tattler (Tringaincanus), 123–24, 132, 134–35, 142, 143
warming. See global warming; heat and warming
water: buffalo, 57–58; in creation myths, 13–14; in Earth's formation, 25; in hurricanes, 157–58, 161; ocean gyres and density of, 174–76; in plants, 204, 211–13; wind impacted by, 155, 169–70. See also oceans
weather: Bible on control of, 257–59, 279, 347n1; controlling, summary of, 279–81; El Niño, 170–71; military and modification of, 271–72, 279; religion for making, 261–62; scientific modification of, 264–71, 349n33; tropical storm, 155–63; in "whirlwind speech," 153. See also geoengineering or planetary engineering; hurricanes; ocean gyres and currents; rain and rain storms; winds
Webster, Noah, 67–69, 304n106

weeds: agricultural development and, 80–81; dating, 80; evolution of, 81–82; in human-dominated ecosystems, 80–82; negative impact of, 81
whirlwinds or vortices, 158–59
"whirlwind speech" and questions: animal domestication in, 35–36; animal preservation and care in, 222, 251, 254–55; on animal reproduction, 140; antiquity of, 8; Behemoth and Leviathan in, 5–6; biblical translations used for, 10–11; on bird migration, 142; in Book of Job, x–xi, xiv, 1–2, 3–8; creation and function account of, 6–9, 103–4, 283–84; Earth system in, 1–2, 3–5, 283–84; God as whirlwind and, 158–59; Job's answers to, 3, 4; rain in, 183–84, 185, 186; religious aspects of, 7–8; science answering, 4–5; weather in, 153
wildlife: DDT in, 84; humans impacted by disease in, 85; reproduction, 89–90. See also specific animals
Willdenow, Carl Ludwig, 195
winds: African, 168–69; atmosphere and circulation of, 163–71; chinook, 154–55, 322n7; doldrums in trade, 166–68; El Niño and, 170–71; Ferrel cell in, 171–72; föhn, 154–55; forces and pressure involved in, 166, 169, 171; geography impacting, 154–55; global belts of, 167; Greek and Roman gods of, 152, 154; Hadley circulation, 166–69; in hurricanes, 155–56, 157–58, 162–63, 322n12; ITCZ in, 167, 168–69; ocean currents and, 171–77; ocean gyres and, 171–77; physical process behind, 154–55; polar cells and, 171; rain and trade, 166, 168–69; science on trade, 163–66; summary of, 181; temperature and, 155, 164, 166, 169–70, 178; trade, 163–71, 167; types of, 154, 322n2; Walker circulation, 169–71;

winds (*continued*)
water's impact on, 155, 169–70; westerly, 171–72, 325n54. *See also* hurricanes
wolf, 43–49
women: animal domestication and nursing by, 43–44, 45; Martu, 63–64; unicorns and young, 37–38
Woolley, Sir Leonard, 70, 73
World War II, 267–71
Wragge, Clement Lindley, 156–57

yak, 56–57
years, 125
YHWH, 31–32
Younger Dryas, 177
Young Woman Seated in a Landscape with a Unicorn (Leonardo da Vinci), 37–38

Zephyr, *152*, 154
zircons, 24–25, 292n31
zoonoses, 85